Advances in Geographical and Environmental Sciences

Advances in Geographical and Environmental Sciences synthesizes series diagnostigation and prognostication of earth environment, incorporating challenging interactive areas within ecological envelope of geosphere, biosphere, hydrosphere, atmosphere and cryosphere. It deals with land use land cover change (LUCC), urbanization, energy flux, land-ocean fluxes, climate, food security, ecohydrology, biodiversity, natural hazards and disasters, human health and their mutual interaction and feedback mechanism in order to contribute towards sustainable future. The geosciences methods range from traditional field techniques and conventional data collection, use of remote sensing and geographical information system, computer aided technique to advance geostatistical and dynamic modeling.

The series integrate past, present and future of geospheric attributes incorporating biophysical and human dimensions in spatio-temporal perspectives. The geosciences, encompassing land-ocean-atmosphere interaction is considered as a vital component in the context of environmental issues, especially in observation and prediction of air and water pollution, global warming and urban heat islands. It is important to communicate the advances in geosciences to increase resilience of society through capacity building for mitigating the impact of natural hazards and disasters. Sustainability of human society depends strongly on the earth environment, and thus the development of geosciences is critical for a better understanding of our living environment, and its sustainable development.

Geoscience also has the responsibility to not confine itself to addressing current problems but it is also developing a framework to address future issues. In order to build a 'Future Earth Model' for understanding and predicting the functioning of the whole climatic system, collaboration of experts in the traditional earth disciplines as well as in ecology, information technology, instrumentation and complex system is essential, through initiatives from human geoscientists. Thus human geoscience is emerging as key policy science for contributing towards sustainability/survivality science together with future earth initiative.

Advances in Geographical and Environmental Sciences series publishes books that contain novel approaches in tackling issues of human geoscience in its broadest sense—books in the series should focus on true progress in a particular area or region. The series includes monographs and edited volumes without any limitations in the page numbers.

Kuppannan Palanisami · Udaya Sekhar Nagothu

India's Water Future in a Changing Climate

Springer

Kuppannan Palanisami
International Water Management Institute
New Delhi, India

Udaya Sekhar Nagothu
Norwegian Institute of Bioeconomy
Research
Ås, Norway

ISSN 2198-3542 ISSN 2198-3550 (electronic)
Advances in Geographical and Environmental Sciences
ISBN 978-981-97-1784-2 ISBN 978-981-97-1785-9 (eBook)
https://doi.org/10.1007/978-981-97-1785-9

This Springer imprint is published by the registered company Springer Nature Singapore Pte Ltd.
The registered company address is: 152 Beach Road, #21-01/04 Gateway East, Singapore 189721, Singapore

Foreword

This book addresses an enormously important subject—the future of water in India in a changing climate. The authors have analysed this complex subject in an exceptionally comprehensive way, examining numerous possible approaches to close what the authors call the "supply–demand gap"—the gap between the demand for water for domestic, agricultural, energy, and industrial purposes and the water resources available for such use. And this gap, as the authors point out, currently stands at 13%—in other words, the amount of water the country consumes, which has been growing steadily in recent decades, exceeds the amount of water replenished from rainfall and other sources by 13%. India, therefore, is currently not living within its hydrologic means, putting the sustainability of its vital water resources at risk. And, as the authors further point out, this gap will grow substantially in the coming years, especially in the context of climate change, if effective action is not taken to address it.

The authors wisely consider a range of supply-side and demand-side options to overcome this gap, so that India may live within its hydrological means and replenish rather than further draw down its water resource endowments. To this end, the authors examine key water issues and challenges that India faces. They first appraise the water sector as such, focusing on the trends in agriculture and irrigation and the drivers of change, especially climate change and its impacts on water supply and demand and on agricultural productivity. They then analyse an array of water management issues and possible solutions, from making groundwater irrigation sustainable and maximizing water reuse to amending current cost-sharing arrangements and irrigation subsidies. The authors conclude by assessing the need to strengthen institutions, policy, and governance and outlining a possible future outlook for India incorporating investment, management, and governance options.

The authors are exceptionally well equipped to address this multifaceted subject. I have had the good fortune of knowing the principal author, Dr. Kuppannan Palanisami, for over 40 years, ever since he was a doctoral student in agricultural economics at Tamil Nadu Agricultural University (TNAU) in Coimbatore, India. Dr. Palanisami has spent a lifetime studying India's water resources, and this book represents the culmination of his life's work in various senior positions, including as Professor and Director of the Water Technology Center at TNAU and subsequently as Principal Researcher at the International Water Management Institute (IWMI), where he is currently Emeritus Scientist. Dr. Palanisami's co-author, Dr. Sekhar Udaya Nagothu, was a colleague of Dr. Palanisami's projects at IWMI and is currently Senior Researcher/Research Professor and Director (Centre for International Development) at the Norwegian Institute of Bioeconomy Research in Ås, Norway.

I commend *India's Water Future in a Changing Climate* to all those who wish to have a better understanding of water management challenges in India and a better appreciation of what might be done to ensure the future sustainability of India's vitally important water resources.

Professor Emeritus Roberto Lenton
Former Founding Executive Director,
Robert Daugherty Water for Food
Global Institute, University
of Nebraska
Director, Sustainable Energy
and Environment Division, UNDP
and Director General, IWMI
University of Nebraska-Lincoln
Lincoln, USA

Preface

Water is essential to ensure economic growth together with food and energy for an increasing population with improved standards of living and quality of life. Demands for water have been steadily increasing over the years in India. On the other hand, water supply is becoming limited due to climate change and frequent extreme weather events. The imbalances in water availability are resulting in water stress, which is compounding either because of scarcity in some areas or excess of water in others. Both water quality and quantity (reliable supply) are becoming a challenge in India. In particular, the increasing water supply–demand gap is attributed mainly to poor water management, weak institutions, poorly maintained irrigation systems, overexploitation of groundwater, and lack of adequate irrigation infrastructure development. The increasing conflicts related to water sharing in interstate rivers is another challenge that needs attention. The looming water crisis, perhaps, is not entirely due to physical scarcity but because of lack of efficient management practices.

India is also identified as one of the highly vulnerable countries to climate change where the shifting pattern of precipitation may affect the spatial and temporal distribution of runoff and increase the frequency and intensity of droughts and floods. The changes in seasonal temperatures could change the crop seasons and crop productivity. Further, the climate change can aggravate other problems such as land degradation, market volatility, rising input costs, food inflation, etc., and is recognized as a potent threat to the sustainability of agriculture and irrigation sectors in the country.

India's rainfed regions are affected by complex climatic challenges, where the most variable and unpredictable environments exist, which render rainfed agriculture a risky proposition. Providing supplemental irrigation is important to stabilize rainfed production, which will have an indirect impact on reducing the burden on irrigated agriculture.

Yet, India's water management needs not to be so dire. With good water management practices, water can be saved and efficiently used with high productivity. Wastewater reuse, hill agriculture improvement, and micro irrigation expansions are some of the areas, which need more attention in the future to address water scarcity.

Further, reworking crop water requirements due to climate change, reworking operation and maintenance costs of multipurpose irrigation projects, and discussion of the success stories at national and international levels are adding strength to this book.

If current water management practices continue, with a free or highly subsidized water supply and excessive groundwater pumping with free electricity, the water future of the country will face crisis. Hence, we need to ensure functioning water institutions and proper water management practices at central and state project levels to address the water problems in India related to quantity, quality, and equity.

The initiative for writing this book started during cross-examination of experts at Krishna Water Disputes Tribunal in 2022. The primary purpose of this book is to document the key issues and challenges in detail and look for possible solutions to address them to enhance the efficiency of water use in irrigated agriculture. Besides, the first author's four decades of experience in water research, extensive reviews, expert discussions, multiple visits across the irrigation projects, etc., helped a lot in compiling this book publication.

We have tried to incorporate most of the possible interventions to address the issues and challenges governing India's water future. The importance of good water governance in managing the water supply and demand aspects is very much emphasized. The book explains the future interventions that need to be considered to sustain the water sector in general and the irrigation sector in particular. With existing knowledge, suitable technology, capacity building, and awareness, water can be better managed so that the country has adequate water not only for current use but also for the future. Hope, the book will be a valuable document to the researchers, water planners, policy makers, and funding agencies interested in India's water future.

We thank Dr. Roberto Lenton, Chair, IWMI Board of Governors for writing the Foreword to this book. We also thank Yosuke Nishida, Editor, Niraja Deshmukh and Satish S.A, Project Co-ordinators, Springer for their sustained efforts in bringing out this publication.

New Delhi, India
Ås, Norway

Kuppannan Palanisami
Udaya Sekhar Nagothu

Acknowledgements

Several experts have shared their views, comments, and also data whenever approached and they are gratefully acknowledged. In particular, I thank Dr. Ruth Meinzen-Dick, Washington, D.C.; Dr. Mark Giordano, Washington, D.C.; Dr. Jan Lundqvist, Stockholm; Dr. Martin Burton, London; Dr. Kei Kajisa, Tokyo; Dr. Upali Amarasinghe, Colombo for their excellent interactions during the course of preparing the book draft.

Many other experts have responded actively with their valuable inputs to the quick survey on water management challenges and possible interventions. Thanks to Dr. Jeremy Bird, London; Dr. Bharat Sharma, Delhi; Mr. Shivakumar, Delhi; Dr. Ramesh Chand, Delhi; Dr. Pramod Agarwal, Delhi; Dr. Alok Sikka, Delhi; Dr. Archisman Mitra, Delhi; Dr. Tushaar Shah, Anand; Dr. Dinesh Marotia, Ranchi; Dr. K. V. Raju, Lucknow; Dr. Suresh Kulkarni, Mumbai; Dr. S. Raman, Mumbai; Dr. Avil Kumar, Hyderabad; Dr. K. Krishana Reddy, Hyderabad; Dr. Suhas Wani, Hyderabad; Er. Regunath Sharma, Hyderabad; Dr. Janakarajan, Chennai; Dr. S. Narayanan, Chennai; Dr. Ravichandran, Chennai; Dr. A. Narayanamoorthy, Karaikudi; Dr. R. Balasubramanian, Coimbatore; Dr. C. Mayilswami, Coimbatore; Dr. S. Pazhanivelan, Coimbatore and Dr. Arjamadutta Sarangi, Bhubaneswar.

Thanks to Dr. Ravinder Malik former World Bank and IWMI water expert and also my colleague at IWMI who reviewed some of the draft chapters and shared his valuable suggestions.

Dr. Martin Burton, Team Leader ADB, India project with whom I worked helped in editing the sections relating to SIMP and sharing his ideas. His support is greatly acknowledged. Thanks to Dr. Suresh Kumar, Director, CARDS, TNAU for sharing the irrigation and related data.

I thank Dr. Rengalakshmi, MSSRF; Mr. Ajit Jain, Dr. Narayanan, Dr. Soman, and Er. Jadahv, Jain Irrigation for providing research support.

I thank the NABARD, Mumbai for providing permission to use my published tank policy report for preparing the book chapter. The copyright permissions given by IWMI, GGRC, NAIP, WTC, TNAU, Allied publishers, Dr. Raman, Dr. Chauhan, Dr. Bharat Sharma, and Dr. Yellamanda Reddy are gratefully acknowledged.

I thank Dr. Arivelarasan who was fully involved with me since the start of the book draft and other research fellows—Dr. Umagowri, Dr. Kavitha, Dr. S. Senthilkumar, who helped me in the data analysis and editing.

I personally thank Dr. V. Raja, Surgeon and my good friend who in fact suggested me to initiate the book work at a time when it is demanded and his sustained encouragements helped me to complete it in time. The support extended by my friends (Kubendran, Ranganathan, Balu, Madhu, Sounder, Maruthachalam, Surilivel) are gratefully acknowledged.

Lastly, the support from NIBIO, Norway is duly acknowledged.

Kuppannan Palanisami

Contents

About the Authors

Dr. Kuppannan Palanisami is an international water expert with 40 years of experience in irrigated agriculture, water economics, technology adoption, water planning, and climate change. He was a visiting professor at the University of Minnesota, USA, from 1982 to 1984 and again in 1990; a visiting fellow at Khon Kaen University, Thailand, from 1984 to 1985; and also worked at the International Rice Research Institute (IRRI), Philippines, from 1986 to 1988. He was a visiting researcher at Waseda University, Japan, in 2000, and at the Research Institute for Humanity and Nature, Japan, in 2004. He worked as an expert consultant for the United Nations Food and Agriculture Organization (FAO) for four East African countries in 1997 and the International Fund for Agricultural Development (IFAD) project in seven Asian countries in 2003. He was Director of the Water Technology Centre and CARDS at Tamil Nadu Agri University from 2000 to 2008 and served as a consultant to the World Bank and Asian Development Bank. Worked as Principal Researcher at the International Water Management Institute (IWMI) from 2008 to 2014 and since then continuing as Emeritus Scientist at IWMI. He has written 20 books on water and climate and published about 120 papers in national and international journals.

Dr. Udaya Sekhar Nagothu has a master's degree in agriculture, another master's degree in natural resource management and sustainable agriculture, and a Ph.D. in development studies. He was the director of the Norwegian Institute for Agriculture and Environmental Research from January 2008 to June 2015. Since July 2015, he has been a senior researcher, research professor, and the director of the Centre for International Development, the Norwegian Institute of Bioeconomy Research, Norway. He has coordinated several large interdisciplinary projects focusing on climate change and its impacts on agriculture, water, forestry, food security, and livelihoods in many countries. He specializes in the multi-actor approach; stakeholder participation; institutional and socio-economic analysis; and applied interdisciplinary research on climate change vulnerability impacts, adaptation and mitigation, water, food, and nutrition security.

Chapter 1
Agriculture and Irrigation in India—Trends and Turning Points

Abstract India's ultimate irrigation potential is about 139.9 million hectares (m·ha) and the irrigation potential created (up to the Eleventh Plan period) is about 113.53 m·ha (81.16%). The total annual replenishable water supply is about 1122 billion cubic metres (bcm) and the annual reliable supply will be only 744 bcm. The future water demand is projected as 910 bcm in 2025 and 1072 bcm in 2050. Major problems in Indian water sector include low irrigation project efficiency (36%), over-exploitation of groundwater (up to 30% including critical and semi-critical areas), declining tank irrigation (about 50%), less coverage under micro irrigation (17.4%), poor technology adoption (8–10%), low wastewater treatments and reuse (26%), reduction in water supplies and increasing crop demand due to climate change, less focus on water management in hill ecosystems, use of old crop water requirement data for water planning, poor cost recovery, inappropriate subsidy calculations, weak institutional and governance structure. Appropriate interventions to sustain the irrigation sector are discussed in this chapter.

Keywords Source-wise irrigated area · Production of crops · Trend in area · Production · Yield of crops · Water potential · Water sector issues · Hill agriculture · Water management technologies · Water quality · Institutions · Governance · Water sector interventions · Phases of water sector

1.1 Introduction

India is primarily an agrarian-based economy. The projections for the future indicate that the population of India may stabilize at 1.6 to 1.7 billion by 2050 (ICAR 2015) and would require about 450 m tonnes of food grains annually, at the present level of food consumption. It is necessary to provide irrigation to at least 130 m ha for food crops alone and 160 m ha for all crops put together to be able to meet the demands of the country in 2050, which can ensure food security (Sarangi et al. 2016). Hence agriculture and irrigation development are the two main areas, which need focus

K. Palanisami and U. S. Nagothu, *India's Water Future in a Changing Climate*,
Advances in Geographical and Environmental Sciences,
https://doi.org/10.1007/978-981-97-1785-9_1

in terms of investment planning, development, and governance. What follows is a brief discussion of the challenges and the need for revitalization of agriculture and irrigation sectors.

1.2 Agriculture Sector in India

Agriculture sector plays a vital role in India's growing economy where 54.6 per cent of the total workforce is engaged in agriculture and allied sector activities (Census 2011). It accounts for 18.6% of India's Gross Value Added (GVA) at current prices during 2021–22. The total geographical area of the country is 328.7 m ha, out of which 139.9 m ha is the reported net sown area and 211.36 m ha is the gross cropped area with a cropping intensity of 141.6%. The net area sown works out to be 42.6% of the total geographical area. The net irrigated area is 75.46 m ha and accounts for 53.40% of the net area sown (MoA and FW 2022). The remaining area is under rainfed agriculture, which is totally dependent on monsoon rainfall.

India's rainfed regions are characterized by complex climatic challenges, where the most variable and unpredictable environments exist, which render rainfed agriculture a risky proposition. The actual rainfall in some places deviates from 20 to 50 percent from normal and the coefficient of variation in rainfall is 34% (Kashyap and Agarwal 2021). Small variations in the timing and the quantity of rainfall occurrence have a significant impact on agricultural output (Sanjeevaiah et al. 2021). Thus, the uncertainty of rainfall, increased frequency of droughts, and hail storms are making rainfed farmers more vulnerable resulting in low productivity, low cropping intensity, increasing cost of cultivation, very low adoption of modern technology, inadequate institutional credit and public investment, low capital formation, and high incidence of rural poverty. Over the years, the rainfed area has decreased from 107.75 m ha during 1972–73 to 64.44 m ha during 2019–20 (i.e., 40% reduction) (Deshpande 2022, MoA and FW 2022). The details of the gross and net irrigated area, area under major crops, and change in crop area are given in Tables 1.1, 1.2 and 1.3, respectively.

The net irrigated area has increased from 20.85 m. ha in 1950–51 to 75.46 m ha in 2019–20, an increase of 262 per cent in seven decades. The gross irrigated area has gone up from 22.56 m ha in 1950–51 to 112.23 m ha in 2019–20, which is an increase of more than 397 per cent. During 1950–51, only 17 per cent of cropped areas got the benefit of irrigation, which has increased to 53.10 per cent during 2018–19. Among the crops, the area under food grains has the largest per cent under irrigation, followed by oilseeds. Among the food grains, rice accounts for the major share followed by wheat and pulses. The details of the irrigated area, production, and yield of major irrigated crops are given in Figs. 1.1, 1.2 and 1.3. The compound growth rate (CGR) of irrigated area has shown that fruits and vegetables have grown at 3.75% annually followed by maize, rice, cotton, and wheat with 2.61%, 1.95%, 1.88%, and 1.66%, respectively (Fig. 1.4). The trend in irrigated crops is given in Table 1.4. It is observed that the per cent of rice irrigated area has increased from 38% in 1970–71 to about 63% in 2018–19 indicating the increasing irrigation demand for rice. The projected

Table 1.1 Gross and net irrigated area in India (m ha)

Year	Net irrigated area	Gross irrigated area	Net area sown	Gross cropped area	Percentage of gross irrigated area over gross cropped area
1950–51	20.85	22.56	118.75	131.89	17.11
1960–61	24.66	27.98	133.20	152.77	18.31
1970–71	31.10	38.20	140.86	165.79	23.04
1980–81	38.72	49.78	140.29	172.63	28.83
1990–91	48.02	63.20	142.87	185.74	34.03
2000–01	55.20	76.19	141.34	185.34	41.11
2010–11	63.67	88.94	141.37	197.68	44.99
2018–19	71.55	102.67	138.44	197.32	52.03
2019–20	75.46	112.23	139.90	211.36	53.10

Source Ministry of Agriculture and Farmers Welfare (2023)

area under irrigated crops in 2050 will be about 117 m ha (Amarasinghe et al., 2007), which means that the demand for water will be significantly increasing from the current levels if efficiency is not improved.

1.3 Changing Dynamics of Irrigated Agriculture in India

The current scenario reveals that agricultural growth can be attained through improving irrigation water productivity rather than land productivity. However, there are challenges that need to be addressed in irrigator sector. The way forward could be through technological advancements or innovations with institutional support in irrigation sector, such as drip irrigation through subsidized loans and incentives for drip equipment, hose reel irrigation equipment, sprinkler irrigation, solar irrigation in various dimensions, underground drainage in agricultural land, rental markets for irrigation are necessary for uplifting irrigation to the next level, which brings systemic change in the agriculture–water–energy nexus. This should be simultaneously accompanied by improvements in irrigation infrastructure at primary (main canal systems), secondary (distributary canal systems), and tertiary levels (field channels). Below, is a list of challenges that the agriculture and irrigation sectors currently face and will continue in the future:

(i) Technical progress in agriculture may be constant or decrease due to biophysical limits of production.
(ii) Future irrigated area expansion may be restricted because of physical limits and conflicting inter-sectoral demands.

Table 1.2 Area under major crops grown in India, 2019–20 (000'ha)

Major crops	Gross irrigated area	Gross unirrigated area	Gross cropped area	Percentage of area under irrigation
Rice	31,265	16,855	48,120	64.97
Jowar	544	4354	4898	11.11
Bajra	1152	6426	7578	15.20
Maize	2851	6882	9733	29.29
Wheat	34,012	1505	35,517	95.76
Barley	438	107	545	80.37
Total Cereals & Millets	70,445	37,538	107,983	65.24
Total pulses	6100	20,307	26,407	23.10
Total food grains	76,545	57,845	134,390	56.96
Sugarcane	5056	186	5242	96.45
Condiments & Spices	3263	907	4170	78.25
Fruits & Vegetables (including root crops)	8334	3331	11,665	71.44
Total Food Crops	93,286	62,500	155,786	59.88
Total Oil seeds	9348	19,998	29,346	31.85
Cotton	4763	8543	13,306	35.80
Tobacco	270	137	407	66.34
Fodder crops	2736	5067	7803	35.06
Other Non-Food Crops	1825	1098	2923	62.44
Total Non-Food Crops	18,943	36,630	55,573	34.09
Total	**112,229**	**99,130**	**211,359**	**53.10**

Source Ministry of Agriculture and Farmers Welfare (2023)

(iii) Environmental safeguards and regulations may constrain the expansion of irrigated agriculture in all locations and put limits to irrigation intensification as well. Deteriorating freshwater quality and increasing levels of pollution may further constrain the productivity of irrigated agriculture. Wastewater reuse is much limited contributing for the environmental pollution particularly in urban and peri-urban areas.

(iv) Hill agriculture, which is not a focus at present will be important in the future to maximize the water use efficiency and reduce poverty in these fragile regions.

Table 1.3 Change in area under irrigation for major crops in India (%)

Major crops	1970–71	2019–20	Percentage change
Rice	38.36	64.97	69.37
Jowar	3.64	11.11	205.22
Bajra	3.98	15.2	281.91
Maize	15.86	29.29	84.68
Wheat	54.25	95.76	76.52
Barley	51.96	80.37	54.68
Total Cereals& Millets	27.6	65.24	136.38
Total pulses	8.75	23.1	164.00
Total food grains	24.11	56.96	136.25
Sugarcane	72.38	96.45	33.26
Condiments & Spices	35.35	78.25	121.36
Fruits & Vegetables (including root crops)	26.91	71.44	165.48
Total Food Crops	25.44	59.88	135.38
Total Oil seeds	7.39	31.85	330.99
Cotton	17.34	35.8	106.46
Tobacco	23.67	66.34	180.27
Fodder crops	19.31	35.06	81.56
Other Non-Food Crops	48.72	62.44	28.16
Total Non-Food Crops	13.08	34.09	160.63
Total	**23.04**	**53.1**	**130.47**

Source Ministry of Agriculture and Farmers Welfare (2022)

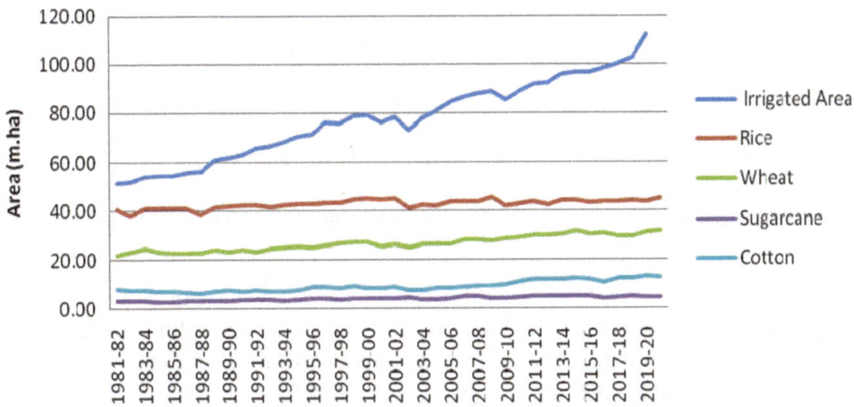

Fig. 1.1 Area of major irrigated crops in India. *Source* Ministry of Agriculture and Farmers Welfare (2022)

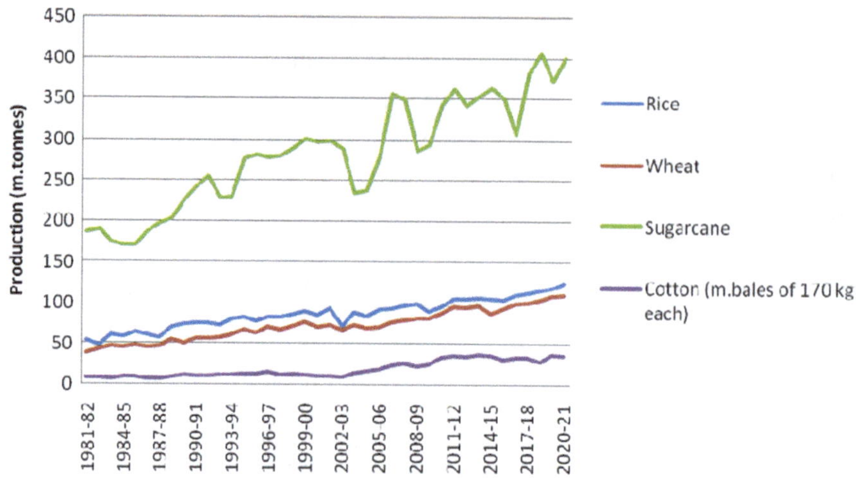

Fig. 1.2 Production of major irrigated crops in India. *Source* Ministry of Agriculture and Farmers Welfare (2022)

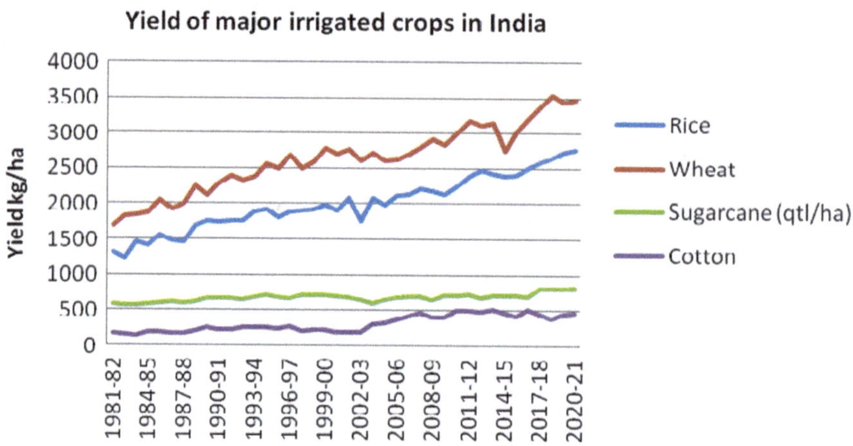

Fig. 1.3 Yield of major irrigated crops in India. *Source* Ministry of Agriculture and Farmers Welfare (2022)

(v) Climate change is likely to change the productivity of agricultural systems due to variations in the quantum of rainfall (mainly resulting in frequent droughts and less flooding periods) and increasing temperature.

(vi) Poor operation and maintenance of the existing irrigation schemes over the years has resulted in deterioration of irrigation infrastructure and uncertain water supplies for irrigation resulting in low levels of water use efficiency, which needs to be improved.

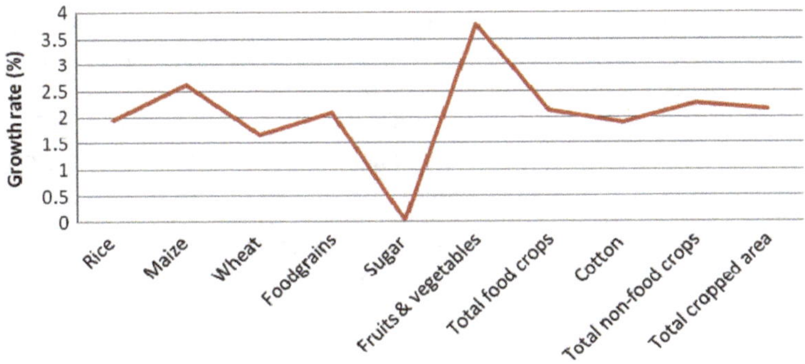

Fig. 1.4 Compound growth rate (CGR) of irrigated area under major crops in India

Table 1.4 Trend in irrigated area under different crops in India (000'ha)

Year	1970–71	1980–81	1990–91	2000–01	2010–11	2018–19
Rice	14,339 (38.36)	16,364 (40.67)	19,470 (45.55)	24,337 (54.37)	25,486 (58.83)	28,473 (62.69)
Jowar	614 (3.64)	768 (4.68)	795 (5.62)	826 (8.33)	641 (8.70)	424 (10.10)
Bajra	533 (3.98)	645 (5.53)	547 (5.10)	783 (7.81)	776 (8.02)	681 (9.40)
Maize	929 (15.86)	1215 (20.14)	1162 (19.72)	1483 (21.79)	2065 (24.54)	2350 (26.28)
Ragi	319 (12.89)	295 (11.78)	212 (9.88)	141 (7.78)	72 (5.76)	94 (9.86)
Wheat	9924 (54.25)	15,553 (69.98)	19,511 (81.14)	22,798 (88.37)	27,654 (92.18)	30,116 (95.34)
Barley	1328 (51.96)	910 (50.58)	530 (54.55)	529 (67.00)	496 (69.91)	358 (74.14)

Source Ministry of Agriculture and Farmers Welfare (2021)
Note Figures in parentheses denote percentage of irrigated area to the cropped area under each crop

(vii) Inefficient water allocation and water use in crop production have resulted in decreasing crop productivity and income per unit of water used.
(viii) Groundwater depletion is increasing in locations where more than 60 per cent of the irrigated area is covered by groundwater source. More than 80 per cent of the addition to net irrigated area has been accounted for by groundwater use.
(ix) Poor technology adoption and ineffective marketing practices will further hamper the farmers' income and their livelihood enhancement.

(x) Institutional and other barriers (including water sharing conflicts between sectors and states) for adoption of more advanced irrigation technologies will also hamper the progress of irrigated agriculture.

(xi) Furthermore, lack of awareness and implementation of the targeted government policies and programmes in the agriculture sector will further slowdown the progress in irrigated agriculture.

While the above-mentioned challenges and changing dynamics of irrigated agriculture may vary between regions and irrigation schemes in India, their net impact is likely to influence the overall agricultural productivity, production, and food security for the growing population in India.

1.4 Irrigation Sector in India

Irrigation sector is fundamental to India's economic development, ensuring food security and poverty alleviation due to increased crop yields and cropped areas. Indirectly it has the potential to increase employment opportunities, higher calorie intakes with better nutrition levels. There are also stability effects because of reduced reliance on rainfall, and if properly managed, it lowers the variance of output and reduces the adverse consequences of drought.

The ultimate irrigation potential of the country is about 139.9 m ha with a contribution of 54.32% by surface water sources and 45.74% by groundwater sources. The average annual rainfall of the country is about 1283 mm and the mean annual utilizable water is about 1122 bcm, wherein, the surface water accounts for 690 bcm and the groundwater 432 bcm (Table 1.5). However, the supply is highly erratic and variable as about 50 per cent of annual rainfall falls in less than one month and 90 per cent of river flows occur in less than four months of the year. Hence, the ability of the current and ageing infrastructure to buffer this variability is rather low thus making it difficult to meet the increasing water demand. Therefore, it is reasonable to assume that India's accessible and reliable water supply will amount to only about 744 bcm which is about 2/3 of the annual utilizable water (Water Resources Group 2009).

(A) **Irrigation Potential Created**

Given the ultimate irrigation potential (UIP) of 139.89 m ha, about 113.53 m ha(81.16%) was created till the end of Eleventh Plan period in India. The Irrigation Potential Created (IPC) in the country, which stood at 12.9 m ha in 1951 had risen to 113.53 m ha by the end of Eleventh Plan period and the increase in the case of major and medium irrigation (MMI) projects was from 9.7 m ha in 1951 to 47.97 m ha, respectively, at the end of Eleventh Plan. However, in many of the states such as Madhya Pradesh, Andhra Pradesh, etc., there is still scope for utilizing the available potentials, which can contribute to food and nutrition security.

Table 1.5 Details of water resources of India, 2019–20

Sl. no	Item	Quantity
1	Average annual rainfall (1985–2015)	1105 mm
2	Annual rainfall (2020)	1283 mm
3	Mean annual natural runoff	1999.2 bcm
4	Total annual utilizable water	1122 bcm
5	Estimated utilizable surface water potential	690 bcm
6	Total replenishable groundwater resources	432bcm
7	Net groundwater availability	411 bcm
8	Ultimate irrigation potential	139.9 m ha
9	From surface water	76 m ha
10	From groundwater	64 m ha
11	Per capita water availability	1729 m^3

Source CWC (2022)

Out of the net irrigated area of 75.45 m ha, canals account for 23.27%, wells 62.50%, and tanks 2.90%. Gross irrigated area over gross cropped area accounts for 53.1% and the net irrigated area over net area sown is 53.9% (Table 1.6).

Source-wise utilization of irrigation water in different states shows that irrigation is mainly concentrated in the northern region of the country (Table 1.7). The state of Uttar Pradesh is having the largest area under irrigation followed by Madhya Pradesh and Rajasthan. The state of Andhra Pradesh is having the largest area under tank irrigation. Tube wells and other wells form the main source of irrigation in Uttar Pradesh, which is also having the largest area under government managed canal systems.

(B) Sources of Irrigation

Various sources of irrigation in India are canals, tanks, tube wells, and other wells, among which tube wells and wells together accounting for about 63.93% of total irrigated area in 2018–19, which has increased from 28.67% in 1950–51. Contribution by canal irrigation has decreased from 39.78% in 1950–51 to 22.96% in 2018–19 followed by tank irrigated area, which has declined from 17.33% in 1950–51 to 2.33% in 2018–19 (Table 1.8 and Fig. 1.5).

(C) Canal Irrigation

Since 1950–51, considerable importance was attached to the provision of canal irrigation through major and medium irrigation (MMI) projects. Canal irrigated area has increased from 8.3 m ha in 1950–51 to 16.43 m ha during 2018–19. Likewise, well irrigated area has increased from 5.98 m ha in 1950–51 to 45.75 m ha in 2018–19. The gap between IPC and IPU has increased over the years from 10% during the first five-year plan to 27.15% (25.6 m ha) at the end of the Eleventh Plan indicating the problems faced by the canal irrigation. Region-wise comparison (by including

Table 1.6 Details of irrigated area in India, 2019–20

Sl. no	Sources	Area (000'ha)
1	Government Canals	17,399
2	Private Canals	162
3	Total Canals (1 + 2)	17,561
4	Tanks	2185
5	Tubewells	36,287
6	Other wells	10,876
7	Other Sources	8547
	Net Irrigated Area (3 + 4 + 5 + 6 + 7)	75,456
	Gross Irrigated Area	112,229
	% of Gross irrigated area over gross cropped area	53.10
	% of Net irrigated area over net area sown	53.93

Source Ministry of Agriculture and Farmers Welfare (2021)

the major states benefitted from MMI schemes) indicates the gap is high in the northern region (67%) followed by the western region (40%). The southern region has the lowest gap of about 24%. The major states account for about 96 per cent of UIP and about 96 per cent of the country's IPC in MMI projects. The main challenges behind the slow decline in canal irrigation across India are ageing of irrigation systems, hydrological changes, violation of crop pattern, low level of participation by irrigation communities, poor cost recovery, inadequate operation and maintenance budget, failure of public-sector agencies to deliver irrigation water as planned, inefficient water conveyance, and inequitable distribution of water (Bjornlund et al. 2020; Chauhan and Shobha 2023).

(D) **Well Irrigation**

Well irrigation in 2018–19 accounted for nearly 64% of the total irrigated area as compared to 28.67 per cent in 1950–51 (Table 1.9). But at the same time, groundwater extraction has accounted for 62–63% of the potential and 60% of the wells in India, which is contributing to a decline in groundwater table. About 14% of groundwater blocks are overexploited while 4% and 12% are at critical and semi-critical stages, respectively (CGWB 2022). Current overexploitation will affect the future water supplies due to increasing well failures. Weak regulations result in the expansion of groundwater irrigation through the drilling of additional wells even in areas where groundwater potential is low. Power subsidies for pumping groundwater further aggravate the situation in water-stressed areas where many states affected by the depletion of groundwater still provide free or heavily subsidized electricity (including solar pumps) for pumping groundwater (Shiferaw 2021). In the present era, the frequency of groundwater vulnerability has been significantly increased due to declining water tables, poor recharge pattern, and high rainfall variability. This is in fact one of the biggest challenges to groundwater dependant irrigation in India.

Table 1.7 Source-wise and State-wise net irrigated area in India, 2018–19 (000' ha)

State	From government canals	From private canals	Tanks	Tube wells and other wells	Other sources	Total
Andhra Pradesh	1303	0	232	1145	116	2796
Arunachal Pradesh	0	0	0	0	55	55
Assam	89	10	9	84	173	366
Bihar	965	0	59	1979	111	3115
Chhattisgarh	899	0	30	575	62	1565
Goa	4	0	8	3	1	16
Gujarat	771	0	45	3303	114	4233
Haryana	1206	0	0	2066	0	3273
Himachal Pradesh	4	0	0	29	84	117
Jammu & Kashmir	183	114	7	8	17	329
Jharkhand	4	0	66	103	58	231
Karnataka	1193	0	138	2136	565	4032
Kerala	83	1	49	167	105	404
Madhya Pradesh	1917	0	346	7718	1368	11,349
Maharashtra	1047	0	0	2098	0	3145
Manipur	0	0	0	0	72	72
Meghalaya	75	25	0	0	0	101
Mizoram	2	14	0	0	0	16
Nagaland	0	0	0	0	118	118
Odisha	0	0	0	0	1092	1092
Punjab	1167	0	0	2944	0	4111
Rajasthan	2017	0	35	6069	162	8283
Sikkim	0	0	0	0	15	15
Tamil Nadu	636	0	322	1603	5	2565
Telangana	425	0	240	1447	98	2210
Tripura	6	0	2	8	70	87
Uttarakhand	79	1	0	222	20	323
Uttar Pradesh	2183	0	80	12,012	117	14,392
West Bengal	0	0	0	0	3107	3107
Andaman & Nicobar Islands	0	0	0	0	0	0

(continued)

Table 1.7 (continued)

State	From government canals	From private canals	Tanks	Tube wells and other wells	Other sources	Total
Chandigarh	0	0	0	0	0	0
Dadra & Nagar Haveli	0	0	0	1	0	2
Daman & Diu	0	0	0	0	0	0
Delhi	2	0	0	19	1	22
Lakshadweep	0	0	0	0	0	0
Puducherry	4	0	0	9	0	13
Total	16,264	165	1668	45,748	7706	71,555

Source Ministry of Agriculture and Farmers Welfare (2021)

Table 1.8 Area and percentage share of different irrigation sources (m ha)

Source of Irrigation	1950–51		1999–00		2018–19	
	Area	%	Area	%	Area	%
Canals	8.30	39.78	17.45	36.34	16.43	22.96
Tanks	3.61	17.33	2.94	6.13	1.67	2.33
Wells	5.98	28.67	24.69	51.42	45.75	63.93
Other sources	2.97	14.23	2.93	6.11	7.71	10.77
Total	20.85	100.00	48.02	100.00	71.55	100.00

Source Ministry of Agriculture and Farmers Welfare (2021)

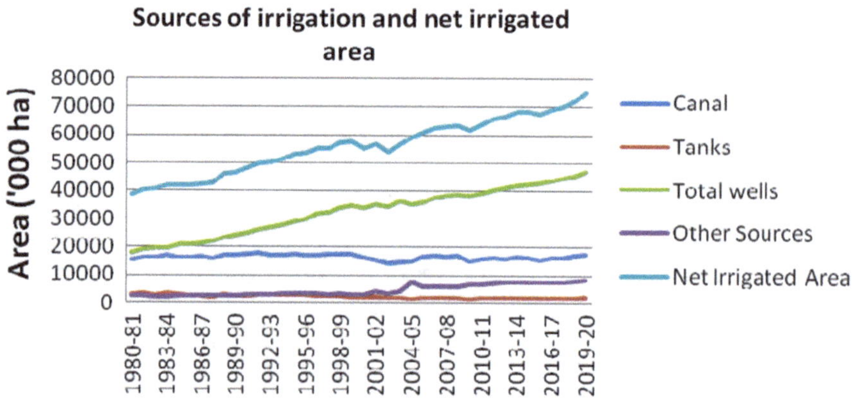

Fig. 1.5 Sources of irrigation and net irrigated area

According to World Bank's recent report, groundwater has been undervalued, overexploited, and mismanaged. Growing depletion levels reduce buffering capacity in the face of higher temperatures and more variable precipitation associated with climate change (Rodella et al. 2023).

(E) **Tank Irrigation**

In terms of irrigated area, the tank irrigated area in India has declined from 3.3 m ha in 1953 to 1.67 m ha in 2018 (about 50% reduction). The share of tank irrigated area to the total irrigated area has also reduced from 15.6% in 1953 to 2.3% in 2018. Studies (Palanisami and Easter 2000) have identified many reasons for this decline, such as illegal encroachment in the tank foreshore, accumulation of silt in the tank basin bed, poor water management, inadequate O&M budget, declining rainfall pattern, poor participation by farmers in tank management, and lack of effective tank management policy at government level. As a result of continuous decline in tank irrigation, the productivity of land under tanks is getting low compared to other sources of irrigation and hence the yield gaps under the tank irrigation are increasing over the years (Palanisami 2022). Tank rehabilitation measures are being taken up by some state governments to improve tank irrigation systems, but they are not adequate enough to improve the tank performance.

(F) **Water Use Efficiency and Technology Adoption**

A CWC study had indicated that the average efficiency of the major and medium projects is only about 28%. Among the efficiencies, average conveyance efficiency is about 69% and on-farm application efficiency is only about 55% indicating the need for improving the efficiency by 60% in the future (CWC 2017). Even though improved soil, crop, and water management technologies are available, the adoption level is only 8–10% that too with high variations in adoption between regions and crops (Palanisami et al. 2015). It is high time that climate-smart technologies, which could sustain the future of agriculture and irrigation sectors are identified for up-scaling and main-steaming in a systematic manner. What is required is a transformative shift in investments in climate-smart technologies that will enhance water security and agricultural productivity in India.

(G) **Micro Irrigation**

The use of drip irrigation systems gives an average water saving of up to 47% and average electricity saving of up to 37% for various crops which results in higher yields under different agro-climatic conditions when compared to conventional surface irrigation methods. The total potential of micro irrigation in India is estimated at around 72 m ha (Narayanamoorthy 2006). However, currently (2019–20) the coverage of micro irrigation is only 12.51 m ha which accounts for only about 17.4%. Among the drip and sprinkler coverage, it is about 28.2% and 12.9%, respectively. The major issues in up-scaling the MI systems are: (a) high initial costs, (b) inefficiency subsidy distribution mechanisms, (c) poor technical know-how of the farmers, and (d) poor maintenance of the systems. Further, effective MI implementation models are lacking in most of the states, and subsidy disbursement is facing problems, which may hinder the up-scaling of the MI. Hence there is a need to relook at models like GGRC, which has proven to be successful.

(H) **Watershed Development and the Challenges**

Watersheds are mainly an integral part of the rainfed ecosystems. They are drainage units, delineated by topographic features of the landscape. Since the start of the watershed programmes several decades ago, several watersheds across the country have been treated successfully. For example, the Integrated Watershed Management Programme (IWMP) has covered an area of about 39.07 m ha from 2009–10 to 2014–15. Further, through the Watershed Development Component (WDC) of Pradhan Mantri Krishi Sinchayee Yojana (PMKSY) scheme, an area of 1.64 m ha has been brought under additional protective irrigation (2015–16) (MoRL 2022). However, over the years the watershed programmes have faced a lot of challenges at the organizational and managerial levels (Palanisami and Sureshkumar 2005; Palanisami et al. 2009). Major challenges were: (a) difficulties in the impact assessment of the management interventions, due to the complex nature of interactions between upstream (causes) and downstream (impacts), (b) conflicts in areas of decision-making, where the watershed management activities are linked with the stakeholders and the overall land-use systems are linked with administrative units, (c) upstream and downstream conflicts, where, there are disagreement between upstream and downstream stakeholders in deciding the type of management interventions, (d) neglect of post-project maintenance where particularly many of the small and medium watersheds are neglected after the management interventions are completed resulting in silting up of the structures such as check dams and percolation ponds. Under such situations, the groundwater recharge is much limited and evaporation losses are higher from the surface water bodies, and (e) conflicts between tank management and watershed interventions, where in many cases watershed activities are introduced in the catchment areas of the tanks resulting in poor inflows to the tanks. Hence there is a need for redesigning the watershed interventions through a holistic approach avoiding conflict with the existing irrigation systems (Palanisami 2022).

(I) **Hill Agriculture**

Agriculture in the hill region of India has considerable potential to grow and contribute to the overall economic growth and livelihoods of the people. India has about 16.2% of total geographical area under the Indian Himalayan Region covering 10 states. Typical characteristics of the hill agriculture include rainfed farming, land fragmentation, dispersed holdings, and terraced farming, where failure of crop in one location due to micro-climatic conditions may not affect the rest of the holdings.

Trends in area and yield of major crops in the hill regions have clearly shown that area and yield increase over the past decade is not significant indicating the subsistence nature of hill agriculture. Most of the region is expected to have a 4% decline in yields. In the context of water management in the hilly regions, the key factor that must be addressed is sustainability. This is particularly apparent in designing and constructing the irrigation structures that suit the cropping systems of the hilly environments. Hence, efficient utilization and management of available rainwater in the hill regions are the core issues if the cropping intensity and production are to be enhanced for ensuring the food security and livelihood options of the households in the region as much of the enormous water resources in hilly areas remain unutilized. The integrated watershed management concept, where community-based sustainable natural resource management is the central theme in hill areas, not only takes a holistic view of the fast-depleting water resources but also helps in augmenting the resources through runoff management and harvesting techniques thereby ensuring groundwater/spring recharge in the hills.

(J) **Water Pricing and Cost Recovery**

In India, irrigation water pricing is mostly based on crop area due to the impracticality of volumetric-based charging and failed to realize water pricing policy targets regardless of different pricing structures (Palanisami et al. 2015). According to the Irrigation Commission of India, reasonable irrigation water rates should be 5% of gross income for food crops and 12% for cash crops but actual receipts have only been about 2.9%. Percentage recovery of working expenses of irrigation projects via irrigation water charges (gross receipt) has been consistently poor over time nationwide (<20% recovery) (CWC 2017). This is due to a lack of: (a) farmers' participation in the irrigation project, (b) transparency between farmers and irrigation management authorities, (c) transparency in water fee collection, (d) agro-power subsidies, (e) rampant water piracy, and (f) poor irrigation service delivery (Chaudhuri and Roy 2019). Also the irrigation subsidy calculations are upward biased in multipurpose projects where other uses like domestic, industry, fisheries, and power generation are not priced proportionate to the water used and benefits generated (Palanisami et al. 2012).

(K) **Weak Crop Water Requirements Data**

As agriculture sectors account for more than 70% of the total water use, a better understanding of agriculture water use particularly in developing countries like India is necessary to manage and utilize the irrigation projects more effectively. In the recent years, there is a growing need to derive the irrigation water requirement of crops using scientific methods incorporating climate variables as the current water allocations from irrigation projects are mainly based on aggregate crop water requirements adjusted for project efficiencies. These estimates were derived from limited field data and broad assumptions, which couldn't represent the actual crop water requirements. Hence, such allocation can be biased resulting in low water use efficiencies. As an improvement over these estimates, consumptive crop water requirements need to be worked out based on climatological considerations for all the irrigation projects in the country. Even though, estimation of crop water demand is becoming complicated as agriculture sector faces location-specific issues in terms of varying weather parameters in a given time frame, it is possible to estimate and validate the CWR estimation procedure, as the needed data for such model estimation is now available, and further the estimation methods are also scientific. Hence, estimation of crop water demand using scientific approach with strong database is crucial while analysing the performance of the various water supply systems.

(L) **Water Quality and Waste Water**

In India, the use of untreated wastewater or poor quality water in agriculture is gaining importance only in the recent years but without any regulation resulting in human health and soil health issues (Raychaudhuri et al. 2014). The wastewater reuse mainly depends on the quality of the effluent available, the crop to be irrigated, the potential risk to the health of the workers, and the skill of the farmers to handle this type of water. A more comprehensive set of water quality guidelines needs to be created incorporating the present understanding of the occurrence and effects of emerging contaminants.

The recycled wastewater is crucial for ensuring a steady supply of water for various uses in India. Industries primarily obtain their water from surface water (41%) followed by groundwater (35%) and municipal water (24%). Urban and peri-urban areas have a limited supply of it (FICCI 2011). Therefore, wastewater in India needs to be reused. Wastewater reuse if done properly could decrease the demand for freshwater by 20–50% (Government of Andhra Pradesh 2019) depending on the extent of the treatment and reuse possibilities. It is observed that about 35% of India's total population is concentrated in urban centres and the estimated wastewater generation is 72,368 MLD (million litres per day) compared to 39,604 MLD in the rural sector for the year 2020–21. The current installed sewage treatment capacity is 31,841 MLD, but the operational capacity is 26,869 MLD which is only about 26% of the actual quantity of wastewater generated and about 74% of wastewater which is left untreated (CPCB 2022). Therefore, a suitable wastewater reuse policy is a prerequisite for effective wastewater management, which is currently lacking in India.

(M) **Climate Change on Agriculture and Irrigation**

India is also identified as one of the most highly vulnerable countries to climate change (INCCA 2010; Cruz et al. 2007). The projections of the Coupled Model Intercomparison Project (CMIP) for India show that the average climate will be warmer by 1.7 to 2.00 C for 2030s and by 3.3 to 4.80 C for 2080s and the rainfall will increase by 5 to 6 per cent and 6 to 14 per cent for 2030s and 2080s, respectively (Chaturvedi et al. 2012). This may affect the spatial and temporal distribution of runoff, soil moisture, groundwater reserves, etc., and may increase the frequency of droughts and floods. The future climatic change, though, will have its impact globally but likely to be felt severely in developing countries like India (Pathak et al. 2014). As climate change aggravates all other problems such as land degradation, market volatility, rising input costs, and slowing response to added inputs that hinder agricultural growth, it is recognized as a potent threat to the sustainability of agriculture.

(N) Water Institutions, Policy, and Governance

Although the central government provides financial resources to the state governments for implementing national-level projects, water governance in India is mostly decentralized at the state level. The states are responsible for the development and management of water resources within their administrative/physical boundaries (Ahmed and Araral 2019).

Currently, the institutional aspects like weaker legal provisions, conflict resolution mechanisms, and water transfer policies and the state water Acts have mostly failed to effectively protect and manage the water sector in India. The tribunals established to solve the interstate water allocation issues are not effective and take long time to implement the decisions. The development activities in these disputed basins are almost negligible. At the same time, the institutional arrangements governing water resource development, allocation, and management have received increasing policy attention in the recent years (Saleth 2004; Neto 2016). Hence, for better management and effective use of water resources, necessary institutional changes are required (Ashish and Parthasarathy 2021).

The National Agricultural Policy (GOI 2000) and National Water Policy (GOI 2012) clearly mention the importance of using the water resources effectively to maximize the crop and water productivities (PIB 2014). Despite their strengths, there are also weaknesses and threats in implementation such as lack of targets and time frame for various activities. It is argued that India lags in water governance due to the poor capacity of Indian states in handling the water issues, the complexity of the Indian decision-making system, and conflicts between states over water allocation (Araral and Ratra 2019). It is expected that the Government of India needs to work in close coordination with state governments to realize the target benefits where GOI's initiatives can be directly linked to SDG6 focusing on water and sanitation-related goals (RIS 2016).

1.5 India's Future Water Demand and Supply

The water resources potential of India has been assessed by the Central Water Commission (CWC) as 1999 billion cubic metres (bcm). As regards to utilization, only about 1122 bcm of the total annual water potential can be used beneficially. Out of which 690 bcm will be from surface water sources and 432 bcm from groundwater sources (CWC 2020). But this annual water potential cannot be realized in the future due to mismanagement of water resources, climate change, over pumping, and pollution. Hence, India's accessible and reliable supply of water amounts only to 744 bcm (Water Resources Group 2009). On the other hand, the demand for water is considerably increasing due to increasing population, and demands from agriculture, domestic, and industrial use. The Standing subcommittee of MoWR has projected the future water demand in India as 910 bcm in 2025 and 1072 bcm in 2050.

To meet the growing water demand, among other things, the groundwater withdrawal is expected to increase to 365 bcm and 423 bcm by 2025 and 2050, respectively. Hence, the augmentation of groundwater supplies is a key driver of agricultural production and water demand growth in the future. With growing demand for water and depletion of the available water sources, an assured supply of good quality water is becoming a growing concern. Therefore, sustainable development and efficient management of water resources is an increasingly complex challenge wherein the role of demand management strategies is gaining importance in bridging the future supply–demand gap.

1.6 Water Future of India—A Digression

Given the above issues and the challenges the Indian water sector is facing now that will continue in the future, the following chart (Fig. 1.6) explains the interventions that need to be focused on in the future to sustain the water sector in general and irrigation sector in particular.

Three phases are important in the development of agriculture and water sector in India. Phase 1 is mostly about the investment/development, phase 2 is about the management of the water resources, and phase 3 deals with the governance where overall return to investment and management can be maximized. The water governance incorporates a range of political, social, economic, and administrative systems mainly to develop and manage water resources. Thus, water governance is important for critical transition from development mode to management mode. It is clear from the chart that climate change is an emerging challenge in the agriculture and water sectors. Future investments need to focus on three main areas: (a) irrigation sector

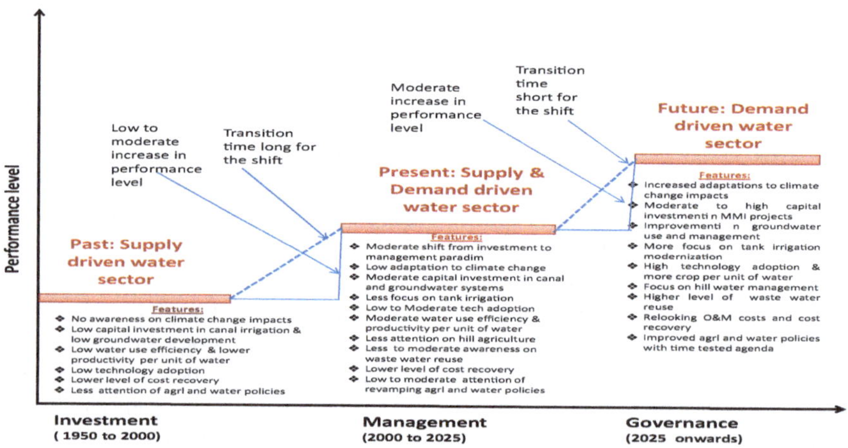

Fig. 1.6 Phases of the water sector in India. *Source* authors' projection

modernization programmes covering surface and groundwater systems to improve the overall water use efficiency, (b) up-scaling the technology adoption to increase the productivity per unit of water, and (c) better governance in agriculture, marketing, and water policies. Hence, a major shift in India's water strategy is needed from the past near exclusive reliance on mere irrigation expansion to a strategy emphasizing improving the performance of irrigation and irrigated agriculture through better governance. In this context, the chapters in this book are planned in such a way that most of the issues/challenges facing India's water future can be adequately addressed.

Box 1.1 Why This Book Is Important

- Discusses the major drivers of water demand & climate change
- Documents the impact of climate change on water sector and the role of climate-smart technologies
- Covers holistically the investment, managementm and governance aspects of water sector
- Addresses hill water management and investment options
- Shows the methodology & re-estimates O&M cost in multiple purpose projects for actual subsidy estimation
- Suggests modernization in a reformed way using Support for Irrigation Modernization Program (SIMP) approach
- Presents a futuristic view of micro irrigation expansion
- Demonstrates the improved crop water requirement estimation using modern approach
- Explains the national and international level success stories relevant to Indian situations

1.7 Outline of the Book

The book's primary purpose is to document the key issues and challenges in detail and prescribe solutions to address the same to enhance the efficiency of water use in irrigated agriculture, as irrigation sector alone accounts for 84% of the total water use in the country. Further, the issues related to irrigated sector are directly linked to issues related to water sector. Issues related to other sectors viz., domestic, industry, and other sectors are accounting for a relatively less share of total water use in the country now and in the future as well and hence the issues related to irrigation are mostly focused in this book. Among other sectors, domestic sector can play a major role in the future in view of the increasing population and the corresponding water needs and one chapter is exclusively focused on wastewater reuse. In addition, hill agriculture is getting importance in the future because of the potential for exploiting the rainwater harvesting and supplemental irrigation to sustain the crop yields and production. Hence, the book is aptly titled as India's water future in a changing climate. The importance of this book is given in Box 1.1.

Table 1.9 provides a list of challenges, the possible solutions that will be dealt in different chapters of the book.

Table 1.9 Key issues, possible solutions, and the book chapters covering them

No	Key issues or challenges	Possible water future solutions/strategies	Chapter
Part I: Water sector			
1	• Rainfed agriculture has risks and uncertainties and more reliance on irrigated agriculture in the future • Irrigated agriculture faces low productivity and low profitability issues where climate change impact is visible	• Focus areas in strengthening irrigated agriculture and adoption of climate-smart practices	1 Agriculture and irrigation-trends and turning points
2	• Drivers such as changing climate differ over the years. Several drivers including climate change affect the water supply, water demand, and agricultural productivity	• Strategies to address the drivers linking the future water supply and demand • Strategies to address the climate change impact linking agricultural productivity and adaptation mechanisms	2 Water future-drivers of change
Part II: Water management and solutions			
3	• The gap between potential created and utilized increases over the years • Poor project efficiencies affect the water allocation and management • Water management technology adoption is very low • Poor cost recovery	• Canal irrigation modernization through structural and non-structural interventions • Technologies to improve the overall performance of the major and medium irrigation projects with capacity building for better adoption and to improve the overall project efficiency • Introduction of micro irrigation in canal commands facing continuous water shortages	3 Stabilizing the canal irrigation

(continued)

Table 1.9 (continued)

No	Key issues or challenges	Possible water future solutions/strategies	Chapter
4	• Overexploitation of groundwater and well failure • Increased cost of well investment and energy for pumping • Poor recharge options/ strategies • Water–energy nexus	• Water–energy nexus, watershed programmes, and artificial recharge of groundwater aquifers • Making micro irrigation mandatory in groundwater over-exploited regions • Reducing subsidies in critical areas where groundwater extraction is no longer feasible • Use of solar power for irrigation and conjunctive use of surface and groundwater	4 Sustaining groundwater irrigation
5	• Declining tank irrigated area over the years due to siltation and encroachment • Farmers abandonment of tank-based agriculture • Ineffective tank-based institutions	• Tank modernization with a participatory approach • Providing groundwater supplementation to stabilize the tank water • Crop diversification and enhanced multiple uses from tank systems	5 Reviving tank irrigation
6	• Quantum of wastewater generated from domestic and other uses has increased over the years contributing to pollution and water quality problems • Low level of wastewater reuse • Lack of effective wastewater reuse policy	• Monitoring of water quality changes and the pollution load in irrigation water • Treatment of wastewater for reuse in agriculture and other uses. Intensification research on the effective use of treated wastewater for different crops • Development and implementation of wastewater treatment and reuse business models involving public and private sector participation • Framing state-wise wastewater reuse policies for implementation. Increasing awareness of water users	6 Maximizing wastewater reuse

(continued)

Table 1.9 (continued)

No	Key issues or challenges	Possible water future solutions/strategies	Chapter
7	• Lack of appropriate and affordable technologies/ practices • Low water productivity per unit of water • Low level of technology adoption	• Cafeteria of irrigation technologies with their economics • Strategies for easy access to and adoption of improved soil, crop, and water management technologies • Strong capacity building and training programmes • Framing policies for implementation of effective technology adoption and up-scaling • Digital agriculture	7 Enhancing water productivity and climate-smart technologies
8	• Current data on water requirement of crops are outdated • Water allocation to crops is not need-based requirements	• Use of improved/scientific method of re-estimation of crop water requirements with climate variables • Reworking delta for all crops • Adoption of the improved method in future planning	8 Re-estimating crop water requirements
9	• Poor adoption of micro irrigation resulting in low area coverage under micro irrigation • Lack of knowledge in use of the micro irrigation systems and their maintenance • Ineffective subsidy implementation programmes	• Creating awareness about operation and maintenance of the micro irrigation systems including drip fertigation • Low-cost micro irrigation designs for different crops • Suitable subsidy implementation and micro irrigation expansion policies	9 Up-scaling micro irrigation
10	• Hill agriculture in general and hill water management in particular are neglected due to their locations and other socio-economic characteristics of the region • Climate change affects hill agriculture	• Creating awareness on climate-smart hill water management practices • Implementation of suitable implementation policies including capacity building programmes	10 Expanding hill agriwater management

(continued)

Table 1.9 (continued)

No	Key issues or challenges	Possible water future solutions/strategies	Chapter
11	• O&M charges in multipurpose irrigation projects are high as other uses are not accounted properly • Subsidy calculations in these projects are biased	• Reworking the O&M costs related to multipurpose irrigation projects giving due weightage to other uses or benefits such as domestic, industry use, power, and fisheries • Reworking subsidy and cost recovery	11 Reworking on cost sharing and irrigation subsidies
12	• Lack of awareness about success stories of irrigation project operation, maintenance, and capacity building hinder the best investment options • Application of advanced tools/methodologies in studying project performance is limited	• Preparation of the list of success stories and case studies of improved water sector investment, management, and performance using both national and international cases • Introduce training and capacity building programmes to stakeholders to follow up with the implementation of success stories/models	12 Learning from success stories and best practices in water management
Part III: Policy			
13	• The existing water institutions, water policies, and governing mechanisms are inadequate to cope with the future water sector investment and management in the country	• National Water Policy needs to be refocused in terms of its relevance and applicability with the inclusion of other relevant areas • The existing water institutions need to be strengthened in terms of their roles and responsibilities • A revamped governance mechanism be established/ adopted	13 Strengthening institutions, policies, and governance
14	• In total, a comprehensive water future outlook is lacking in the irrigation sector of India	• A water futures outlook for India incorporating successful options on water sector investment, management, and governance	14 Water future—A way forward

References

Ahmed M, Araral (2019) Water governance in India: evidence on water law, policy, and administration from eight Indian States. Wat 11:2071. https://doi.org/10.3390/w11102071.

Amanullah, Khalid S, Imran, Khan HA, Arif M, Altawaha AR, Adnan M, Fahad S, Shah A, Parmar B (2020) Effects of climate change on irrigation water quality. In: Fahad S, Hasanuzzaman M, Alam M, Ullah H, Saeed M, Khan IA, Adnan (ed) Environment, climate, plant and vegetation growth. Springer, Cham. https://doi.org/10.1007/978-3-030-49732-3_6

Amarasinghe UA, Shah T, Turral, H Anand BK (2007) India's water future to 2025–2050: business-as-usual scenario and deviations. Research Report 123, International Water Management Institute, Colombo. https://iwmi.cgiar.org/Publications/IWMI_Research_Reports/PDF/PUB123/RR123.pdf

Araral E, Ratra S (2016) Water governance in India and China: comparison of water law, policy and administration. Wat Pol 18:14–31

Ashish C, Parthasarathy R (2021) Understanding water institutions and their impact on the performance of the water sector in India. Wat Pol 23:466–486

Bjornlund V, Bjornlund H, Van Rooyen AF (2020) Exploring the factors causing the poor performance of most irrigation schemes in post-independence sub-Saharan Africa. Inter J Wat Res Dev 36(sup1):S54–S101. https://doi.org/10.1080/07900627.2020.1808448

Census (2011) Government of India. https://censusindia.gov.in/census.website/data/census-tables

Central Water Commission (CWC) (2022) Annual Report (2021–22) Department of Water Resources, River Development and Ganga Rejuvenation, Ministry of Jalshakti. New Delhi. https://cwc.gov.in/sites/default/files/annual-report-21-22.pdf

Central Ground Water Board (2022). http://cgwb.gov.in/documents/2022-11-11-GWRA%202022.pdf

Chaturvedi RK, Joshi J, Mathangi Jayaraman G, Bala, Ravindranath NH (2012) Multi-model climate change projections for India under representative concentration Pathways. Curr Sci 103(7):1–12

Chaudhuri S, Roy M (2019) Irrigation water pricing in India as a means to conserve water resources: challenges and potential future opportunities. Envir Conser 46(1):99–102. https://doi.org/10.1017/S037689291800036X

Chauhan MK, Shobha R (2023) Rehabilitation of canal irrigation schemes in India: a qualitative analysis. Wat Poli 25(1):59. https://doi.org/10.2166/wp.2022.237

Chowdhury K, Behera B (2018) Is declining groundwater levels linked with the discontinuity of traditional water harvesting systems (tank irrigation)? empirical evidence from West Bengal, India. Ground Sustain Dev 7:185–194. https://doi.org/10.1016/j.gsd.2018.05.007

CPCB (2022) Status of water supply, wastewater generation and treatment in class I cities and class II towns of India. Series: CUPS/70/2009–10. Central Pollution Control Board, India

Cruz RV, Harasawa H, Lal M, Wu S, Anokhin Y, Punsalmaa B, Honda Y, Jafari M, Li C, Ninh NH (2007) Asia, climate change 2007: impacts, adaptation and vulnerability. In: Parry ML, Canziani OF, Palutikof JP, Van der Linden PJ, Hanson CE (eds) Contribution of working group ii to the fourth assessment report of the intergovernmental panel on climate change. Cambridge University Press, Cambridge, United Kingdom, pp 469–506

Central Water Commission (CWC) (2017) Pricing of water in public system in India. Information systems organisation—Water planning and projects wing. Retrieved April 27, 2020. http://cwc.gov.in/sites/default/files/Pricing%20of%20Water%20in%20Public%20System%20in%20India%202017.pdf

Deshpande RS (2022). Under the shadow of development: rainfed agriculture and droughts in agricultural development of India. NABARD Research and Policy Series No. 2. Department of Economic Analysis and Research. https://www.nabard.org/auth/writereaddata/tender/2007223429Paper-2-%20Rainfed-Agriculture-Dr.-Deshpande.pdf

Federation of Indian Chambers of Commerce and Industry –FICCI (2011) Water use in Indian industry survey, FICCI Water Mission, New Delhi

GOI (2000) National Agricultural Policy, Ministry of Agriculture, New Delhi

GOI (2012) National Water Policy, Ministry of Water Resources, New Delhi. https://nwm.gov.in/sites/default/files/nwp20025617515534.pdf

Government of Andhra Pradesh (2019) Action plan for usage of treated waste water from sewage treatment plants. Municipal Administration and Urban Development Department, Andhra Pradesh, India

Halder S, Roy MB, Roy PK (2020) Analysis of groundwater level trend and groundwater drought using Standard Groundwater Level Index: a case study of an eastern river basin of West Bengal India. SN Appl Sci 2:507. https://doi.org/10.1007/s42452-020-2302-6

https://wdcpmksy.dolr.gov.in/aboutus

ICAR (2015) Vision, 2050, Government of India. https://icar.org.in/files/Vision-2050-ICAR.pdf

IIT (2022) Water reuse-a vital step towards water value, Roorkee

Indian Network for Climate Change Assessment (INCCA) (2010) Greenhouse gas emissions. Ministry of Environment and Forests, Government of India

Joshi PK, Jha AK, Wani SP, Laxmi J, Shiyani RL (2005) Meta-analysis to assess impact of watershed program and people's participation.Research Report 8, Comprehensive Assessment of watershed management in agriculture, International Crops Research Institute for the Semi-Arid Tropics and Asian Development Bank, pp 21

Joshi PK, Pangare V, Shiferaw B, Wani SP, Bouma J, Scott C (2004) Socioeconomic and policy research on watershed management in India: Synthesis of past experiences and needs for future research. Global Theme on Agroecosystems Report no.7. Patancheru 502324, Andhra Pradesh, India: International Crops Research Institute for the Semi-Arid Tropics, pp 88

Mali SS, Shirsath PB, Islam AA (2021) High-resolution assessment of climate change impact on water footprints of cereal production in India. Sci Rep 11:8715. https://doi.org/10.1038/s41598-021-88223-6.

Minhas PS, Jayanta KS, Dotaniya ML, Abhijit S, Madhumonti S (2022) Wastewater irrigation in India: current status, impacts and response options. Sci of the Total Environment 808:152001

Ministry of Agriculture and Farmers Welfare (2021) Land Use Statistics at a Glance, 2010–11 to 2018–19, Directorate of Economics and Statistics, Government of India. https://eands.dacnet.nic.in/LUS_2018-19/LAND%20USE%20STATISTICS%20AT%20A%20GLANCE%202009–10%20to%202018–19.pdf

Ministry of Agriculture and Farmers Welfare (MoA and FW) (2022) Annual Report 2022–23, Ministry of agriculture and farmers welfare, Government of India. https://agricoop.nic.in/Documents/annual_report_english_2022_23.pdf

Ministry of Agriculture and Farmers Welfare (2022) Agricultural Statistics at a Glance 2021. Directorate of Economics and Statistics. Government of India. https://agricoop.nic.in/Documents/CWWGDATA/Agricultural_Statistics_at_a_Glance_2021_1.pdf

Ministry of Agriculture and Farmers Welfare (2023) Land use statistics at a glance, 2010–11 to 2019–20 Directorate of Economics and Statistics, Government of India. https://eands.dacnet.nic.in/LUS_2019-20/Cover%20page.pdf

Ministry of Rural Development (MoRL) (2022) Notes on Demands for Grants, (2023–2024) 305 No. 88/Department of Land Resources. https://www.indiabudget.gov.in/doc/eb/sbe88.pdf

Ministry of Jal Shakti (2023) Government of India. https://pib.gov.in/PressReleaseIframePage.aspx?PRID=1895757

Narayanamoorthy A (2006) Potential for drip and sprinkler irrigation in India, Gokhale Institute for Politics and Economics, Pune

Narayanamoorthy A (2022) Is investing in canal irrigation a bad idea? Hindu Business Line. The writer is a former Full-Time Member (Official), Commission for Agricultural Costs and Prices, New Delhi. https://www.thehindubusinessline.com/opinion/is-investing-in-canal-irrigation-a-bad-idea/article65802538.ece

NareshK APK, Swaroopa R, Surabhi J, Rani S, Nitin C (2011) Impact of climate change on crop productivity in Western Ghats, coastal and northeastern regions of India. Curr Sci 101(3):332–341

Neto S (2016) Water governance in an urban age. Utili Pol 43:32–41

Palanisami K, Easter KW (2000) Tank Irrigation in the 21st century-what next? Discovery Publications, New Delhi

Palanisami K, Suresh Kumar D (2005) Impact assessment of watershed development – issues, methods and experiences. Associated Publishing Company, New Delhi

Palanisami K, SureshKumar D, Wani SP (2009) A Manual on impact assessment of watersheds. global theme on agro-ecosystems, Report no. 53, ICRISAT, Patancheru, Hyderabad

Palanisami K (2022) Tank irrigation in India: future management strategies and investment options. NABARD Research and Policy Series No. 10/2022. Department of Economic Analysis and Research. National Bank for Agriculture and Rural Development, Mumbai. https://www.nabard.org/auth/writereaddata/tender/2911220521paper-10-tank-irrigation-in-india-by-dr-palanisami.pdf

Palanisami K, Kakumanu KR, Malik RPS (2015) Water pricing experiences in India: Emerging issues. In: Dinar A, Pochat V, Albiac-Murillo J (eds) Water pricing experiences and innovations. Global Issues in Water Policy. Springer, Cham, p 9

Palanisami K, Raman S, Mohan K (2012) Micro irrigation- economics and outreach. MacMillan Publishers India, New Delhi

Palanisami K, Shiferaw B, Joshi PK, Nedumaran S, Suhas P, Wani (2011) Impact of watershed projects in India: application of various approaches and methods. Chapter 11

Integrated Watershed Management in Rainfed Agriculture, pp 349. https://doi.org/10.13140/2.1.4930.3684

Pathak H, Pramanik P, Khanna M, Kumar A (2014) Climate change and water availability in Indian agriculture: impacts and adaptation. Ind J Agri Sci 84 (6):671–679

Press Information Bureau (PIB) (2014) National Water Policy 2012. Ministry of Water Resources. Government of India

Rodella A-S, Zaveri E, Dilip; Bertone, Francois Maurice Denis Charles. (2023) The Hidden Wealth of Nations: The Economics of Groundwater in Times of Climate Change. World Bank, Washington, D.C.

Raychaudhuri M, Raychaudhuri S, Jena SK, Ashwani Kumar, Srivastava RC (2014) WQI to Monitor Water Quality for Irrigation and Potable Use. DWM Bulletin No. 71 Directorate of Water Management, Bhubaneswar, India, p 43

Research and Information System for Developing Countries (RIS) (2016) India and sustainable development goals: the way forward. http://ris.org.in/sdg/india-and-sustainable-development-goals-way-forward

Saleth MR (2004) Strategic analysis of water institutions in India: application of a new research paradigm. Research Report 79. Colombo, Sri Lanka: International Water Management Institute

Sarangi A, Rajender Kumar, Kaur R (2016) Irrigation management, Agriculture World. 20. https://issuu.com/krishijagran1/docs/aw_final_25-10-2016/20

Sanjeevaiah SH, RudrappaKS LMT, Huggi L, Hanumanthaiah MM, VenkatappaSD LingegowdaN, Sreeman SM (2021) Understanding the temporal variability of rainfall for estimating agro-climatic onset of cropping season over south interior Karnataka. India. Agron. 11:1135. https://doi.org/10.3390/agronomy11061135

Shiferaw B (2021) Addressing groundwater depletion: lessons from India, the world's largest user of groundwater. Independent evaluation Group (IEG). https://ieg.worldbankgroup.org/blog/addressing-groundwater-depletion-lessons-india-worlds-largest-user-groundwater

The Economic Times (2020) Retrieved on July 10, 2020. https://economictimes.indiatimes.com/news/economy/agriculture/govt-on-track-to-doubling-farmers-income-by-2024-agriculture-ministry-official/articleshow/76897730.cms?from=mdr

Chapter 2
Water Future—Drivers of Changing Climate and Impacts

Abstract The key drivers of change in water demand are climate change, population growth, and growing water scarcity. Other drivers include changing diets and food demand, competition between sectors, deteriorating water quality, and falling groundwater levels. The gross irrigated area in India will increase from 76 to 117 M ha in 2050. The gross groundwater irrigated area will increase to 60 M ha by 2025 and to 70 M ha by 2050. The total food requirement will be between 382 and 449 million tonnes by 2050. Annual irrigation requirement will increase from 4.63 to 20.46% in 2050 and climate change will reduce major crop yields by 4.5 to 9%. The net irrigation requirements are expected to increase by another 5% and gross irrigation requirements by 10%. It is expected that the efficiency of groundwater irrigation would increase to 75% by 2050 and surface irrigation projects would also increase to about 60%. Efficient and integrated water management and implementation strategies are expected to bridge the future water supply–demand gap.

Keywords Drivers of change · Future water supply and demand · Trends in area · Production · Productivity of crops · National Action Plan on Climate Change (NAPCC) · Climate change and water · Sahi-Fasal climate change programme · National Water Mission

2.1 Introduction

Drivers relating to climate change refer to the elements that directly or indirectly contribute to the growing greenhouse gas (GHG) emissions. The term "driver" is used here mainly to indicate an association to provide insights on what constitutes overall changes in global GHG emissions. The main drivers of GHG emissions include consumption, international trade, population, economic growth, structural change to a service economy, and energy consumption and they are interlinked with each other (IPCC 2014). In the water sector, both climate and non-climate drivers influence the water resources in general and irrigated agriculture in particular. Fluctuations in temperature and rainfall variability are expected to be the important drivers

K. Palanisami and U. S. Nagothu, *India's Water Future in a Changing Climate*, Advances in Geographical and Environmental Sciences,
https://doi.org/10.1007/978-981-97-1785-9_2

of change in natural and managed systems such as water resources. Heat extremes that are currently witnessed in several parts of the world will be even more a critical driver in the future for irrigation sector to manage. Physical and biological responses to changing temperatures are often better understood than responses to other climate parameters because the anthropogenic signal is easier to detect for temperature than for other parameters. For example, mean temperature and the seasonal cycle in temperature over relatively large spatial areas could show the clearest signals of change in the observed climate. Precipitation has much larger spatial and temporal variability than temperature thus making it difficult to quantify the impact with accuracy (IPCC 2001).

Non-climate drivers mainly include increasing urbanization and pollution which can influence both directly and indirectly through their effects on climate variables. Socio-economic processes such as land-use change (e.g., forestry to agriculture; agriculture to urban area), land-cover modification (e.g., ecosystem degradation or restoration), and increasing demand for food and water also affect the performance of different ecosystems over the years (Blanco et al. 2014) (Table 2.1).

Table 2.1 Direct and indirect effects of non-climate drivers

Non-climate driver	Examples	Direct effects on systems	Indirect effects on climate
Land-use change	Conversion of forest to agriculture (e.g., Lepers et al. 2004)	Declines in wildlife habitat, biodiversity loss, increased soil erosion, etc.,	Change in albedo, evapotranspiration, water, and heat balances
	Urbanization and transportation	Ecosystem fragmentation, biodiversity loss, deterioration of air quality, increased runoff and water pollution	Change in albedo, urban heat, local precipitation, wind, precipitation, evaporation
	Afforestation	Restoration or establishment of tree cover, soil improvements	Change in albedo, water and energy balances, potential carbon sequestration
Land-cover modification	Ecosystem degradation (desertification)	Reduction in ecosystem services, reduction in biomass, biodiversity loss	Changes in micro-climate
Pollution	Tropospheric ozone, toxic waste, oil spills, pesticides increased emissions	Reduction in breeding success and biodiversity, species mortality, health impairment, melting of snow and ice	Direct and indirect aerosol effects on albedo, temperature, and precipitation

Note albedo normally indicates the fraction of the solar radiation reflected by a surface and expressed as a percentage; aerosol refers the mixture of very small solid or liquid particles that can spread in the air; troposphere means the lowest part of the atmosphere, from the surface to about 10 km in altitude at mid-latitudes where clouds and weather phenomena occur

2.2 Drivers of Change

The following drivers of change are considered important while estimating the water demand (Tables 2.2 and 2.3). The projection of these drivers of change for 2030 and 2050 clearly indicates the magnitude of their influence on water demand in general and irrigated water demand in particular. The chapter provides a brief discussion of the drivers of change that will provide a better understanding of the linkages and their impacts (Figs. 2.1, 2.2, 2.3, 2.4, 2.5 and 2.6).

Population

There are a number of estimates and projections of future population growth in India. There are also short- and long-term population forecasts by Natarajan (1993), United Nations (1995, Revised estimates), Visaria and Visaria (1996), and Upali et al. (2007). It is expected that the growth in urban population will be more than rural population and will result in more domestic water demand due to high per capita water use. For example, if the urban population increases to 60% of the total population by 2050, as against the current 51%, the total irrigation and domestic water demand could possibly increase by another 10 Bm3 which will be about 2% of the total water demand (Upali et al. 2007) (Fig. 2.1).

Area Under Crops

The net sown area in the future is expected to remain at the present level of 142 million hectares (M ha) (Upali et al. 2007) due to constraints in expanding the net sown area. However, with growing food demand, the irrigation expansion is likely to continue and the gross irrigated area is expected to increase from 76 to 117 M ha during 2000–2050 (Fig. 2.2). Groundwater irrigation will be the major driver of irrigation expansion and the gross groundwater irrigated area is projected to increase to 60 M ha by 2025, and 70 M ha by 2050 which is possible with initiatives like artificial recharge.

Crop Productivity

The increase in crop yields over the years offers more scope for meeting the increasing demand for food and the resulting increase in irrigation water demand. It is expected that the increase in overall irrigated crop yield for major crops will be 3.4 t/ha and 4.0 t/ha in 2030 and 2050, respectively. It is possible to reduce the future irrigation water demand if the present level of increase in crop yield is maintained.

Food Demand

India produced 176.4 million tonnes of food grains during 1990–91 including 162.1 million tonnes of cereals and 14.3 million tonnes of pulses. During 2000–01, the total food grain production was 199 million tonnes, which has increased to 244 million tonnes by 2010–11 and further to 310.74 million tonnes during 2020–21, indicating the correlation between population growth and food grain production to meet the growing food demand.

Table 2.2 Driver of change and water demand, India in 2030 and 2050

No	Drivers of change	Unit	2030	2050
1	Population	Million	1389	1583
	Urban	Million	514	807
	Rural	Million	875	776
	Total**	Million	1286–1333	1346–1581
	Urban**	Million	476–603	646–971
	Rural**	Million	810–730	700–610
2	Area under crops (Irrigated)			
	Food crops	Million ha	95	99
	Rice	Million ha	32	33
	Wheat	Million ha	35	36
	Maize	Million ha	3	3
	Food grains	Million ha	78	81
	Sugarcane	Million ha	5	5
	Cotton	Million ha	5	5
	Non-food crops	Million ha	19	20
	Net sown area**	Million ha	144	145
	Irri area to GCA	%	45–53	52–64
	Irri food crops area to GIA	%	70	70
3	Production			
	Grains	Million tonnes	303	315
	Rice	Million tonnes	88	93
	Wheat	Million tonnes	120	127
	Maize	Million tonnes	9	10
	Food grains	Million tonnes	186	199
	Sugar	Million tonnes	405	402
	Cotton	Million tonnes	2	2
4	Productivity of crops			
	Rice	Kg/ha	2747	2796
	Wheat	Kg/ha	3480	3561
	Maize	Kg/ha	3039	3106
	Food grains	Kg/ha	2376	2444
	Sugarcane	Kg/ha	80,173	79,529
	Cotton	Kg/ha	446	428
	Irri crops**	t/ha	3.4	4.0
5	Demand			
	Rice	Million tonnes	109	117
	Wheat	Million tonnes	91	102

(continued)

Table 2.2 (continued)

No	Drivers of change	Unit	2030	2050
	Food grains	Million tonnes	291	377
	Sugarcane	Million tonnes	42	55
	Cotton	Million tonnes	4	6
	Food grain**	Million tonnes	280–291	382–449
6	Climate change*			
	Temperature variability	^{0}C	2.67	3.27
	Rainfall variability	mm/day	0.10	0.23
7	Irrigation efficiency			
	Surface water	%	3–55 50**	42–60 60**
	Groundwater	%	70 72**	75 75**
8	Depth or GIR			
	Groundwater	ha m	0.51	0.49
	Surface water	ha m	0.73	0.61

Source Population, food demand, and water demand are sourced from Upali et al. (2007). * *Source* Assessment of climate change over the Indian region, Ministry of Earth Science, GOI (2020).***Source* NCIWRD. Others are authors' estimates.

Table 2.3 India Water demand projections (BCM)*

S.NO	Sectors	MOWR		NCIWRD	
		2025	2050	2025	2050
1	Irrigation	910	1072	611	807
2	Drinking water	73	102	62	111
3	Industry	23	63	67	81
4	Energy	15	130	33	70
5	Other	72	80	70	111
	Total	1093	1447	843	1180

NCIWRD: National Commission on Integrated Water Resources Development
BCM: Billion Cubic Metres. *high demand estimates
MOWR: Ministry of Water Resources
Source: CWC, Basin Planning Directorate, XI Plan Document. *Report of the Standing Sub-Committee on "Assessment of Availability & Requirement of Water for Diverse Uses-2000"*.

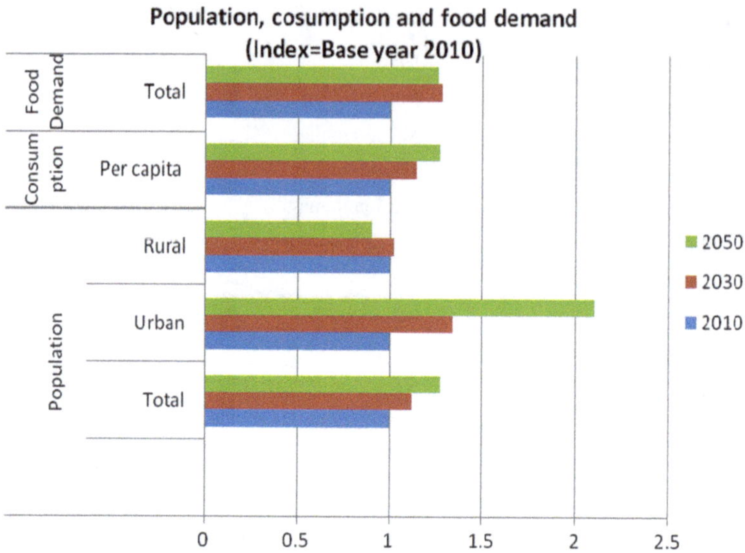

Fig. 2.1 Projected population and food demand, India

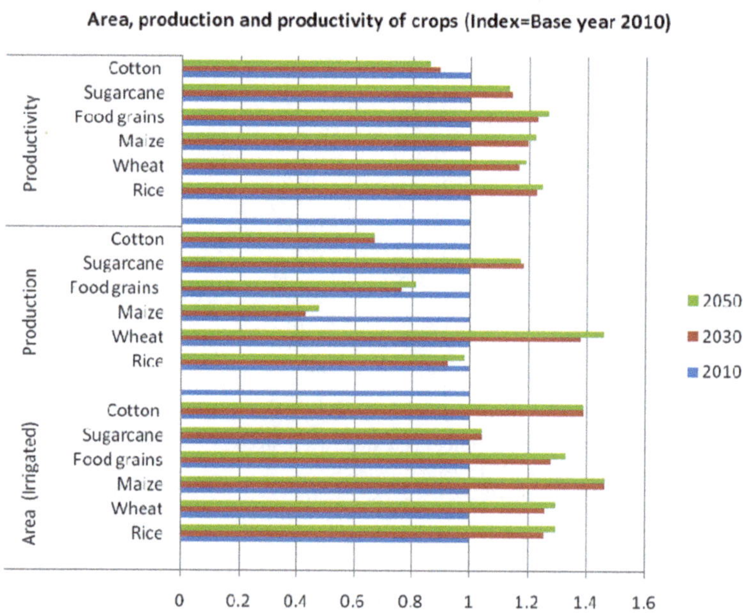

Fig. 2.2 Trend in area, production, and productivity of crops, India

The projected food grain requirements will be roughly 218 and 284 kg per head per year for the years 2030 and 2050, respectively. Using the population projections and per capita food requirements, the total food requirement for the country has been estimated between 382 and 449 million tonnes by 2050 (Fig. 2.3). The major drivers are the trends in rural and urban populations, levels of nutritional intakes (calorie supply) from crop and animal products, per capita consumption of different crop categories, and the feed conversion ratios. According to a study made by IWMI (Upali et al. 2007), the BAU scenario projects an average calorie supply of 3,000 kcal/person/day by 2050. However, the average calorie supply based on the NCIWRD was over 4,000 kcal/per person/day. Feed requirement, losses in storage & transportation, seed requirement, and carryover for years of monsoon failure in the country will be about 12.5% of the food grain production (NCIWRED, 1999). In the recent years, diet pattern has changed more towards milk, eggs, and livestock products and demand for cereals may not increase as rapidly as expected. Hence, the per capita consumption of rice, maize, and other coarse cereals is expected to decrease by 9% to 47% from 2000 to 2050 (Upali et al. 2007).

Self-sufficiency in food production at national level will be the driving force in the estimation of future water requirement. It is also expected that much of the additional grain requirement will be produced under irrigated conditions where the ratio of food and non-food productions in terms of area sown will be 70:30 and for the irrigated and unirrigated cropped areas, it will be 54:46.

Depth and Efficiency of Irrigation

Allocation of water to irrigation project(s) is decided based on the pre-determined values or assumptions. The Net Irrigation Requirement (NIR) varies with the crop consumptive use, effective rainfall during the crop growing period, and type of irrigation. The Gross Irrigation Requirement (GIR) or "Delta", i.e., depth of irrigation

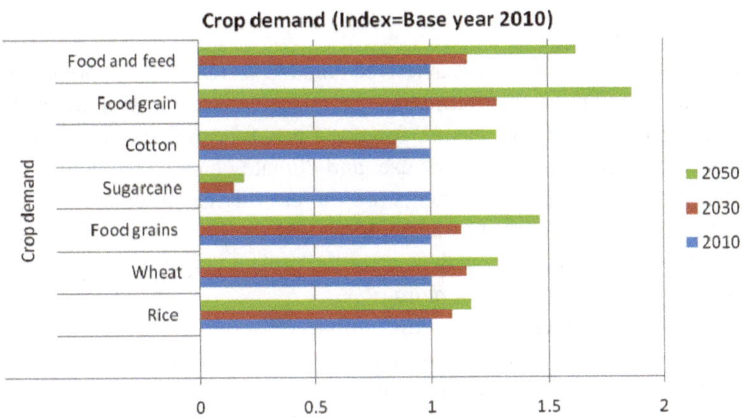

Fig. 2.3 Trend in food demand by crops, India

at canal head is a function of Net Irrigation Requirement (NIR) and efficiency of the irrigation project (incorporating both conveyance and application efficiencies). The Irrigation Commission (1972) assumed an average value of delta (or irrigation depth) as 0.76 m at national level for surface and groundwater combined. Subsequently, the National Commission on Agriculture (1976) assumed an average delta of 0.65 ha m of groundwater to irrigate a cropped hectare and a delta of 0.9 ha m if the source is surface water, as conveyance losses are higher in the latter. The two together give an average delta of 0.8 ha m per cropped hectare. It was further envisaged that with more scientific and economical use of water in the future, the average water depth for irrigation (delta) could be reduced to 0.7 ha m. Using these values, future water demands can be projected. The data on current water use by the crops is higher than what is really needed based on improved consumptive use studies (Palanisami et al. 2015). It is understood that under the climate change regime, the actual crop water requirements might have changed due to changes in weather parameters but most of the projects still follow the old crop water requirement data or *delta* as the basis for water allocation among the projects. For example, in the Krishna Water Disputes Tribunal, project-specific crop water requirements using a scientific approach were worked out. The results indicated wide differences between the data based on current water allocation and the data derived using a scientific approach (Palanisami 2019).

Different efficiencies are being adopted over the years to estimate mainly the water requirement for irrigation. As such, based on current experiences, 35–40% efficiency in surface water, and 65–70% efficiency in groundwater will be a fair approximation (Fig. 2.4). It is expected that with the increasing economic value of water and increasing demand for water supplies, improvement in overall irrigation efficiency will most likely take place in the future. It is expected that the efficiency of groundwater irrigation will increase to 75% by 2050 from the present level of 65%. The efficiency in surface irrigation projects would also increase to about 60% from the present level of 30–40%.

Water Demand

Water requirement for irrigation is a derived demand which includes both consumptive and non-consumptive requirements. The major determinants of water demand are increasing population, increasing food and non-food requirements, increasing area under crops, efficiency of water use, and climate change. The major parameters of the crop irrigation requirements are the crop irrigated area, crop calendar, crop coefficients, potential evapotranspiration, and the occurrence of rainfall. The net irrigation water requirement adjusted for the project irrigation efficiencies gives the gross irrigation requirement or *delta*.

It is estimated that water demand will continue to rise and could present significant challenges for the future. Different estimates have shown that the future demand will be increasing in all sectors and the irrigation sector alone will account for about 72% of the total water demand in 2025 and 68% in 2050 (CWC, 2000). Surface water supplies account for about 57% of the demand and the rest by groundwater supplies (CWC, 2000). A study by the International Water Management Institute (Upali et al., 2007) on the water demand and supply indicated that the Business As Usual (BAU)

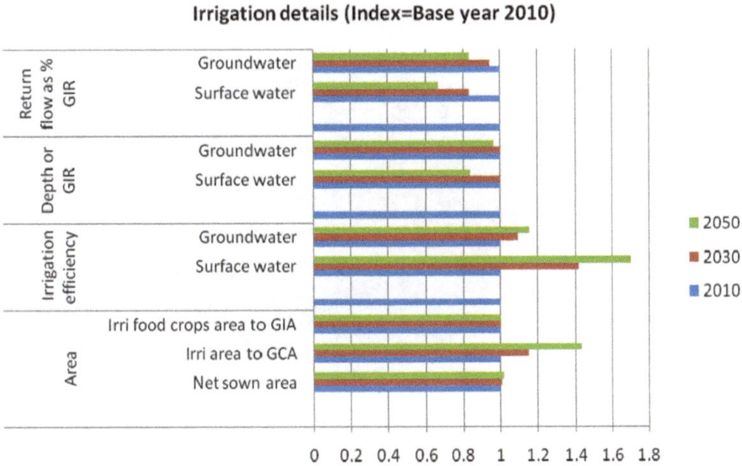

Fig. 2.4 Trend in irrigated area, irrigation depth, and irrigation efficiency, India

scenario projects an increase in the total water demand from 680 BCM in 2000 to 833 BCM by 2025, and to 900 BCM by 2050. The projected irrigation water demand by the year 2050 will be around 628 km3 and 807 km3, respectively, under low and high demand assumptions (Upali et al., 2007). The estimates by government agencies as given in Table 2.3 show comparatively a higher water demand and the difference in estimates is mainly due to the assumption made on future food grain demand where the IWMI study assumed a decreasing food grain demand and the NCIWRD study assumed a significant growth in food grain consumption. The water demand by various sectors is given in Figs. 2.5 and 2.6 for 2025 and 2050, respectively.

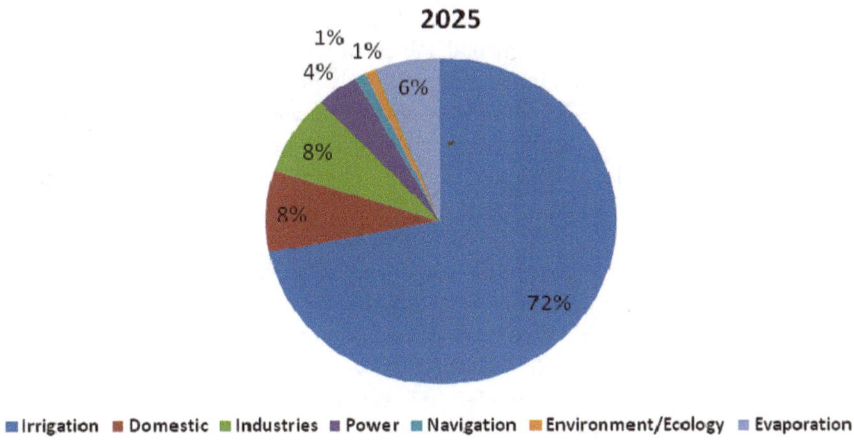

Fig. 2.5 Water demand by uses, 2025, India

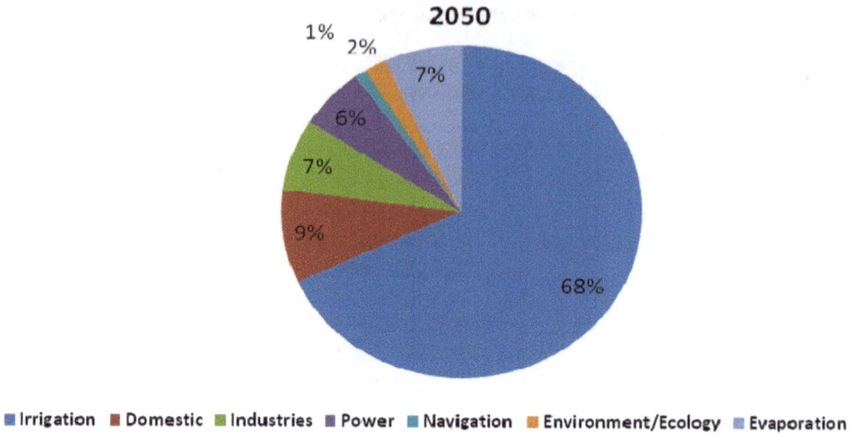

Fig. 2.6 Water demand by uses, 2050, India

2.3 Climate Change and Climate Change Scenarios for Indian Sub-continent

The impending climate change caused by the greenhouse gas (GHG) emissions is now becoming an established fact. Studies conducted on the impact of climate change on hydro-climatic factors are grouped into two types: "empirical statistical analyses" of past trends in observed variables like precipitation, temperature, soil moisture, stream flow, and so on; and "model-based projections" of future trends in these variables. However, these are not always in sync (Srinivasan 2019). Mostly, mathematical models such as General Circulation Models (GCM) are used to simulate the behaviour of the atmosphere with different scenarios for various levels of GHG emissions and it is possible to predict changes in the hydrologic cycle.

According to the Intergovernmental Panel on Climate Change (IPCC) (IPCC 2001), increased precipitation intensity, particularly in the summer monsoon season could result in more flood-prone areas in temperate and tropical regions of Asia. Further, countries such as India whose water use is more than 20% of the total water resources available are expected to experience moderate to severe water stress, particularly during the drought periods and the surface runoff is expected to reduce significantly.

The *Fifth Assessment Report* of the IPCC has further shown that climate change may affect the short-term variability of water resources through increased frequency and intensity of droughts and floods or induce long-term changes in mean renewable water supply (IPCC, 2014). Notably, the following changes are expected due to climate change: a) more intense rainfall in fewer spells could result in more intense floods and droughts; b) the shift of the rainfall towards winter; and c) significant reduction in the glacier's mass resulting in increased flows in the initial few decades and reduced flows thereafter where annual average river runoff to increase by 10–40%

in high latitudes, but decrease by 10–30% in some dry regions at mid-latitudes and in the dry tropics (Milly et al 2005). As expected, many of the areas that are already facing water shortages will have less water supplies in the future thus impacting surface and groundwater availability for irrigation (Mall et al 2006).

Most of the Indian landmass below the Ganges plain is likely to experience a 0.5–1 °C rise in average temperature during 2020–2029 and then 3.5–4.5 °C rise during 2090–2099. Many parts of peninsular India, especially Western Ghats will experience a 5–10% increase in total precipitation (IPCC 2001, 2014) and this increase will be accompanied by greater temporal variability. Throughout the sub-continent, more wet days will contribute more to total precipitation. It is also observed that more of India's precipitation may be received in fewer than 100 h of hailstorms—and half in less than 30 h—as has been the case during recent decades. This would mean that higher precipitation intensity and a larger number of dry days in a year will be a possibility. It is also expected that the increased frequency of extremely wet rainy seasons will also result in more runoff (Gosain and Rao, 2007). The IPCC Fifth Assessment Report (IPCC, 2014) thus projects a general increase in seasonal mean rainfall over India, but an increase in intense rainfall events over central India and longer dry spells over other regions. In Table 2.4, a summary of the projected climate change is provided.

2.4 Climate Change and Surface Water Supplies

Surface runoff would show much variation across the river basins as well as sub-basins of India where an increase in precipitation in the Mahanadi, Brahmaputra, Ganges, Godavari, and Cauvery is expected. In the remaining basins, a decrease in precipitation is expected whereas Sabarmati and Luni basins will have drastic decrease in precipitation and consequent decrease of total runoff. Overall, the impacts will be in terms of severe droughts and intensive floods in various parts of the country due to climate change (Gosain and Rao 2003). Most of the river basins in the country will have the impact of climate change at various levels in the twenty-first century (Table 2.5).

In addition to increased temperature and rainfall variability, other factors also affect the water resources in the country. For example, changes in the cropping pattern and land-use pattern, overexploitation of groundwater resources, and drainage issues will influence the water availability. For example, in the Gangetic basin, a reduction in the Ganges discharge by 60% over 25 years will lead to about 50% drop in water availability in surface water resources, and lowering of the groundwater table (Adel 2002). Irrigation water demand will be more sensitive to climate change as field-level changes in the weather parameters may influence the field-level water demand such as increased crop evapotranspiration. Rehana and Majumdar (2013) indicated that the water requirements of most of the crops in Bhadra command are likely to increase because of the increased evapotranspiration. A study by Mohan and Ramasundaram (2014) has shown that compared to 2010 level, annual irrigation requirement will

Table 2.4 Projected climate change during the twenty-first-century over India

Region	Temperature	Rainfall	References
All India	Rise in temperature by 2 °C and 3.5 °C	Increase by 7 and 14%	Kumar (2009)
All India	Increase in winter temperature by 1–4 °C. increase in CO2 concentration	Increase in rainfall by 20%	Bhaskaran et al. (1995)
		Increase in heavy rainy days during the summer and an increase in inter-annual variability	
All India	Average temperature change 2.33 to 4.78 °C, with a doubling in CO2 concentration	Increase in frequency of heavy rainfall events	Lonergan (1998)
All India	Annual mean surface temperature change between 3.5 and 5.5 °C by the end of the century	7 to 10% Increase in annual mean rainfall	Lal et al. (2001)
		5–25% decline in winter rainfall, 10–15% increase in summer rainfall	
		30% increase rainfall by 2050 in Northwest India	
	More warming in winter	Higher rainfall in Western semi-arid regions of India	
		Decrease in winter rainfall between 10 and 20% over central India by 2050	
All India	In the northern region, the maximum temperature may exceed 4 °C	Decrease in number of rainy days in western and central parts by more than 15 days; foothills of the Himalayas and in north-east India, number of rainy days increase by 5–10 days	Rupakumar et al. (1992)
	A general increase in minimum temperature up to 4 °C all over the country	Increase in rain intensity by 1–4 mm/day, except for small areas in northwest India, where rainfall intensities may decrease by 1 mm/day	
Godavari basin	Rising temperature by 1.5 °C to 3.5 °C	Increasing variability from 13.6 to 17.8%	Palanisami and Reddy (2012)
Krishna Basin	Rising temperature by 2.05 °C to 4.1 °C	1 Increasing variability from 13.6 to 17.8%	Palanisami and Reddy (2012)
Cauvery basin	Rising temperature by 1.93 °C to 4.03 °C	Increasing variability from 13.6 to 17.8%	Palanisami and Reddy (2012)

Source Mall et al. (2006), Palanisami and Reddy (2012), Palanisami (2019)

increase from 4.63% to 20.46% in 2050 under different climate change emissions scenarios (Table 2.6).

Impact on Groundwater

The increasing inter-annual variability of the monsoon rains will affect groundwater recharge and could threaten the sustainability of the groundwater resource, as in

Table 2.5 Impact of climate change on water resources, India

Basin	Impact	Reference
Ganges and Brahmaputra	Increased risk of flooding in lower reaches of Ganges and Brahmaputra during July–September	Seidel et al. (2000)
River basins in East and North-east India	Most of the basins vulnerable towards climate change in terms of water availability during twenty-first century	Rehana and Majumdar (2013)
River basin in Central India—Narmadha (Madhya Pradesh)	Considerable decrease in extreme flood events and higher water availability for irrigation purposes	Sudheer (2016)
River basins in South India		
(a) Upper Bhavani	The annual runoff decreases by 5% under low dry season rainfall and runoff increases by 17% under high annual rainfall	Wilk and Hughes (2002)
(b) Tungabhadra	Increase in the precipitation and runoff, decrease in the actual evapotranspiration over the sub-basins	Meenu et al. (2013)
(c) Malaprabha	The monthly precipitation increases in the future	Anandhi et al. (2008)
(d) Krishna, Ganges, and Godavari	Increased trend in rainfall over India during 2071–2100 except some parts in northwest and southern India Increase in extreme rainfall and rainfall intensity in all three river basins during 2071–2100	Central Water Commission, and National Institute of Hydrology (2008)
Indian sub-continent	Increased intensity of floods and severity of droughts in several parts of the country. Acute water scarcity in Sabarmati and Luni river basins	MoEF (2004)
	Severe drought-event frequency in central north-eastern, peninsular, and west central India	Ojha et al. (2013)
	Occurrence of extreme dry spells will increase over central, southeast coastal, eastern, and some parts of north-eastern India	Ghosh (2016)

Table 2.6 Change in irrigation water requirement with reference to the year 2010

No	Scenario	Annual irrigation requirement relative to 2010 (%)	
		2020	2050
1	A2	24.73	14.17
2	A1B	5.99	4.64
3	B1	-0.91	20.45

Note The A2 scenarios are of a more divided world. The A2 scenarios are characterized by continuously increasing population and high emissions. A1B - A balanced emphasis on all energy sources. The B1 scenarios are of a world more integrated, and more ecologically friendly. The B1 scenarios refer to the rapid economic growth, reductions in material intensity, and the introduction of resource-efficient technologies

many locations water levels have gone deep beyond the level where pumping is not economical (Shah 2009). Further, the shifting pattern of rainfall affected the spatial and temporal distribution of runoff and increased the frequency of droughts resulting in poor groundwater recharge. Lower groundwater tables and the resulting increase in the energy required to pump water will make irrigation more expensive for small and marginal farmers.

The slow decline in snowmelt runoff could seriously affect the groundwater recharge causing shortage in water availability in states like Punjab, Haryana, Uttar Pradesh, and Rajasthan. In several locations increased rainfall intensity may also lead to higher runoff resulting in reduced recharge (Ministry of Environment and Forest, 2004). Depending upon the spatial as well as temporal distribution of monsoon season rainfall with increasing trend (+10% to + 12% of the normal) in parts of northern Andhra Pradesh, and northwestern India (Rajasthan) may possibly increase the quantum of groundwater recharge. However, for eastern Madhya Pradesh and some parts of Gujarat and Kerala, an increase in temperature and decline in rainfall (−6% to −8% of the normal) may result in reduced recharge and affect groundwater availability thus making these regions as India's groundwater hotspots (Shah 2009).

Water Quality

Changing climate will also have significant impacts on the quality of available water. Melting glaciers will increase the risks of floods during the rainy season and reduce dry season water supplies. Melting ice and thermal expansion of oceans will result in sea level rise and will lead to deteriorating groundwater quality due to saltwater contamination. Already water-stressed regions of Sabarmati basin, Krishna basin (particularly the Tungabhadra sub-basin), and the Cauvery and Pennar basins may be vulnerable to increased temperatures and salinity intrusions due to rising sea levels. In addition, water quality in many rivers has reached critical levels, mainly due to discharges of inadequately treated domestic sewage, inadequate controls on the discharges of industrial waste disposals, loss and destruction of catchment areas, deforestation, and poor agricultural practices.

2.5 Impact of Climate Change on Agriculture

Climate change is unlikely to have a significant impact on municipal and industrial demands, but it may substantially affect irrigation withdrawals. In municipal and industrial sectors, it is likely that non-climatic drivers will continue to have significant effects on the demand for water. Irrigation withdrawals, however, are more climatically determined where a change in rainfall and temperatures would result in an increase in irrigation demands (IPCC 2001).

Further, in addition to the reduction in water availability, higher temperatures and changing precipitation patterns could severely affect the production patterns of different crops by reducing the crop productivity. Among the crops, wheat yields are predicted to fall by 5–10% with every increase of 1 °C temperature and overall crop yields could decrease by up to 30% in South Asia by the mid-twenty-first century (IPCC 2001). India could experience a 40% decline in overall agricultural productivity by the 2080s (IPCC 2007). It is projected that climate change over the period 2010–2039 reduces major crop yields by 4.5 to 9% and the long-run (2070–2099) impact will be a reduction in yields by 25% or more in the absence of long-run adaptation (Guiteras 2009). Overall, kharif (monsoon) season crops will have more risk of flood as well as droughts and rabi and summer crops will have enhanced ET resulting in more irrigation demand. Further, climate change is not only affecting the yield but also increasing the variability in yield among major crops in several river basins of India (Palanisami et al. 2014). The crop gross margin also will be reduced due to changes in temperature and rainfall during the crop season (Table 2.7) (Palanisami and Reddy 2012).

At macro level, using the temperature and declining rainfall during 1970–2015, the Economic Survey 2018/19 observed that during the years when rainfall levels drop 100 mm below average, farmer incomes would fall by 15% during the kharif and seven per cent during the rabi crops seasons. Climate change could reduce annual agricultural incomes in the range of 15 to 18% for irrigated areas and up to 20–25% for unirrigated areas. It stresses that climate change will increase farmer uncertainty

Table 2.7 Climate change impact on crop gross margin in selected river basins, India

Details	Godavari basin			Krishna basin		
	Current	Mid century	End century	Current	Mid century	End century
Gross margin	1362.7	828.7	740.1	2297	1789	1400
Total reduction		534	622.6		508	697
Reduction due to temperature		551.9	625.3		419	549
Reduction due to rainfall		-17.9	2.5		89	148
% reduction		39.2	45.6		22.1	30.3

Note Mid century = 2021–2050, End century = 2071–2100
Source Palanisami and Reddy 2012

and calls for effective use of technologies and crop insurance programmes to make farming more resilient and sustainable (Lahiry 2018).

Even though the models used in the estimation might have limitations of data and assumptions, the overall inference is that most of the surface water storages will have increased runoff only during periods of heavy rains but at the same time will have increased evaporation from large open surfaces of reservoirs due to increase in mean temperature.

There is reasonable consistency in temperature projections under climate change whereas predicting the possible changes in the Indian monsoon will be a major challenge (Saha et al. 2014). In the absence of reliable historical records of climate, water, and human activity, models rely mostly on conventional data sources that are over-simplified. For instance, a modelling study by Gosain et al. (2006) did not incorporate the structures like dams and diversions where the translation of rainfall into runoff and recharge is difficult because of lack of reliable database as watersheds have been so drastically altered in the recent years due to several developmental programmes (Srinivasan, 2019).

The impact of increased ET can also be reflected in the water footprint (WF) which is a reciprocal of the water productivity. For example, the total WF of the cereal crops in India will change in the range of − 3.2 to 6.3% in the future. Although the national-level green and blue WF is projected to change marginally, distinct trends are observed for kharif and rabi crops (Mali et al. 2021).

Considering the increasing inter-annual variability of rainfall in India, more in-depth studies covering different basins, aquifers, and agro-climatic regions of India are needed to assess the sensitivity of the basins to climate change. Further, the development of more reliable "damage due to flood" and "damage due to drought" relationships will help to finetune the already made inferences on the impact of climate change on water resources.

2.6 National Missions and National Action Plan on Climate Change

Large-scale planning would be clearly required for adaptation measures for climate change impacts if catastrophic human misery is to be avoided (GOI 2004). The National Action Plan on Climate Change (NAPCC) launched by GOI in 2008 has established a national strategy to address climate change impacts, mitigation, and adaptation. The National Water Mission (NWM) which is one of the eight missions stated within the framework of NAPCC also emphasize the need to improve the overall water use efficiency in all aspects.

Sahi-Fasal Climate Change Programme

The objectives of the National Water Mission (NWM) related to the conservation of water include "focused attention to vulnerable areas including over-exploited areas" and "increasing water use efficiency by 20 percent". In keeping with these objectives NWM launched the "Sahi-Fasal" campaign in November 2019 (GOI 2019), an initiative to encourage farmers in water-stressed areas to grow crops which are: (i) less intensive in water use; (ii) more productive per unit of water consumed; (iii) economically remunerative; (iv) healthy and nutritious; (v) suited to the agro-climatic-hydro characteristics of the area; and (vi) are environmental-friendly. Key steps formulated towards promoting Sahi-Fasal include creating awareness among farmers on alternative crops which consume less water; better irrigation practices; micro irrigation; and soil moisture conservation. In addition, steps are taken to support policy makers in framing enabling policies including effective pricing of inputs (water and electricity); improved procurement and marketing for these alternate crops; and creating storage facilities.

2.7 Climate Change and Water—The Way Forward

India has about 20 river basins with varying levels of water supply and most river basins are water stressed and several of them are considered as closed basins. In addition to the growing irrigation demand, increasing demand from a growing population, coupled with economic activity, adds pressure on the already stressed water resources in the country. With the country already experiencing water stress, there is a need to augment water supply both in water-rich regions lacking infrastructure and effectively manage water demand in water-scarce regions.

The adoption of climate-smart agriculture (CSA) is considered important to mitigate the negative impact of climate change. Innovative strategies to enhance the resilience in farming are gaining importance which include mainly the modernized irrigation infrastructure, improved water management practices, resilient crop varieties, adoption of sustainable farming practices, development of advanced early warning systems for forecasting extreme weather events, introduction of effective agricultural insurance mechanisms, strong research and development, business models with public–private partnership through targeted policy interventions, etc. The policy measures include prioritizing and scaling-up CSA technologies and precision agriculture and adopting market-based measures like carbon finance to promote greenhouse gas mitigation and increase farmers' income. Furthermore, revamping of agricultural support policies towards sustainability is also needed (Gulati and Thangaraj 2023).

It is expected that most irrigated areas in India therefore would require more in the future and global net irrigation requirements would increase relative to the situation without climate change by 3.5–5% by 2025, and 6–8% by 2075 (Doll and Siebert 2001). Considering the impact of climate change, i.e., the increased ET demands for several crops, variation in rainfall distribution, and the resulting changes in effective rainfall contribution, it is assumed that the net irrigation requirements may increase by another 6% and gross irrigation requirements by 10%. Efficient and integrated water management and implementation strategies outlined above have to be examined and planned accordingly. The subsequent chapters in the book will be able to address the key issues and various implementation and management strategies in such a way that they could be able to address the impact of climate change in a manner deemed to be necessary.

References

Adel MM (2002) Man-made climatic changes in the Ganges basin. Int J Climatol 22:993–1016

Anandhi A, Srinivas VV, Nanjundiah RS, Kumar ND (2008) Downscaling precipitation to river basin in India for IPCC SRES scenarios using support vector machine. Int J Climatol 28(3):401–420. https://doi.org/10.1002/joc.1529

Bhaskaran B, Mitchell JFB, Lavery JR, Lal M (1995) Climatic response of the Indian subcontinent to doubled CO2 concentrations. Int J Climatol 15:873–892

Blanco G, Gerlagh R, Suh S, Barrett J, De Coninck HC, Diaz Morejon CF, Mathur R, Nakicenovic N, Ofosu Ahenkora A, Pan J, Pathak H, Rice J, Richels R, Smith SJ, Stern DI, Toth FL, Zhou P (2014) Drivers, trends and mitigation in: Climate Change 2014: mitigation of climate change. Contribution of Working Group III to the Fifth Assessment Report of the Intergovernmental Panel on Climate Change [Edenhofer O, Pichs-Madruga R, Sokona Y, Farahani E, Kadner S, Seyboth K, Adler A, Baum I, Brunner S, Eickemeier P, Kriemann B, Savolainen J, Schlömer, Stechow C, Zwickel T, Minx JC (eds.) Cambridge University Press, Cambridge, United Kingdom and New York, USA.

Central Water Commission (CWC) (2000) Report of the standing sub-committee on assessment of availability and requirement of water for diverse uses. XI Plan Document, Basin Planning Directorate, New Delhi.

Central Water Commission and national Institute of Hydrology (2008) Preliminary consolidated report on effect of climate change on water resources, Ministry of Water Resources, Government of India, New Delhi

Doll P, Siebert S (2001) Global modelling of irrigation water requirement. University of Kassel, Kassel, Germany

Ghosh S, Vittal H, Sharma T, Karmakar S, Kasiviswanathan KS, Dhanesh Y, Sudheer KP, Gunthe SS (2016) Indian summer monsoon rainfall: implications of contrasting trends trends in the spatial variability of means and extremes.PLOS ONE 1(7): e0158670 https://doi.org/10.1371/journal.pone.0158670

GOI (1976) National commission on agriculture. Ministry of Agriculture, New Delhi.

GOI (2004) National environment policy. Ministry of Environment and Forests, New Delhi. p. 38.

GOI (2019) Sahi Fasal. National Water Mission, New Delhi.

GoI (2020) Assessment of climate change over the Indian region, Ministry of Earth Science,

Gosain AK, Rao S (2003) Impact assessment of climate change on water resources of two river systems in India. Jalvig Same 22:21

Gosain AK, Rao S, Basuray D (2006) Climate change impact assessment on hydrology of Indian river basins. Curr Sci 90:346–353

Guiteras R (2009) The impact of climate change on Indian agriculture. Report, University of Maryland, USA

Gulati A, Thangaraj P (2023) El niño and weather vagaries: deconstructing historical trends. Agri-Food Trends and Analytics Bulletin (AF-TAB) 3(1)

Intergovernmental Panel on Climate Change (IPCC) (2014) Climate change 2014: impacts, adaptation, and vulnerability (1). Cambridge University Press, Cambridge and New York

IPCC (2001) Impacts, adaptation and vulnerability: contribution of working group II to the third assessment report of the IPCC. Cambridge University Press, Cambridge, UK

IPCC (2007) Summary for Policy-makers, Climate Change. Mitigation. Contribution of Working Group III to the Fourth Assessment Report of the IPCC. Cambridge University Press, Cambridge, UK.
Irrigation commission (1972) Report of the Irrigation Commission (I), Ministry of Irrigation and Power, Government of India
Kumar KS (2009) Climate sensitivity of Indian agriculture. Working paper 43. Madras School of Economics. Chennai.
Lahiry SC (2018) Climate change and its impact on water resources. India Water Portal.
Lal M, Nozawa T, Emori S, Harasawa H, Takahashi K, Kimoto M, Abe-Ouchi A, Nakajima T, Numaguti A (2001) Future climate change: implications for Indian summer monsoon and its variability. Curr Sci 81(9):1196–1207
Lepers E, Lambin EF, Janetos AC, DeFries R, Achard F, Ramankutty N, Scholes RJ (2004) A synthesis of information on rapid land-cover change for the period 1981–2000. Bioscience 55:115–124
Lonergan S (1998) Climate Warming and India (eds) World Bank Technical Paper No. 402, Washington DC
Mali SS, Shirsath PB, Islam A (2021) A high-resolution assessment of climate change impact on water footprints of cereal production in India. Sci Rep 11:8715. https://doi.org/ https://doi.org/10.1038/s41598-021-88223-6
Mall RK, Gupta A, Singh R, Singh RS, Rathore LS (2006) Water resources and climate change: an Indian perspective. Curr Sci 90(12):1610–1626
Meenu R, Rehana S, Mujumdar PP (2013) Assessment of hydrologic impacts of climate change in Tunga-Bhadra river basin, India with HEC, HMS and SDSM. Hydrol Process 27(11):1572–1589https://doi.org/10.1002/hyp.9220
Milly PCD, Dunne KA, Vecchia AV (2005) Global pattern of trends in streamflow and water availability in a changing climate. Nature 438(7066):347–350. https://doi.org/10.1038/nature04312
Ministry of Water Resources (1999) National Commission for Integrated Water Resources Development, Ministry of Water Resources, p 542.
MoEF (2004) India's initial national communications to the United Nations framework convention on climate change. Ministry of Environment and Forest (MoEF), Government of India, New Delhi
Mohan S, Ramasundaram N (2014) Climate change and its impact on irrigation water requirements on temporal scale. Irri Drain 03(01).
Nagothu, Udaya Sekhar, Gosain AK, Palanisami K (2012) Water and climate change: an integrated approach to address adaptation challenges. Macmillan Publishers India Ltd - ISBN 935–059–059-X
Natarajan KS (I993) Population projections for India. Proceedings of the 17th Conference of the Indian Association for Study of Population, Annamalai University, Tamil Nadu.
NCIWRDP (1999) Integrated water resources development – a plan for action. Report of the National Commission for Integrated Water Resources Development (NCIWRDP), Ministry of Water Resources, New Delhi, India
Ojha R, Kumar ND, Sharma A, Mehrotra R (2013) Assessing severe drought and wet events over India in a future climate using a nested bias correction approach. J Hydrol Eng 18(7):760–772. https://doi.org/10.1061/(ASCE)HE.1943-5584.0000585
Palanisami K (2019) Estimation of project based crop water requirements in Krishna basin. Affidavit, Submitted to the Krishna Water Disputes Tribunal, New Delhi.
Palanisami K, Krishna Reddy (2012) Project report of the collaborative research project 'Climate Change, Water, Agriculture, Food Security and livelihoods in India', The Norwegian Institute for Agriculture and Environmental Research (BIOFORSK) and IWMI, Hyderabad.
Palanisami K, Suresh Kumar D, Malik RPS, Raman S, Kar G, Mohan K (2015) Managing water management research: analysis of four decades of research and outreach programmes in India.Econ Polit Wkly 1(26&27) June 27.

PalanisamiK RCR, SureshkumarD MRPS (2014) Enhancing the crop yield through capacity building programs: application of double difference method for evaluation of drip capacity building program in Tamil Nadu State. India. Agric Sci 5(1):33–42. https://doi.org/10.4236/as.2014.51003

Rehana S, Mujumdar PP (2013) Regional impacts of climate change on irrigation water demands. Hydrol Process 27(20):2918–2933

Rupakumar K, Pant GB, Parthasarathy B, Sontakke NA (1992) Spatial and sub seasonal patterns of the long term trends of Indian summer monsoon rainfall. Int J Climatol 12:257–268

Saha A, Ghosh S, Sahana AS, Rao EP (2014) Failure of CMIP5 climate models in simulating post-1950 decreasing trend of Indian monsoon. Geo Res Lett 41(20):7323–7330

Seidel K, Martinec J, Baugmgartner MF (2000). Modelling runoff and impact of climate change in large Himalayan basins. In: Majumdar PP, Tiwari VM (2019) (eds) International conference on integrated water resources management pp 19–21, Quoted in Water futures of india-status of science and technology, Indian institute of science. Bangalore.

Shah T (2009) Climate change and groundwater: India's opportunities for mitigation and adaptation. Environ Res Lett 4:13

Srinivasan V (2019) Climate adaptation in the water sector in India. In Navroz K. Dubash (ed). India in a Warming World: Integrating Climate Change and Development . Published to Oxford Scholarship DOI: https://doi.org/10.1093/oso/9780199498734.001.0001. Print ISBN-13: 9780199498734.

Sudheer KP (2016) Impact of climate change on water resources in Madhya Pradesh- an assessment report. Retrieved from https://www.devalt.org/images/L2_projectPdfs/ (16)CCimpactwaterresources.pdf?oi

United Nations (1995) World population prospects. The (1994) Revision. UN, New York

Upali A, Shah T, Turral H, Anand BK (2007) India's water future to 2025–2050: business-as-usual scenario and deviations. Research Report 123, International Water Management Institute, Colombo.

Visaaria L, Visaria P (1996) Prospective population growth and policy options for India, 1991–2001. The Population Council, New York

Wilk J, Hughes DA (2002) Simulating the impacts of land-use and climate change on water resource availability for a large south Indian catchment. Hydrol Sci J 47(1):19–30. https://doi.org/10.1080/02626660209492904

Chapter 3
Stabilizing the Canal Irrigation

Abstract Irrigation sector in India is known for its *Build–Neglect–Rebuild* approach. Overall, the major and medium irrigation (MMI) projects account for about 73% of the irrigation potential already created and about 82% of the ultimate irrigation potential (UIP). The gap between the created and utilized potential is increasing over the years, which is about 25.6 M ha (27.15%) at the end of the Eleventh Plan. The average water use efficiency of the canal irrigation projects in the country is only about 36%. The measures to improve the performance of irrigation schemes include improving the overall water use efficiency, investment in command area development including more pressurized (micro) irrigation systems and participatory irrigation management, modernization of the existing schemes, inter-basin transfers in feasible locations, etc. A recent development in the area of canal irrigation modernization in India is the establishment of Support for Irrigation Modernization Program (SIMP) which is a natural progression by the ADB and the GoI and will support the states in preparing future irrigation modernization programmes for MMI schemes.

Keywords Potential created and utilized · Cost of potential created and utilized · Canal irrigation performance gaps · Canal modernization · Project efficiencies · Support for Irrigation Modernization Program (SIMP) · Irrigation modernization plans

3.1 Introduction

Large-scale canal irrigation began to develop since the past one hundred years, when large-scale capital investments in irrigation were made possible, based initially upon economic interests in irrigated agriculture for exporting goods under British rule. Later, after independence, irrigation development was made primarily to boost food production to ensure food security in the country. However, the irrigation sector had several challenges that discouraged the prioritization of canal irrigation for some period. On one hand, the increasing costs of construction, availability of land, and

K. Palanisami and U. S. Nagothu, *India's Water Future in a Changing Climate*, Advances in Geographical and Environmental Sciences,
https://doi.org/10.1007/978-981-97-1785-9_3

disputes over sharing water rights between neighbouring states constrained the investments in new major and medium irrigation (MMI) projects, whereas, on the other, poor maintenance and performance of the projects had reflected in poor utilization of the created potential. This also discouraged the government's interest to further invest and develop in canal irrigation. However, in the recent years, there is growing pressure on MMI schemes to meet both the increasing food demand, and the need to release water for other uses as non-agriculture demands are increasing over the years. In this context, MMI canal projects need to address among others, the four key challenges to sustain the future of canal irrigation in India.

i. Low utilization of the potential created: Potential created is not fully utilized resulting in cost overruns and inefficient water use. In many cases, there is overuse of water, inequitable water distribution, and low income to farmers.
ii. Low water use efficiency: In most of the MMI projects, the water use efficiency is in the range of 30–40% resulting in lower productivity per unit of land and water.
iii. Growing conflicts at interstate water sharing: Many river basins are having interstate water sharing issues. Long-standing water sharing conflicts affect both the productivity and income level of the farmers as performance improvements are not addressed due to standing water sharing conflicts.
iv. Lack of prioritization of investments in canal irrigation modernization

3.2 Status of the Gap Between Irrigation Potential Created and Utilized

One of the indicators of the irrigation project performance is the gap between the irrigation potential created and the potential utilized. The Ultimate Irrigation Potential (UIP) estimated for India by the CWC is 139.89 million ha (M ha), of which 58.47 M ha (42%) is from major and medium irrigation sources and 81.43 M ha (58%) is from minor irrigation sources (including groundwater and surface water sources like tanks). Out of this, about 113.53 M ha (81%) of irrigation potential was created towards the end of the Eleventh, which included 47.97 M ha (42%) from major and medium irrigation sources and 65.56 M ha (58%) from minor irrigation sources (CWC 2020). Table 3.1 provides state-wise irrigation potential created (IPC) through the MMI projects. According to the Ministry of Water Resources, the gap between the created and utilized potential is increased over the years, which is about 25.6 M ha (27.15%) at the end of the Eleventh Plan compared to 10% gap observed during the first plan period (Figs. 3.1 and 3.2).

Overall, MMI projects accounted for about 73% of potential already created and about 82% of the UIP of 58.47 M ha. There are variations among the states where seven states (i.e., Jammu &Kashmir, Gujarat, Karnataka, Rajasthan, Chhatisgarh, Tamil Nadu, and Maharashtra) have created more than their UIP. On the other hand, gaps between the potential created and utilized are also increasing in these states compared to others. In the case of Assam and Madhya Pradesh, only about half of

Table 3.1 State-wise irrigation potential through the major & medium irrigation projects ('000 ha)

Name of state/UT	UIP of MMI projects	UIP of minor projects	Total UIP	IPC up to XI plan		
				MMI	MI	Total
Andhra Pradesh/ Telangana	5000.00	6260.00	11,260.00	4803.73	3340.55	8144.28
Arunachal Pradesh	0.00	168.00	168.00	1.20	132.24	133.44
Assam	970.00	1900.00	2870.00	455.96	1016.82	1472.78
Bihar	5223.50	5663.50	10,887.00	3054.46	5924.78	8979.24
Chhattisgarh	1146.93	571.00	1717.93	1269.32	842.30	2111.62
Goa	62.00	54.00	116.00	55.55	25.92	81.47
Gujarat	3000.00	3103.00	6103.00	3679.09	2071.97	5751.06
Haryana	3000.00	1512.00	4512.00	2206.29	1637.67	3843.96
Himachal Pradesh	50.00	303.00	353.00	30.45	186.22	216.67
Jharkhand	1276.50	1108.00	2384.50	530.71	534.20	1064.91
Jammu Kashmir	250.00	1183.50	1433.50	325.61	745.66	1071.27
Karnataka	2500.00	3474.00	5974.00	2965.83	1704.17	4670.00
Kerala	1000.00	1679.00	2679.00	715.69	763.65	1479.34
Madhya Pradesh	4853.07	11,361.00	16,214.07	2506.43	2534.34	5040.77
Maharashtra	4100.00	4852.00	8952.00	4128.71	3185.60	7314.31
Manipur	135.00	469.00	604.00	158.50	120.69	279.19
Meghalaya	20.00	148.00	168.00		77.77	77.77
Mizoram	0.00	70.00	70.00		51.74	51.74
Nagaland	10.00	75.00	85.00		124.51	124.51
Orissa	3600.00	5203.00	8803.00	2147.36	1887.43	4034.79
Punjab	3000.00	2967.00	5967.00	2684.39	3497.71	6182.10
Rajasthan	2750.00	2378.00	5128.00	3167.13	2487.76	5654.89
Sikkim	20.00	50.00	70.00		42.74	42.74
Tamil Nadu	1500.00	4032.00	5532.00	1578.27	2331.99	3910.26
Tripura	100.00	181.00	281.00	29.25	161.86	191.11
Uttar Pradesh	12,154.00	17,481.00	29,635.00	9288.09	25,320.13	34,608.22
Uttarakhand	346.00	518.00	864.00	288.98	585.34	874.32
West Bengal	2300.00	4618.00	6918.00	1901.41	4159.68	6061.09

(continued)

Table 3.1 (continued)

Name of state/UT	UIP of MMI projects	UIP of minor projects	Total UIP	IPC up to XI plan		
				MMI	MI	Total
Union Territories	98.00	46.00	144.00	0.00	61.93	61.94
Total	**58,465.00**	**81,428.00**	**139,893.00**	**47,972.41**	**65,557.36**	**113,529.78**

Note Major and medium irrigation schemes are classified based on their cultivable command area. Major irrigation schemes are those with a CCA greater than 10,000 ha, and medium irrigation schemes are those with a CCA greater than 2,000 ha and less than 10,000 ha
Source CWC (2020)

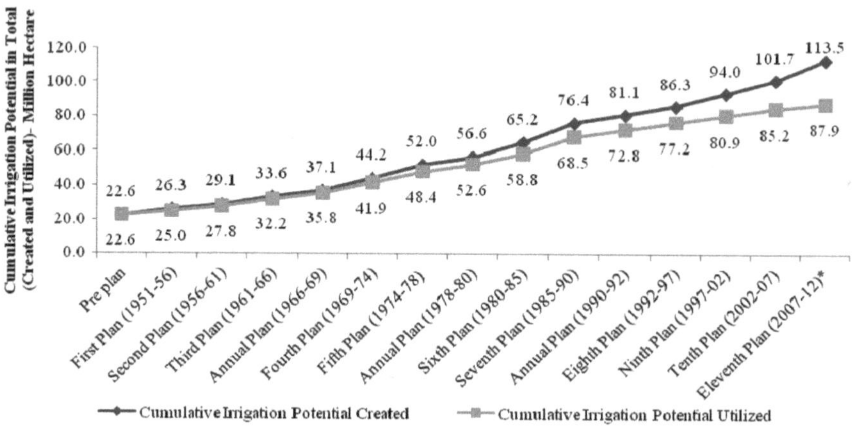

Fig. 3.1 Gap between IPC and IPU in India: Plan-wise. *Source* Data from Central Water Commission for various years

the UIPs have been created. Region-wise comparison (by including the major states that benefitted from MMI schemes) indicates the gap is high in the northern region (67%) followed by the western region (40%). Southern region has the lowest gap of about 24% (Table 3.2). These major states account for about 96% of UIP and about 96% of the country's IPC in MMI projects.

Major reasons for the increasing gap between irrigation potential created and utilized are:

(a) The differences in the management of irrigation systems between different regions. For example, the northern systems of Punjab, Haryana, and Western UP were typically designed and built to provide irrigation through run-of-the-river canals as protection against rainfall uncertainties during crop season and mostly they run with very low irrigation intensities. The southern systems were mostly similar to other regions of India, typically designed for much higher intensities and cover crops with heavy water requirements, such as paddy, sugarcane, and banana. The system of water distribution was based on "water duty" or

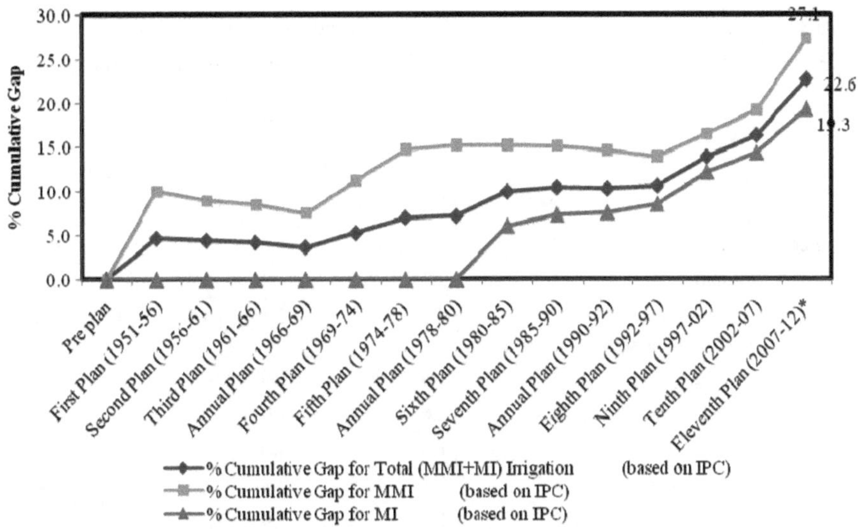

Fig. 3.2 Plan-wise cumulative Gap for MMI, MI, and Total Irrigation: All India. *Source* Data from Central Water Commission for various years

"localization". Hence, there was always a gap in the utilization of the created potential which varied among the regions (Palanisami 1984). There were some modifications in the 90 s and beyond with needed changes in cropping systems and project management.

(b) Variation or deviation in the crop area in the commands (area under irrigation) where farmers started to cultivate the crops of their choice (mostly higher water demanding crops). Thus there were significant deviations from the planned crop pattern in the project area resulting in comparatively lesser area covered than planned in the project documents. Imbalances in the development of the command areas resulting in a skewed distribution of wet crops where the irrigation institutions are weak to enforce the norms for adopting the planned crop pattern.

(c) Lack of actual data on IPU. As such there is a considerable lag between the data availability, its publication, and the reference periods. The data from the major agencies like erstwhile MoWR, Directorate of Economics & Statistics (DES) of Ministry of Agriculture and Farmers' Welfare (MoA&FW), and the CWC (Water and Related Statistics 2015) showed the irrigation potential utilized by MMI projects only up to 2006–07. The MI Census and the NSSO surveys also indicated that the canal irrigated area was on the decline in spite of the massive investments. However, research papers based purely on remote sensing showed higher coverage under irrigation (Dheeravath et al. 2010).

(d) Unexpected delays in completion of the irrigation project infrastructure and poor water delivery mechanisms. Delays in the acquisition of the land for water distribution were mainly due to the unwillingness of the landowners to provide

Table 3.2 Region-wise and State-wise irrigation potential and cost per ha in India

Region	States	Irrigation potential (M. ha)			Capital cost in Rs lakhs/ha (2017–18 prices)		Working expenses in Rs thousand/ha (2017–18 prices)	
		IPC	IPU	% Gap	IPC	IPU	IPC	IPU
Northern	Haryana	2.2	1.9	14.2	6.4	7.3	0.2	0.3
	Punjab	2.7	2.5	6.5	3.2	3.4	3.0	3.1
	Uttar Pradesh + Uttarakhand	9.6	8.0	16.7	2.1	3.5	3.0	3.8
	Average	4.7	1.6	66.9	3.9	4.7	2.1	2.4
Eastern	West Bengal	1.9	1.6	17.2	1.0	1.3	1.9	2.1
	Odisha	2.1	1.9	12.5	4.0	5.0	1.5	1.5
	Bihar + Jharkhand	3.6	2.1	42.5	0.8	1.3	1.5	2.4
	Average	2.5	1.8	27.8	1.9	2.6	1.6	2.0
Southern	Andhra Pradesh	4.8	3.2	32.5	10.6	12.7	14.9	16.4
	Karnataka	3.0	2.3	21.4	5.5	8.7	0.6	0.7
	Tamil Nadu	1.6	1.6	1.4	1.2	1.2	4.6	4.6
	Average	3.1	2.4	23.7	5.8	7.5	6.7	7.2
Western	Gujarat	3.7	1.9	49.1	4.4	6.2	4.3	4.9
	Madhya Pradesh + Chhattisgarh	3.8	2.1	43.8	2.7	3.6	2.5	3.5
	Maharashtra	4.1	2.3	44.0	13.5	20.4	5.1	7.6
	Rajasthan	3.2	2.5	20.2	1.5	2.1	5.0	5.5
	Average	3.7	2.2	40.1	5.5	8.1	4.2	5.4

land for field channel construction. Normally, more money is spent than the planned budget on the construction of the main dam and the main and branch canals, which leaves limited funds for the water delivery system at the outlet level.

(e) Poor cost recovery from irrigation projects, as the water charges are very low and not revised in several states over the years. The cost recovery as such accounted for less than 8% of the working expenses. The maintenance expenditure as per cent of capital investment was less than 1% of the capital cost causing deferred and delayed maintenance. The deferred maintenance leads to faster degradation, resulting in the degeneration of the whole irrigation system, inefficiency, and inequities in water supplies between head and tail reaches of the canal systems.

(f) Increased groundwater supplementation in the canal systems indicating the growing uncertainties in canal water supplies. However, the data on canal irrigated area also included the area supplemented by groundwater thus making biased estimates on the IPC–IPU gap.

3.3 Costs of the IPC and IPU

Irrigation investment normally increases among the projects and the regions, as the cost overrun is very common in irrigation projects. Unless the created potential is fully used, the cost overrun will be reflected in the unit cost of the project, both in terms of potential created and utilized. Higher the IPC–IPU gap, higher will be capital cost as well as working expenses embedded in the irrigation projects. Thus the per ha costs of irrigation potential created and utilized will give signals on how best these systems should be given priority for future improvement.

Capital Cost

Capital cost was worked out using the cost data available from 2002–03 to 2013–14. Figure 3.3 shows the state-wise per ha costs of potential created and utilized during 2002–03 to 2013–14. In the case of capital cost, the picture is slightly different where the eastern region has the lowest capital cost followed by the northern region. The higher cost in the southern and western regions might be due to large size of the projects and the cost escalation due to delayed completion of the capital works. The capital cost of potential utilized is about 1.5 times higher than the potential created in the western region.

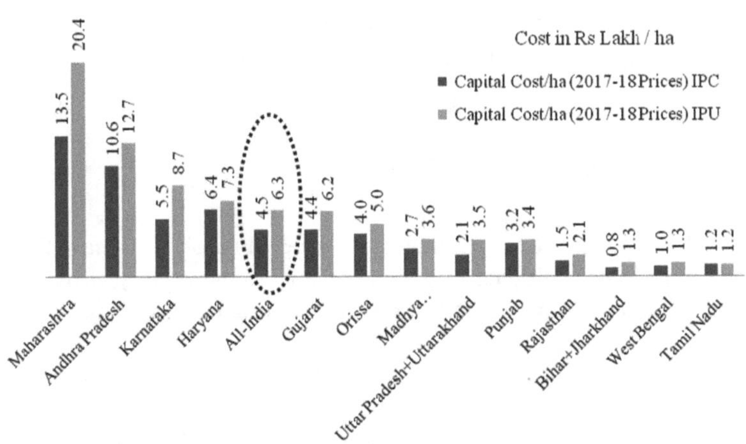

Fig. 3.3 State-wise per ha capital cost: 2002–03 to 2013–14 in 2017–18 Prices. *Source* Sharma and Banerjee (2020)

Fig. 3.4 State-wise per ha working expenses: 2002–03 to 2013–14 in 2017–18 prices. *Source* Sharma and Banerjee (2020)

Working Expenses

Working expenses as a percentage of capital expenditure have been 11% per ha of IPC and 9.8% per ha of IPU. In the case of states, it varied from as low as 0.4% in Haryana to 47.7% in Tamil Nadu for IPC. The working expenses of the potential utilized were higher by 1.3 times than the potential created in the western region followed by the eastern region and northern region indicating the financial implications of increasing gap between the potential created and utilized (Fig. 3.4).

3.4 Bridging the IPC–IPU Gap

Even though State Irrigation departments and the GOI departments have the potential in the creation of additional irrigation potential in the country, their priorities, limitation of funds and capacity, and lack of targeted training of staff make it difficult to realize it.

Despite the efforts made by the GOI by launching the Accelerated Irrigation Benefit Programme (AIBP) in 1996–97 to provide Central Loan Assistance (CLA) to the MMI projects to speed up the pending works, the progress was not satisfactory. Hence, the following factors must be considered while making efforts to re-engineer the MMI projects as management entities (Sharma and Banerjee 2020). This way, the paradigm shift from irrigation development to irrigation management can be achieved.

a. Funds for the selected projects should be released only when the major constraints (for example, acquisition of the land from the farmers for main canals as well as distributary canals and field channels) have been addressed and responsibility and accountability in fund management ensured.

b. Creation of irrigation potential should be allowed only when there are no gaps in the main/branch canals of the irrigation systems where all the associated minors and distributaries have been completed.
c. Integration with command area development (CAD) where funding of CAD and completion of the activities should be in unison with the main irrigation system development. All the projects may be considered as complete only when corresponding CAD works are also simultaneously completed to ensure water deliveries.
d. Canal modernization programme should be planned in such a way that it should include those interventions that could help minimize the IPC–IPU gap and increase WUE as well.
e. Introduction of underground pipeline (UGPL) systems and canal-based micro irrigation. Wherever, land acquisition for the construction of the irrigation networks is an issue, construction of UGPL will have added benefits such as achieving utilized potential quickly at a reasonable cost and promoting more irrigation cooperatives for better water distribution (Shah et al. 2010). UGPL has been successfully tried in the sandy and undulating and several other regions in Punjab, some projects in Rajasthan, and now in Sardar Sarovar Project in Gujarat. Cost of UGPL is about two times the average cost of Rs. 22,000/ha with open channels, but the savings in water, land, and higher productivity can justify the cost (Sharma and Banerjee 2020).
f. Prioritizing investment options: The capital cost of creation and utilization of MMI in the eastern region (eastern Uttar Pradesh, Bihar, Jharkhand, Chhattisgarh, West Bengal, Odisha, Assam) will be only about one-fourth of the All India averages and the working expenses for operation and maintenance of the systems are only about one-third of the all India averages. Hence, priority can be given to the eastern region so that the IPC–IPU gap can be bridged. Also, such prioritization will have added benefits of flood control in the region and thus the returns to irrigation investment will be comparatively high in this region. Future canal modernization programmes can consider this option.
g. Use of information and communication technology (ICT) should be considered. The low water use efficiency in the MMI projects can also be attributed to poor information on actual water demand both at reservoir and at field level. There is a wide communication gap between the irrigation department, water users associations, and the farmers about timing and quantity of water release. ICT can be used to build a good information feedback system on when to release water and how much adjusting to the localized rainfall pattern. Further, GIS and RS tools can be easily applied to quantify the area irrigated, stages of crop, and irrigation water requirement.

3.5 Performance of the Canal Irrigation Projects

Irrigation sector in India is known for its ***Build–Neglect–Rebuild*** approach. But once a project is constructed it is of paramount importance that the project serves the purpose for which it was constructed. In all the cases, a good management approach targeting higher performance is missing. Further, most of the existing MMI schemes in the country were planned and developed as protective irrigation systems to supplement rainfall. Over the years, changing climate, population growth, water and food demands made these MMI schemes as productive irrigation systems, ensuring an assured crop with stable income to farmers. However, due to problems in effective water distribution, always inequity existed in water availability between head and tail reaches of the canal system. There are hardly any mechanisms in place to get feedback from the farmers on whether water is being delivered to them or not in terms of quality and quantity. This results in poor accountability in irrigation projects and farmers supplementing canal irrigation with other private sources like groundwater which in turn increased the cost of cultivation (Hiremath et al. 2016).

Looking at the existing performance of the MMI projects, most of these MMI projects are underperforming thus indicating the need for revitalization to meet the increasing future water and food demands. An analysis of the 35 MMI project efficiencies in the country by the Central Water Commission (CWC) had revealed that the average water use efficiency is only about 36%. Out of the 35 MMI projects, only one project had the efficiency of more than 60%, 19 projects (54%) had the efficiency between 36 and 60%, and 14 projects (40%) had the efficiency of less than 36%. Among the efficiencies, average conveyance efficiency is about 69% and on-farm application efficiency is only about 55% (Table 3.3). Out of the 35 projects, about 67% of the medium irrigation projects and 52% of the major irrigation projects showed efficiency above the mean efficiency levels (Figs. 3.5 and 3.6). The major trigger in improving the overall efficiency will be the on-farm application efficiencies. This means primarily that soil, crop, and water management technologies and improved practices are important in improving the on-farm application efficiency. Depending upon the soil and crops, the water management practices vary much. As such current adoption of these technologies or improved practices is only about 12% indicating the scope for intensifying the adoption of these technologies at farm level (Palanisami et al. 2014). Considering the option of targeting low hanging fruits, priority for future investment can be given for those projects with a current mean efficiency level of above 36% so that the return to investment will be quicker and comparatively higher.

Table 3.3 Water use efficiency of major and medium irrigation projects, India

S.no	Name of project	Major/ Medium	State	Culturable command area (hectare)	Conveyance efficiency (%)	On-farm application efficiency (%)	Overall project water use efficiency(%)
1	Koil Sagar Project	Medium	A.P	11,700	83	75	62
2	Bhairavanithippa Project	Medium	A.P	4856	86	67	58
3	Augmentation Canal Project	Major	Haryana	85,443	79	72	57
4	Matatila Dam Project	Major	U.P	179,880	68	80	54
5	Eastern sone canal	Major	Bihar	126,055	73	73	53
6	Vamsadhara Project	Major	A.P	82,087	91	58	53
7	Durgawathi Irrigation Project	Major	Bihar	21,110	81	65	53
8	Dholabaha Dam Project	Medium	Punjab	2600	74	71	53
9	Kamla Irrigation Project	Major	Bihar	28,331	70	74	52
10	Naugarh Dam Irrigation Project	Major	U.P	64,221	71	70	50
11	Ahraura Dam Irrigation Project	Medium	U.P	14,964	70	70	49
12	Tungabhadra High Level Canal	Major	A.P	45,800	81	58	47
13	Godavari Delta System (Sir Arthur Cotton Barrage)	Major	A.P	410,108	83	54	45
14	Sri Ram Sagar Project	Major	A.P	371,054	78	57	45
15	Rajolibanda Diversion Scheme	Major	A.P	35,410	82	51	42
16	East Baigul Reservoir Project	Medium	U.P	16,605	64	65	42
17	Krishna Delta System (Prakasam Barrage)	Major	A.P	529,000	87	46	40
18	Nizam Sagar Project	Major	A.P	93,659	87	45	39
19	Walmiki Sarovar Project	Major	U.P	6271	62	62	38
20	Pili Dam Project	Medium	U.P	4044	58	65	38

(continued)

Table 3.3 (continued)

S.no	Name of project	Major/ Medium	State	Culturable command area (hectare)	Conveyance efficiency (%)	On-farm application efficiency (%)	Overall project water use efficiency(%)
21	Ranjit Sagar Dam Project	Major	Punjab	300,000	51	65	33
22	Tungabhadra Low Level Canal	Major	A.P	61,163	72	45	32
23	*Saran Canal Irrigation Project	Major	Bihar	360,000	53	69	30
24	Upper Morhar	Major	Bihar	5725	77	40	30
25	Kurnool–Cuddapah Canal System	Major	A.P	65,465	62	45	28
26	Gandipalem Project	Medium	A.P	6478	73	38	28
27	Gajuladinne (Sanjeevaiah Sagar)	Medium	A.P	10,300	57	45	26
28	Nagarjuna Sagar Project	Major	A.P	889,000	56	39	22
29	Somasila Project	Major	A.P	54,650	56	32	18
30	Kaddam Project	Major	A.P	27,519	51	36	18
31	Srisailam Project	Major	A.P	59,900	50	34	17
32	Narayanapuram Project	Medium	A.P	15,855	47	32	15
33	Yeleru Project	Major	A.P	27,240	50	28	14
34	Nagal Lift Project	Major	Haryana	35,721	48	27	13
35	Upper Manair Project	Medium	A.P	6984	–	–	–
Weighted average value					69	55	36

Source CWC (2016)

3.6 Interstate Water Sharing and Disputes

Poor water management and steadily increasing water demands create pressure for efficient water allocation and management. In addition, increasing conflicts on water allocations in the interstate rivers is alarming in the recent years. All the major rivers in the country are interstate, and water management is basically a state subject on which the Centre has very limited control. A major challenge facing many states is the absence of permanent and efficient dispute resolution mechanisms for interstate rivers. The Interstate Water Disputes Act of 1956 allows the states to discuss and resolve the conflicts before engaging in adjudication and subsequently, ad hoc tribunals are established whenever conflicts between two states cannot be resolved by mutual discussion. These interstate water tribunals often contribute to long-drawn

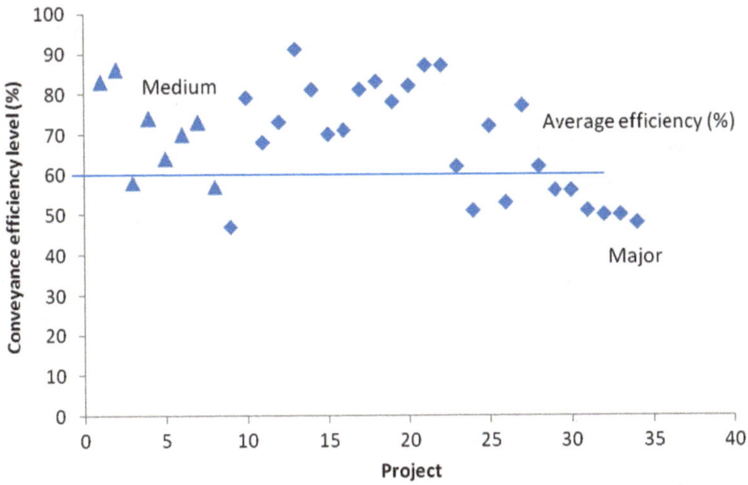

Fig. 3.5 Conveyance efficiency of projects in India (%)

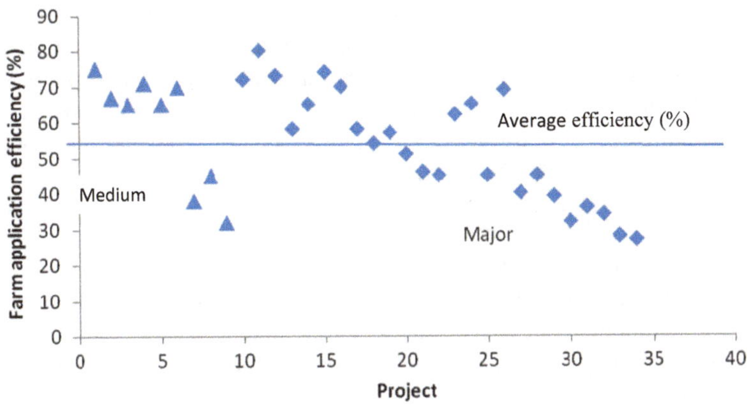

Fig. 3.6 On-farm application efficiency of projects in India (%)

negotiation processes between the conflicting states and in many cases resulted in hardening of the situations. Further, states are reluctant to accept its verdicts for one reason or another due to growing conflicts with added political pressures.

The tribunal awards are non-binding to the states to follow. There is no particular institution or agency that is responsible to implement the tribunal decisions. Further, the long pending tribunal's awards may not be relevant due to the delays, and changing situations from the time when the tribunal started its initial hearings. In several cases, the political situation in the states also changes over time thus making the situation more complex. Water has now assumed the role of a political weapon for interstate water disputes because of the emotional nature of the water problems. It has been observed that due to prevailing conflicts and court cases in sharing the interstate waters, the party states are not taking sincere efforts in improving the water use efficiency in the canal irrigation projects and also understate the water use data. This inefficiency in water use in the interstate river projects could pose a major threat to the main objective of improving the efficiency, productivity, and income per unit of water. In the midst of persistent water sharing conflicts, the water management research and technology adoption are given less priority in these project commands. Hence, the framework for water management research and improving water use efficiencies in the interstate river projects need a transformative change in the coming years. A sustained political will is essential to take some hard decisions in the future, either at the central or state levels.

3.7 Prioritizing Canal Modernization

As discussed in the previous chapter, some of the key drivers of change in water demand are climate change, population growth, and growing water scarcity. But there are others, including, changing diets and food demand, competition between sectors, deteriorating water quality, and falling groundwater levels that affect the water demand. The MMI projects therefore require modernization and not rehabilitation as rehabilitation merely means restoration to the original conditions. Irrigation project modernization means improving the efficiency of water use through added management interventions. Modernization could incorporate modern technological advances that could offer significant opportunities for addressing the challenges faced by the irrigation sector.

This will primarily utilize remote sensing to carry out basin-wide water accounting, crop pattern and estimation of actual evapotranspiration, modern information and communication systems to access real-time data for decision-making at system level and farmer level. The doppler-based flow monitoring in the canal system is shown in Box 3.1.

Several states have been implementing irrigation modernization programmes, which include mainly lining of the canal system, improvement of canal networks, drainage structures, and installation of automated gate operation. Modernization of irrigation systems incorporates not only strengthening canal infrastructure but also upgrading operations and management of the canal system for better water delivery to farmers both in the head and tail reaches of the system, and further improves water use efficiency and crop productivity. In recent years, improved agricultural production and marketing components have also added as part of the modernization programme. However, most of the projects even after the interventions are completed, have not improved their overall performance in terms of equity in water distribution, water use efficiency, crop yields, and farm income.

Box 3.1. Flow Monitoring System in Ghataprabha Project, Karnataka

Online Doppler-Based Flow Monitoring System: A Doppler-based flow monitoring system is installed in the GLBC canal network having first measuring point at Hidkal Dam monitoring live reservoir levels and capacity. This system covers 50 strategic monitoring points along the main canal and distributaries. The total command area of GLBC is 1,61,887ha covered under 150 distributaries. However 75% of water is conveyed through 45 large distributaries. Hence this system is able to continuously monitor 75% of flow in the command area along with 5 major milestones on the main canal. The sensor units, based on side looking Doppler, are powered by solar panels installed nearly. Flow monitoring data is transmitted through GSM/GPRS network to central server located at KNNL Office in Bangalore. Processed information is communicated to concerned officials and water users in the form of SMS as well as online. About 10% of water loss is minimized through precession control and detection.

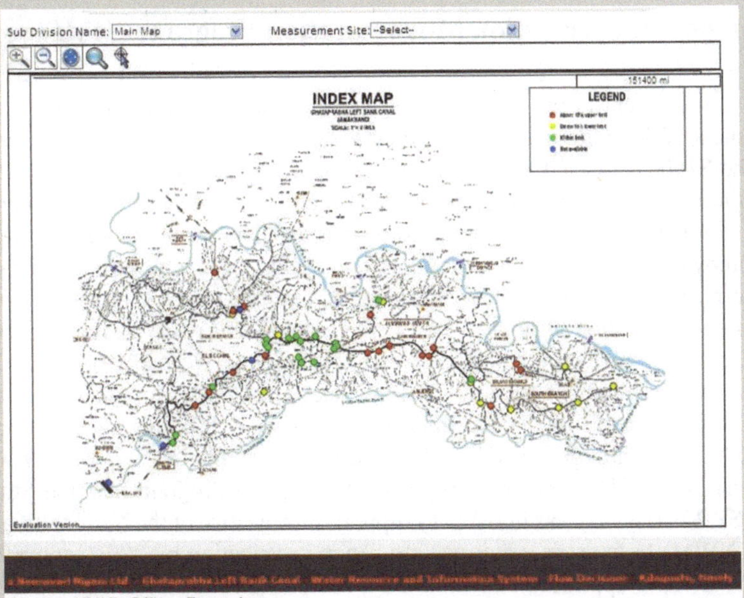

Source: KNNL Office, Bangalore

There are examples of good programmes supported by GoI and International Finance Institutions (IFIs) to emulate. In April 2012, MOWR, RD&GR through CWC with an objective to improve safety and operational performance of selected dams, along with institutional strengthening with system wide management approach, embarked upon the Dam Rehabilitation and Improvement Project with World Bank assistance at a cost of INR 3,466 crores. The DRIP has been successful in bringing together dam owners, engineers, scientists, academicians, industries, the World Bank as well as renowned dam safety professionals. In addition to rehabilitation of dams, other important activities included developing guidelines, O&M manuals, Emergency Action Plans, developing a web-based asset management tool, conducting dam safety conferences for knowledge assimilation, national and international training programmes, technical exposure visits, etc. The Dam Safety Bill of 2018 provides an institutional mechanism for the safety of dams.

Another good example of a national programme is the ongoing National Hydrology Project (NHP) implemented by the GoI with financial support from the World Bank. The MoWR and RD&GR have introduced NHP as a central sector scheme, where the Ministry fully funds the establishment of a water resources information system. The project cost is equally divided between the Government of India (50%) and the World Bank (50%). For implementing NHP, DoWR, RD&GR has set up a permanent National Project Management Unit which is responsible for overall project planning, management, and coordination.

3.8 Support for Irrigation Modernization Program (SIMP)

A recent development in the area of canal irrigation modernization in India is the establishment of Support for Irrigation Modernization Program (SIMP). The overall goal of the SIMP is to establish a system to support the states in preparing future irrigation modernization programmes for MMI schemes that sustain and enhance system performance by improving water use efficiency, land, crop, and water productivity, thereby contributing to raising and sustaining farmer incomes. In fact, irrigation modernization refers to the process of upgrading infrastructure, operations, and management of irrigation systems to sustain the water delivery service requirements of farmers and optimize production and water productivity (ADB 2019).

The Asian Development Bank (ADB) has supported the government over the past decade to develop strategies to improve India's water use efficiency and crop water productivity through several studies and projects. The first major study was the National Water Use Efficiency Improvement Support Program Scoping Study (NWUEISP 2014) which was initiated to support the objective of the National Water Mission (NWM) to improve water use efficiency in irrigation by 20%. Innovative tools such as rapid appraisal, benchmarking, remote sensing, surface and energy balance algorithm for land (SEBAL), etc., are suggested.

The SIMP is a natural progression by the ADB and the GoI using the study results and projects outlined above with a view to building on the experience gained and expanding the modernization of MMI schemes throughout India in a more systematic and streamlined manner.

3.9 Irrigation Modernization Plans (IMPs)

To produce a comprehensive irrigation modernization plan (IMP) a detailed study is required of each irrigation scheme selected for modernization. This plan will be prepared on behalf of the State. Once a State has gained experience in the process, the State through the State Irrigation Modernization Unit may choose to carry out its own IMP preparation. Key features of the modernization plan among other things will include: (a) current and future predicted water demand and water balance for the scheme, (b) condition and performance of the irrigation and drainage water delivery services, (c) current agricultural practices, opportunities, and pathways for improvement, (d) current and future potential practices related to conjunctive use and management of surface and groundwater resources, (e) current and future potential practices related to the use, management, and costs of energy for pumping, (f) costs and benefits of modernization interventions.

3.10 Assessing and Monitoring Scheme Performance

The first step before implementing the modernization of an MMI scheme is to assess its current performance and constraints using rapid performance assessment techniques, incorporating FAO's Rapid Appraisal Process (RAP) to provide a detailed assessment of issues influencing the management, operation, and maintenance of I&D schemes. The key performance indicators (KPIs) from the RAP include external and internal indicators where the external indicators compare input and output of an irrigation system to examine the overall performance. These indicators are expressions of various forms of efficiency, for example, water use efficiency, crop yield, crop water productivity, and increased income, economics of investment. But they do not provide any details on what internal processes lead to these outputs and what should be done to improve the performance. They, however, could be used for comparing the performance of different irrigation projects, nationally or internationally. Tools of the performance assessment will include mainly partial budgets, cost curves, production functions, and double-difference method. Once these external indicators are computed, they are used as a benchmark for monitoring the impacts of modernization covering several key components.

3.11 Improving Future of Canal Irrigation—The Way Forward

To address the present and future water and food security, the GOI (CWC 2020) has identified the following focus areas in irrigation.

- Improving the system's overall water use efficiency.
- Adoption of the piped distribution system in place of open canal system mainly to reduce conveyance losses and improve delivery.
- Investment in pressurized (micro) irrigation systems in canal commands and participatory irrigation management.
- Repair, renovation, and restoration (RRR) of existing water bodies.
- Construction of more minor irrigation structures for regulation and improvement in the management of groundwater.
- Increase groundwater potential by recharge techniques.
- Inter-basin transfer of river water by interlinking of rivers.

In order to bridge the gap between irrigation potential created and utilized, it is important to integrate the development activities with command area development (CAD) so that duplication of works can be avoided. Canal modernization programme should be planned in such a way that it should include those interventions that could help minimize the IPC–IPU gap and increase WUE as well. Prioritizing investment options needs to be examined taking into account both structural and non-structural interventions such as agriculture and marketing. Investment in eastern region will have lower capital and working expenses as compared to other regions in India. The use of information and communication technology (ICT) should form an integral part of canal irrigation modernization. Further, GIS and RS tools must be applied to quantify the area irrigated, stages of crop, and actual irrigation water requirement.

It is also possible to save more water when the WUE of the projects is increased from the current level. For example, moving from 30% efficiency level to achievable 60% efficiency level will result in a saving of about 102 BCM. When the WUE in groundwater is increased from the present level of 55% is increased to 75%, a saving of another 44 BCM is possible (CWC 2014). While doing the effective modernization, it is possible to follow up with higher efficiencies under different water use and methods as indicated by the Central Water Commission (CWC 2014) Table 3.4.

The future of MMI modernization needs to address the major outcomes such as increasing farmer income, optimizing agricultural production, raising water use efficiency, raising productivity of water, and sustaining the physical infrastructure over time. Hence, the focus should be on modernization of the physical system combined with modernization of institutional, MOM, and agricultural aspects including technology adoption, crop diversification, marketing, and crop insurance. With the existing technologies like GIS and mobile-based ICT, it is now possible to assess water needs in different parts of a canal command area and water can be delivered as per the crop demand. In the future, digital agriculture will play a key role in training,

Table 3.4 Suggested irrigation efficiencies under different water use and methods

Water use and methods	Efficiency (%)
Irrigation efficiencies	
Conveyance:	
—Through unlined canal for surface water	55–60*
—Through lined canal for surface water	70–75*
Application for both surface and groundwater:	
—Flood irrigation	65
—Furrow irrigation	80
—Sprinkler	85
—Drip	90
Overall efficiency for surface water system	30–65
Overall efficiency for groundwater system	65–75

* Conveyance efficiency of the canal depends on many factors such as length of the canal, type of soil, material used for lining, etc.

knowledge exchange, and technology adoption by farmers. The sustainability of the modernized systems has to be ensured with asset management plans and MOM cost recovery. A blueprint approach should be avoided and a *cafeteria approach can* be promoted. Since climate change impacts, especially, weather extremes are already visible in agriculture sector, a re-estimation of water requirements at both scheme and farm level for the designed and actual cropping pattern needs to be done for better water allocation decisions. It is expected that when the modernization is completed, a sustained increase in farmers' income should be ensured. The SIMP can be taken as a future model for modernization of the MMI projects in the country.

References

ADB (2019) Support for irrigation modernization program. Concept Note, Asian Development Bank, Manila, Updated 11 November

CWC (2014) Guidelines for improving water use efficiency in irrigation, domestic & industrial sectors. Central Water Commission, New Delhi, India

CWC (2016) Summary report on water use efficiency studies for 35 irrigation projects. Central Water Commission New Delhi, India

CWC (2015)Water and related statistics. Water Resources Information System Directorate, Information System Organization, Water Planning and Project Wing, Central Water Commission, New Delhi, India

CWC (2020) Annual report (2019–20), New Delhi available from web link http://cwc.gov.in/sites/default/files/arcwc2019-20.pdf

Dheeravath V, Thenkabail PS, Chandrakantha G, Noojipady P, Reddy GPO, Biradar CM, Gumma MK, Velpuri M (2010) Irrigated areas of India derived using MODIS 50 m time series for the years 2001–2003. ISPRS J of Photo and Rem Sen 65:42–59

Hiremath BD, Shah P, Chaudhary S (2016) ICT interventions to improve performance of canal irrigation sector in India. ICTD'16, Proceedings of the Eighth International Conference on Information and Communication Technologies and Development, Art. No. 51. Doi:https://doi.org/10.1145/2909609.2909635

Malhotra SP, Raheja SK, David S (1983) A performance monitoring study of the Warabandi system of irrigation management in North-west India. The Ford Foundation, New Delhi

MFLW (2015a). India: innovations for more food with less water task 2 report prepared by Lahmeyer international in association with Lahmeyer international Indiabets consulting services, centre for environment and development, and total management services. Project No. 45072–001, Regional Technical Assistance TA7967-REG, Asian Development Bank, Manila, April

MFLW (2015b) Innovations for more food with less watertask1: final report. report prepared by the international water management institute, Project No. 45072, Regional Technical Assistance TA-7967-REG, Asian Development Bank, Manila, May

NWUEISP (2014) Scoping study for a national water use efficiency improvement support program. Report prepared by Martin Burton and Amarjit Dhingra, Project No. SC 100903 (IND) and SC 100905(IND), Asian Development Bank, Manila

Palanisami. K. 2012. Water Resources Group: Expert Input on Preparing Detailed Project Report for Water Enabled Growth in Karnataka. Water Management Interventions for Narayanpur Right Bank Canal and Ghataprabha Project in Krishna Basin in Karnataka. International Water Management Institute (IWMI) ICRISAT Campus, Patancheru, Hyderabad

Palanisami K, Ranganathan CR, Nagothu US, Kakumanu KR (2014) Climate change and agriculture in India: Studies from selected river basins. pp 101. Routledge, New Delhi

Palanisami K (1984) Irrigation water management- the determinants of canal water distribution in India. Agricole Publishing Academy, New Delhi

Sharma BR, Banerjee P (2020) Gap between irrigation potential created and utilized for surface irrigation in India: reasons, cost, government policies and the way forward. Draft, Indian Council for Research on International Economic Relations, New Delhi

Chapter 4
Sustaining Groundwater Irrigation

Abstract India's annual groundwater extraction is 239.16 bcm compared to the total annual groundwater recharge of 437.60 bcm. Stage of groundwater extraction (as on 2022) is 60%. Out of the total 7089 groundwater assessment units in the country, about 14% are "Over-exploited", 4% are "Critical", and 12% are "Semi-critical" units. Irrigated agriculture accounts for about 87% of total annual groundwater extracted. As on 2020–21, the average failure rate (drying up) of open dugwells is about 56% and borewells is 65% in critical and over-exploited areas indicating the negative externalities in irrigated agriculture. It is predicted that the demand of energy may rise between 3.2 and 4.9% by 2030. At national level, groundwater pumping by electric and diesel pumpsets with an average pumping of 1600 h/year accounts for about 14.38 million tonnes of carbon emissions. Groundwater recharge is affected by widespread impact of climate change. Managed Aquifer Recharge (MAR) is a promising adaptation measure. Water conservation through watershed development programme, adoption of micro irrigation, crop diversification towards less water intensive crops, addressing water–energy nexus and use of solar and wind energy for pumping, regulatory measures, and groundwater governance are suggested in sustaining the groundwater irrigation in the country.

Keywords Groundwater exploitation · Well failure · Subsidized electricity · Carbon foot print · Climate change and groundwater · Water–energy nexus · Micro irrigation · Groundwater recharge · Regulations and acts · Watershed development

4.1 Introduction

Groundwater in India is the primary source of water, which accounts for more than 60% of the irrigated agriculture and 85% of the drinking water. Increasing population, food demand, urbanization, and industrialization need more water, which makes groundwater an important resource. India has achieved food security, which can be mainly attributed to introduction of high yielding varieties, mechanization, irrigation

K. Palanisami and U. S. Nagothu, *India's Water Future in a Changing Climate*,
Advances in Geographical and Environmental Sciences,
https://doi.org/10.1007/978-981-97-1785-9_4

Table 4.1 Dynamic groundwater resources assessment of India, 2022

1	Total annual groundwater recharge	437.60 bcm
2	Annual extractable groundwater resources	398.08 bcm
3	Annual groundwater extraction	239.16 bcm
4	Stage of groundwater extraction	60.08%

Source (Central Ground Water Board 2022)

Table 4.2 Groundwater resources assessment over years

Sl. no	Groundwater resources assessment	2004	2009	2011	2013	2017	2020	2022
1	Annual groundwater recharge (bcm)	433	431	433	447	432	436	438
2	Annual extractable groundwater resource (bcm)	399	396	398	411	393	398	398
3	Annual groundwater extraction (bcm)	231	243	245	253	249	245	239
4	Stage of groundwater extraction (%)	58	61	62	62	63	62	60

Source (Central Ground Water Board 2022)

technologies, and use of plant protection chemicals promoted by the green revolution in the 60 and 70 s. This has made groundwater as an integral part of irrigated agriculture in the country.

Over the last 50 years, the number of borewells has increased from 1 to 20 million and has strongly proven that India is the world's largest user of groundwater (Bekele 2021). In India, the total annual groundwater recharge is 437.60 billion cubic metres (bcm), and annual groundwater extraction is 239.16 bcm (Table 4.1). As per the report of CCWB 2022, the average groundwater extraction is about 60%, which varies across regions. Further, It was observed that the annual groundwater recharge and the annual groundwater extraction has shown a slow but steady increase from 2004 to 2022 period, and the variations between the years might be due to changes in rainfall pattern during these periods (Table 4.2).

4.2 Status of Groundwater Development

Assessment of Groundwater Extraction

Among various sectors, agriculture is the predominant consumer of groundwater, and accounts for about 87% of the total annual groundwater extracted. Consumption by domestic and industrial sectors together accounts for the remaining 13%. In some regions the groundwater used for domestic purpose was observed to be more than 40%. This was in the case of Kerala, Manipur, Meghalaya, Mizoram, Nagaland,

Table 4.3 Categorization of assessment units, 2022 (Blocks/Mandals/Firkas/Taluks)

Sl. no	Category	Number of assessment units		Recharge worthy area		Annual extractable groundwater resource	
		Number	%	Lakh sq.km	%	Bcm	%
1	Safe	4780	67	16.18	66	291.88	73
2	Semi critical	885	12	3.03	12	47	12
3	Critical	260	4	0.77	3	13.02	3
4	Overexploited	1006	14	4.30	17	46.05	12
5	Saline	158	2	0.4	2	NA	NA
	Total	7089		24.69		398.08	

Source (Central Groundwater Board 2022)

Tripura, Jammu and Kashmir, Ladakh, Andaman and Nicobar Island, Delhi, Goa, and Chandigarh where surface water resources for domestic use are comparatively limited.

Categorization of Assessment Units

As per the report of Dynamic Ground Water Resources of India (Central Groundwater Board 2022), out of the total 7089 groundwater assessment units (Blocks/Mandals/Talukas/Firkas) in the country, about 1006 units located in various states (14%) have exceeded the annual replenishable groundwater recharge and hence categorized as "Over-exploited" (Table 4.3). The over-exploited units are predominantly concentrated in the northwestern part of the country including the parts of Punjab, Haryana, Delhi, Gujarat, Rajasthan, and Western Uttar Pradesh, and the southern India including the parts of Karnataka, Tamil Nadu, Telangana, and Andhra Pradesh. Groundwater extraction between 90 and 100% is observed in 260 (4%) assessment units and categorized as "Critical". There are 885 (12%) units categorized as "Semi-critical" where the groundwater extraction is ranging between 70 and 90%. About 4780 (67%) units are categorized under "Safe" units, where the groundwater extraction is less than 70% and the remaining 158 assessment units (2%) are categorized as "Saline" (Table 4.4).

4.3 Climate Change and Groundwater

Climate change is emerging as a global threat, which influences the natural resources use and also significantly affects the social and economic activity of the human being. The average temperature is predicted to rise from 0.5–1 °C during 2020–2029 and 3.5–4.5 °C in 2090–2099 in most of the Indian landmass below the Ganges plain (IPCC 2001). As India receives more of its water from rainfall than snowmelt,

Table 4.4 State-wise categorization of blocks/Mandals/Talukas in India, 2022

Sl no	State/union territories	Total no. of assessed units	Safe		Semi critical		Critical		Overexploited		Saline	
			Nos	%	Nos	%	Nos	%	Nos	%	Nos	%
1	Andhra Pradesh	667	598	89.7	19	2.8	5	0.7	6	0.9	39	5.85
2	Arunachal Pradesh	11	11	100.00								
3	Assam	28	27	96.43	1	3.57						
4	Bihar	535	469	87.66	46	8.60	12	2.24	8	1.50		
5	Chhattisgarh	146	116	79.45	24	16.44	6	4.11				
6	Delhi	34	4	11.76	8	23.53	7	20.59	15	44.12		
7	Goa	12	12	100.00								
8	Gujarat	252	189	75.00	20	7.94	7	2.78	23	9.13	13	5.16
9	Haryana	143	36	25.17	9	6.29	10	6.99	88	61.54		
10	Himachal Pradesh	10	10	100.00								
11	Jharkhand	263	241	91.63	11	4.18	6	2.28	5	1.90		
12	Karnataka	234	139	59.40	35	14.96	11	4.70	49	20.94		
13	Kerala	152	122	80.26	27	17.76	3	1.97				
14	Madhya Pradesh	317	226	71.29	60	18.93	5	1.58	26	8.20		
15	Maharashtra	353	272	77.05	62	17.56	7	1.98	11	3.12	1	0.28
16	Manipur	9	9	100.00								
17	Meghalaya	12	12	100.00								
18	Mizoram	26	26	100.00								
19	Nagaland	11	11	100.00								
20	Odisha	314	300	95.54	8	2.55					6	1.91

(continued)

Table 4.4 (continued)

Sl no	State/union territories	Total no. of assessed units	Safe		Semi critical		Critical		Overexploited		Saline	
			Nos	%	Nos	%	Nos	%	Nos	%	Nos	%
21	Punjab	153	17	11.11	15	9.80	4	2.61	117	76.47		
22	Rajasthan	302	38	12.58	20	6.62	22	7.28	219	72.52	3	0.99
23	Sikkim	6	6	100.00								
24	Tamil Nadu	1166	463	39.71	231	19.81	78	6.69	360	30.87	34	2.92
25	Telangana	594	494	83.00	80	13.60	7	1.20	13	2.20		
26	Tripura	59	59	100.00								
27	Uttar Pradesh	836	557	66.63	169	20.22	47	5.62	63	7.54		
28	Uttarakhand	18	14	77.78	4	22.22						
29	West Bengal	345	232	67.25	31	8.99	22	6.38			60	17.39
30	Andaman and Nicobar	36	35	97.22							1	2.78
31	Chandigarh	1			1	100.00						
32	Dadra and Nagar Haveli	1							1	100.00		
	Daman and Diu	2							2	100.00		
33	Jammu and Kashmir	20	19	95.00	1	5.00						
34	Ladakh	8	7	87.50	1	12.50						
35	Lakshadweep	9	7	77.78	2	22.22						

(continued)

Table 4.4 (continued)

Sl no	State/union territories	Total no. of assessed units	Safe		Semi critical		Critical		Overexploited		Saline	
			Nos	%	Nos	%	Nos	%	Nos	%	Nos	%
36	Puducherry	4	2	50.00			1	25.00			1	25.00
	Grand total	7089	4780	67.43	885	12.48	260	3.67	1006	14.19	158	2.23

Note Blocks—Bihar, Jharkhand, Odisha, Chhattisgarh, Madhya Pradesh, Uttar Pradesh, Uttarakhand, Rajasthan, Haryana, Punjab, Kerala, Manipur, Mizoram, Tripura, West Bengal; Taluks—Karnataka, Goa, Maharashtra, GujaratMandals—Andhra Pradesh, Telangana District—Arunachal Pradesh, Himachal Pradesh, Assam, Meghalaya, Nagaland, Sikkim, Jammu and Kashmir Valley, Ladakh, Dadra and Nagar Haveli, Daman and Diu, Islands—Andaman and Nicobar, Lakshadweep; Firka—Tamil Nadu; Region—Puducherry, UT—Chandigarh, Tehsil—Delhi.

Source (Central Ground Water Board 2022)

variation in rainfall due to climate change will result in less soil moisture and higher crop evapotranspiration (ET) demand in summer (Shah 2009).

Further, the extensive impact of climate change prompting groundwater recharge is also observed due to the intensification of precipitation. This "intensification" of precipitation is significant in the tropics, where the most of the world's population is projected to live by 2050 (Gerland et al. 2014). Precipitation (P) and evapotranspiration rates (ET) are determined by climate change and land cover whereas the transmission of water surplus (P-ET) to an underlying aquifer is indicated by underlying soil and geology. The increasing rate of evapotranspiration reduces the water surpluses. It is estimated that evapotranspiration has increased by 10% on an average between 2003 and 2019 (Pascolini-Campbell et al. 2021). Snowmelt occurring in continental northern latitudes, mountainous and polar regions affect water flow regimes that impact groundwater recharge. In the case of temperate regions, less snow accumulation, earlier snowmelt, increased winter precipitation and an intensified frequency of rain-on-snow events are likely to occur (Harpold and Kohler 2017). This causes reduction in seasonal duration and magnitude of the groundwater recharge, which reduces water storage in catchments and increases the severity of extreme summer low flows (Dierauer et al. 2018).

Heavy rainfalls triggered by climate change have the potential to increase groundwater recharge and mobilize pollutants such as chloride and nitrate in the Vadose zone immediately above aquifers particularly in drylands and temperate regions. Further, groundwater is also contaminated by surface runoff, which contains waste and stored chemicals on or near the ground, penetrated into aquifers (WHO 2022).

4.4 Groundwater Development and Externalities

Groundwater is considered as a common pool resource with open access. This leads to overexploitation of groundwater resulting in decrease in water availability and increase in pumping costs, which generate negative externalities. In fragile resource regions, investment in wells by few affordable households excludes other users of groundwater which raises "free-riding" externalities. Contemporary externalities in groundwater extraction normally arising out of groundwater depletion are both stock and cost related. A few individuals with capital access strategically capture all the available groundwater using high tech drilling/pumping technologies, depriving the others, which is known as technological externality. In most of these situations, farmers incur significant financial losses by pumping unsustainable water by means of additional investments in drilling/deepening, increased pumping costs, and reduced crop income (Amami 2018).

Groundwater abstractors are likely to increase abstraction until their marginal private productivity exceeds the marginal private cost of abstraction. However, since the marginal social cost of the groundwater draft is higher than the private cost, significant amount of groundwater will be used than the socially desirable. Volumetric pricing of groundwater is mostly considered as superior method of promoting efficient groundwater but it can work only where the abstractors are few and large and where the culture of a metered water supply has formed deep roots. Furthermore, the groundwater pricing becomes difficult, if the groundwater abstractions are small, numerous, and dispersed (Shah 2014).

A classic case of well failure and externalities can be observed in Coimbatore district, Tamil Nadu state, which is one of the progressive and well irrigated regions in the state. It is seen that the failure rate (drying up) of open dugwells was 56% and borewells was 65% in 2020–21 resulting in both direct costs (foregone well investment costs) and indirect costs (foregone farm income) to the farmers. The total cost of failed wells was about Rs 5436 million for open wells, and Rs 3527 million for borewells indicating the magnitude of groundwater externalities induced by groundwater overexploitation in hard rock regions (Table 4.5).

Furthermore, besides technical factors (i.e., easy availability of pumps, subsidized electricity), the socio-economic factors also prompted the overexploitation process over the period. Social pressure obliged the farmers to remain in agriculture in spite of lesser earnings to maintain their social status within the society. Consequently, the externalities of well failure are internalized to certain extent owing to competitive deepening or drilling of wells. Indeed, social cost (i.e., the cost associated with the failed wells) is also increasing, as addition of one new bore well in the hardrock regions will result in the failure of other wells in the vicinity. In spite of the increasing well failure and declining irrigated area per well, higher prices for the farm products had compensated to some extent for the loss of irrigated area and therefore farmers still continue their investment in digging new wells. In certain situations, the change in cropping pattern towards high-value crops helped farmers to certain extent to bear the cost of externalities that arose. In several cases, farmers shift towards drip irrigation also helped to manage the situation. The cost of well water in hard rock regions is given in Box 4.1

Table 4.5 Wells failure and cost of well failure in Coimbatore District, Tamil Nadu

Sl. no	Details	Open wells		Bore wells	
		1991–92	2020–21	1991–92	2020–21
1	Total number of wells	75,621	49,787	2426	35,000
2	Failed wells (%)		56		65
3	Direct cost of failed wells (Rs million)		4182		2389
4	Indirect cost of failed wells (Rs million)		1254		1138
5	Total cost of failed wells (Rs million)		5436		3527

Note Since the farmers are investing every year new borewells, the average number of borewells in a typical 2 ha farm will be about 4. Further, all the open well farms are having the 2–4 bore wells. Considering this, the total number of borewells in the district was arrived at 35,000 in 2020–21. Accordingly the total number of wells in 2020–21 was around 84,787

4.5 Groundwater–Energy Nexus

Groundwater withdrawal through wells (mainly tube wells) directly depends on energy supply and the association between the two is often known as energy–groundwater–irrigation nexus. The energy–groundwater–irrigation nexus has significantly increased agricultural productivity, food and livelihood security as well as the social and economic development (Shah et al. 2003; Dubash 2007). Furthermore the abstraction of groundwater for irrigation has been closely linked with access to the free or subsidized electricity in the country (Sarkar 2020) and has led to perverse the groundwater–energy nexus. This mounting energy subsidy affected the groundwater resources and state's fiscal health. The emerging challenge in the future of groundwater development will be mainly due to lack of accountability on the part of farmers and utilities owing to surge in the groundwater demand and unmetered energy supply from free or subsidized power. The irrigation–energy nexus in different phases is given in Table 4.6.

Box 4.1 Well investment, groundwater pumping, and cost of irrigation water

Particulars	Tumkur, Karnataka	Coimbatore, Tamil Nadu
Average farm size (acres)	4.2	5.6
Average depth of new wells (ft)	800	1000
Capital cost/well (Rs)	65,000	80,000
Annualized cost (AC) of well (Rs/year)	40,127	49,387
Minimum water pumped (m^3/year)	6374	5980
Maximum water pumped (m^3/year)	9486	8750
Cost of water (Rs/m^3) with minimum water pumped	6.30	8.26
Cost of water (Rs/m^3) with maximum water pumped	4.23	5.64

Consumption of Electricity by Agricultural Sector

Agricultural electricity consumption has increased from 120,209 gigawatt hours (GWh) in 2009–10 to 168,913 GWh in 2014–15 and further to 207,791 GWh in 2018–19 (Fig. 4.1). Lower levels of groundwater table increase the electricity consumption per energized pumpset to draw groundwater from deep down. First, it takes more time for pumpsets to run in order to extract the groundwater; additionally higher power subsidies further aggravate the running time of pumpsets and making them inefficient (Narayanamoorthy, 1997; Palanisami and Suresh Kumar 2003). On the other hand, the higher use of electric power is further aggravated due to free or subsidized electricity tariffs.

Table 4.6 Irrigation–energy nexus in different phases

Phases	Policy initiatives	Challenges
Phase 1: Drive for rural electrification (1950–1965)	Promotion of electricity use in agriculture and increase the irrigated area	Capital for well investment was a major constraint and rate of electrification was rather slow
Phase 2: Increasing rural electrification with the introduction of subsidized power to farmers (1965–1995)	Increased power demand due to the spread of green revolution. Increased cultivation costs and pumping costs, few states introduced flat tariff and some considered free power with political pressure	Increasing water–energy nexus resulted in groundwater overexploitation in several locations. Power utilities faced increased financial burden
Phase 3: Groundwater management with increasing power subsidies (1995–2015)	Strategies and programmes introduced to manage the nexus in states where the groundwater extraction was alarming with increased well investment due to irrigation intensification with high-value crops. Well spacing norms introduced	Increased burden on state electricity departments. Power price increased in non-agri consumers (cross subsidization). Political lobby for maintaining free electricity was strong. No incentives to improve pump efficiency in the rural sector. Adoption level of water saving technologies is rather slow
Phase 4: In addition to the existing electric power use, promotion of solar power (2015 onwards)	In addition to the existing electric power use, both central and state government programmes to popularize solar power for pumping are introduced with subsidies	Irrigation–energy nexus continues. Solar power as an income generating option at farm level is seen. Efforts to convert electric pumpsets to solar for pumping are being taken in several states. Still farmers prefer subsidized/free electric power. Micro irrigation adoption is increasing but further expansion is facing practical difficulties

Source Authors' compilation

It is predicted that the demand of energy may rise between 3.2 and 4.9% CGR by 2030, whereas the aggregate load of pumpsets is also estimated to grow from 76 Giga-watt (GW) to 175–206 GW (5.8–6.9% CAGR). Pumpset usage pattern, intensity, and availability of viable alternatives determine fundamental difference between past and future growth in electricity demand, rather than electricity access or pumpset ownership (Ali 2018).

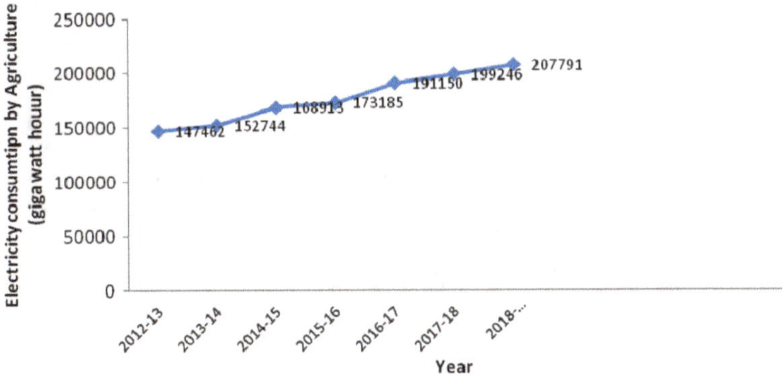

Fig. 4.1 Electricity consumption by agriculture

Cost of Energy in India (Rs/kwh)

It is also important to examine with the rising electric power cost, how the alternative source of energy such as solar and wind energy could be effectively tapped and used to minimize the increasing financial budget on the power utilities. Currently, wind is the major source of renewable energy, accounted about 33% of the total renewable energy, while sun shared 12–15% of the total renewable energy. Furthermore, the costs of wind power in India are also decreasing rapidly, which would further confirm the significance of wind energy. During the financial year 2020, the average cost of state electricity supplied in India was Rs. 6.15 per kWh, which has increased from Rs. 3.4 In 2009 (Fig. 4.2). Conversely, the solar power tariff witnessed a new low of Rs. 2.0/kWh in November 2020. In 2018, India's ministry of new and renewable energy released a solar-wind hybrid policy, which provides a framework to promote grid-connected hybrid energy through set-ups. This would use land and transmission infrastructure efficiently and manage the variability of renewable resources to certain extent.

Carbon Footprint of India's Groundwater Economy

Deep tube wells in the country are operating for about 1,600 h/year (i.e., diesel pumps operate for around 600 h and electric pumps for 800–1,000 h) with about 40% operating efficiency, and 25% or more transmission and distribution losses. About 9.1 kWh electric power is used to lift 1000 m^3/m. In contrast, diesel pumps are even less efficient, nevertheless they uplift water to a smaller head and do not incur the transmission and distribution losses. A litre of diesel provides the equivalent of 10 kWh of energy. In India, groundwater pumping by electric pumps and diesel pumpsets emitted 11.09 million tonnes and 3.29 million tonnes of carbon respectively, which together account for 14.38 million tonnes of carbon emissions. The groundwater irrigation alone accounts for about 16 million tonnes of carbon emissions, which is roughly 4% of India's total carbon emissions. Major concerns of the carbon footprint of India's groundwater economy are: (a) lifting a 1000 m^3/m using electricity emits 5.5 times more carbon than using diesel; and diesel pumps are

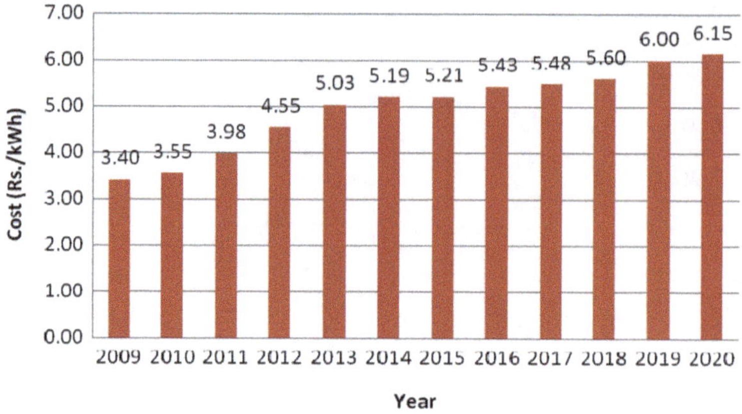

Fig. 4.2 Average electricity cost (Rs./kWh)

concentrated in eastern India enriched with alluvial aquifers; (b) Dynamic head over which groundwater is lifted to higher leads to higher energy use and carbon emission; and diesel pumps become extremely inefficient beyond a depth of 10–15 m, forcing switch to electricity which has a larger carbon footprint anyway. The biggest carbon emitters from groundwater extraction in India are the states of Tamil Nadu, Karnataka, Andhra Pradesh, and Gujarat, which have larger area under deep tubewell irrigation. Even though, the deep tubewells irrigate only 4.1 million of the 31 million ha under electric pumpsets, they account for about 2/3rd of carbon emission (Shah 2009). It is expected that by 2040, India will become the largest CO_2 emitter, and its CO_2 emission is projected to increase by 50%. However, India's per capita CO_2 emissions is still remaining below the global average (India energy outlook, 2021).

Balancing Water–Energy Nexus

Compared to conventional gravity irrigation methods, practising micro irrigation can save around 39 and 58% of water and energy respectively. For example, total energy saving at national level during 2020–21 period works out to be 31,891 lakh KWh where drip irrigation accounts for about 13,900 lakh kwh and sprinkler irrigation accounts for about 17,991 lakh kwh (Ministry of Agriculture and Farmers Welfare 2021) (Table 4.7).

Table 4.7 State-wise area under micro irrigation and energy saving, 2020–21

States	Drip irrigation		Sprinkler irrigation		Total energy saving under micro irrigation (lakh kWh)
	Area under drip irrigation as on 31.3.2021 (ha)	Energy saving (lakh kWh)	Area under sprinkler irrigation as on 31.3.2021 (ha)	Energy saving (lakh kWh)	
Andhra Pradesh	1,388,126	3426.09	519,165	1346.61	4772.70
Arunachal Pradesh	4017	2.52	3494	0.00	2.52
Assam	4208	1.27	16,217	0.53	1.80
Bihar	13,763	18.92	106,979	399.50	418.42
Chhattisgarh	31,311	63.83	331,283	989.82	1053.65
Goa	1386	3.96	1346	3.69	7.65
Gujarat	865,959	1687.50	764,933	1714.48	3401.98
Haryana	40,018	93.08	600,461	2256.88	2349.96
Himachal Pradesh	7934	1.19	6403	2.80	4.00
Jammu & Kashmir	1779		280		
Jharkhand	25,686	25.87	17,713	40.67	66.53
Karnataka	788,981	1764.22	1,304,281	1709.72	3473.94
Kerala	24,168	92.40	9096	28.49	120.89
Madhya Pradesh	330,335	682.69	258,600	761.61	1444.31
Maharashtra	1,349,979	3678.38	576,325	1536.61	5214.99
Manipur	358	0.19	7039	0.12	0.32
Meghalaya	308		307		
Mizoram	5551	7.09	1744	1.74	8.83
Nagaland	3589	0.82	6210	20.52	21.34
Odisha	29,425	75.64	115,396	336.80	412.44
Punjab	36,416	126.42	14,055	49.86	176.28
Rajasthan	287,619	698.04	1,730,876	6209.25	6907.29
Sikkim	6667	22.75	7943	11.35	34.10
Tamil Nadu	805,282	1190.13	348,010	124.79	1314.92
Telangana	203,279	103.82	74,871	21.70	125.52
Tripura	444	0.41	1651	1.61	2.02
Uttar Pradesh	41,273	63.69	227,897	86.77	150.46
Uttarakhand	12,737	2.86	10,300	1.30	4.15

(continued)

Table 4.7 (continued)

States	Drip irrigation		Sprinkler irrigation		Total energy saving under micro irrigation (lakh kWh)
	Area under drip irrigation as on 31.3.2021 (ha)	Energy saving (lakh kWh)	Area under sprinkler irrigation as on 31.3.2021 (ha)	Energy saving (lakh kWh)	
West Bengal	10,347	2.48	92,984	207.36	209.84
All India	6,320,945	13,899.87	7,155,859	17,991.68	31,891.55

Source Ministry of Agriculture and Farmers Welfare, 2021

4.6 Watershed Development Programmes and Groundwater Recharge

The watershed projects were largely aimed to conserve soil and water to raise farm productivity. Conserving soil implies increasing farm productivity and transferring good soils to the next generation. The major objective of watershed development projects was to achieve convergence of investments in irrigation at the field level, expansion of cultivable area under assured irrigation, increase the adoption of precision-irrigation, introduction of sustainable water conservation practices, and recharge of aquifers (PMKSY 2019). The major focus of government-supported watershed projects is to reach a target of 175 m.ha of degraded land throughout India. Over the past several decades, the government-supported watershed development under a range of central schemes viz., the Desert Development Plan (DDP) and the Drought Prone Areas Program (DPAP) mostly covering the arid-zones. Institutions such as the National Rainfed Area Authority (NRAA), State-level Nodal Agencies (SLNA), and District Watershed Development Units (DWDU) were also created for projects covering more than 25,000 ha.

However, for future focus, two types of watershed projects are important: (a) watershed projects that help to increase water supplies to irrigation, particularly in low rainfall areas such as the western and southern India, and (b) watershed projects that are built to develop rainfed agriculture particularly in catchment areas with more assured rainfall such as the hilly regions of central India. Of the two types of watershed projects, those in arid and semi-arid areas have been developed more with government investment as the structures like check dams, ponds, and tanks increased the availability of groundwater through recharge. Construction of these structures was done with collective action and the use of labour under the popular MGNREGA programme. In the hilly catchment areas, the watershed development works were comparatively limited and they are still under droughts with low crop productivity and limited livelihood options to the rural households. These regions can benefit from the decentralized water harvesting structures such as diversion channels and check dams mainly for protective irrigation and groundwater recharge.

As a result of watershed programmes, water conservation has created, on an average of 38 ha additional water storage capacity in a typical 500 ha watershed (Joshi et al. 2008). The increasing water storage capacity further contributed in reducing the rate of runoff and increasing the groundwater recharge. In hard rock regions the groundwater recharge is rather slow and depends upon the quantum of water stored in the ponds and check dams. Normally, wells close to the recharge structures recorded maximum water level rise on 48 day onwards followed by far away (more than 400 m) bore wells recorded maximum water level rise on 91 days after the start of filling the recharge structure. This is supported by minimum water level fluctuations in the nearby wells and maximum fluctuation in distanced wells. Overall, the zone of influence was about 400 m from the recharge structures than normally assumed 1000 m (Palanisami et al. 2009).

4.7 Groundwater Management Options/Initiatives

Groundwater management must be regulated to achieve the goals and objectives of the policies set by the government, or else it may be overexploited by entities and individuals acting on their own without any restrictions (UNESCO 2022). Groundwater management must explore and put into practice the operational decisions and practices that can guide groundwater abstraction and other activities which influence the groundwater and the aquifers.

Conventional approaches implicitly give emphasis on technical options for changing inflow or outflow from aquifers to bring mass balance of water within aquifers, considered as the primary points for management. To increase inflow, attempts to manage aquifers should focus directly on technologies to increase groundwater recharge, which consist of water harvesting structures to stimulate recharge, infiltration galleries, watershed management, and injection wells. Whereas for the outflow, most conventional approaches emphasize the technical interventions, which focus on well spacing and limitations on well depths or pump capacities that directly stabilize the ability to extract groundwater from the aquifers. Conventional approaches also give importance to technologies to increase the water use efficiency by means of irrigation, which includes promotion of drip and sprinklers irrigation. In case of urban water use, it often focuses on changes in household appliances such as washing machines and toilets.

As of now, India's principal tool for groundwater demand management is managing the energy–irrigation nexus with sensitivity, which can minimize the carbon footprint. However, as far as micro irrigation is concerned, energy savings are significant than water savings. Indeed, the water management structures and strategies aim to achieve maximization of both water productivity and energy efficiency under climate change regime (Shah 2009).

Managed Aquifer Recharge (MAR) is one of the promising adaptation measures to reduce hydrological variability and vulnerability to climate change (Palanisami et al. 2008). It involves in developing infrastructure or modifying the landscape to increase groundwater recharge. The MAR measure can be used to increase groundwater levels in over-exploited aquifers subject to declining yields, to store water for future use, to control saltwater intrusion, to prevent land subsidence, and to sustain the ecosystem functions. Further, groundwater impacts and groundwater potential are also important to identify the possibilities for the communities to improve drought monitoring through early warning systems. The International Groundwater Resources Assessment Centre (IGRAC) has adequate experience in MAR research at both regional and local levels, and its activities fall into three main pillars, such as MAR in-situ assessment, MAR information management, and regional mapping of MAR suitability.

Gravity Recovery and Climate Experiment (GRACE)

In the past two decades, the application of Gravity Recovery and Climate Experiment (GRACE) has gained remarkable attention in the study of groundwater in estimating groundwater storage. This GRACE data has been used by Long et al. (2016) to analyse groundwater depletion in northwestern India. The aquifers in the region showed a significant decline in groundwater storage that could threat both the agricultural output and groundwater sustainability. GRACE has contributed to improving the existing knowledge of hydro-climatic processes to facilitate water resources management in India. Further, improving the spatio-temporal resolution of GRACE data and additional hydro-climatic datasets from satellite missions (e.g., GPM, SWOT, and SMAP) may be instrumental in reducing the uncertainties (Soni and Syed 2015). In addition, number of schemes were initiated at state and central government levels to improve groundwater irrigation efficiency and energy use.

Several measures were taken up by the Government of India in the past to replenish/augment groundwater resources such as constitution of Central Ground Water Authority (CGWA) to regulate groundwater development and compilation of a document titled "Master Plan for Artificial Recharge to Ground water in India", which foresees the implementation of nearly 11 million rainwater harvesting and artificial recharge structures to increase the groundwater resources. In this context, the Ministry of Jal Shakti has circulated a Model Bill to all the States/UTs to enact suitable legislation for regulation of groundwater development, which includes provision of rainwater harvesting. For mapping the major aquifers, the CGWB has taken up National Aquifer Mapping & Management Programme (NAQUIM), which resulted in the Dynamic Ground Water Resources Assessment of India—2022. This ensures sustainability of the water resources, and categorizing over-exploited, critical, and semi-critical assessment units in the country. Further, many state governments are also implementing watershed development programmes, which primarily focused on groundwater conservation.

Further, to consolidate water conservation and management initiatives, the Ministry of Jal Shakti has launched "JalKranti Abhiyan" and aimed at consolidating water conservation and management initiatives involves all the stakeholders through a holistic and integrated approach. Similarly, during April 2020, the Atal Bhujal Yojana was implemented to improve groundwater management in the identified water-stressed areas, where emphasis was given on community participation and demand management. Furthermore, the schemes of the Government of India such as Pradhan Mantri Krishi Sinchai Yojana (PMKSY), Ground Water Irrigation (GWI), and Har Khet Ko Pani (HKKP) envisage the creation of additional irrigation potential from groundwater sources (Central Ground Water Board 2022). In 2020, the World Bank began support for the government's programme in over 9,000 water-scarce gram panchayats (village councils) in seven Indian states—from Haryana and Uttar Pradesh in the north, to Rajasthan, Gujarat, and Maharashtra in the west, to Madhya Pradesh in central India, and Karnataka in the south. Over 2,200 villages have drawn up water budgets that show how much groundwater is available, how much is estimated to be recharged, and how much can be set aside for agriculture, by far the largest user of this resource. The plans also focus on getting farmers to adopt high efficiency irrigation practices and broach the more complex issue of getting them to switch to less water intensive crops.

In Haryana, the state government's "Mera Pani, Meri Virasat" programme—My Water, My Heritage is incentivizing farmers to replace water-guzzling paddy with maize, millets, or pulses.

The World Bank has been working in collaboration with the Government of India to improve groundwater management in affected areas. Case studies of World Bank Independent Evaluation Group in Andhra Pradesh, Telangana, and Rajasthan found that in the absence of demand-side action, the supply-side measures like watershed management programmes, tank rehabilitation activities, and aquifer recharging, did not end up with sustainable use. Further, measures like surface water harvesting through farm ponds and check dams, water-efficient irrigation systems (e.g., more efficient drips and sprinklers), and less water intensive crops, should be integrated on the demand side for improved management and reduced depletion.

A World Bank project, which builds small irrigation facilities in the remote rainfed western region of West Bengal has significantly helped the farmers from disadvantaged tribal communities.

A World Bank report (2012) exposed a model that has been adopted in the state of Andhra Pradesh, which is one of the drought-prone areas of India. The state has demonstrated the large-scale success in self-regulation of groundwater use at the cost of USD 2,200 per village per year. As a result, the farm income has been doubled and the groundwater use was close to sustainable levels where in many cases, farmers reduced their water use voluntarily.

The state government of Punjab announced a unique scheme on the 14th of June 2018, which enables the farmers to earn money and save water for each unit for electricity they save. This scheme was targeted for optimum utilization and conservation of water and electricity. The Gujarat government started implementing the Surya Shakti Kisan Yojana scheme in 2018. More than 4000 farmers using Solar Irrigation Pumpsets (SIPs) through Suryashakti Kisan Yojana (SKY) have replaced their electric pumpsets resulting in a net reduction in carbon emissions and irrigation costs (Aditi 2022).

Other major scheme of the Government of India is the ambitious Pradhan Mantri Kisan Urja Surakshaevam Utthan Mahabhiyan (PM KUSUM) scheme to distribute solar pumps targeting 35 lac farmers. The scheme supports to set up stand-alone solar pumps and grid-connected solar irrigation facilities, which aimed to provide clean and reliable power and also doubling the farmer incomes (World Bank 2020).

4.7.1 Emerging Challenges in Groundwater Development and Management

Despite various initiatives taken by the Government of India, state governments, and other national and international organizations, still the groundwater development and management in the country faces challenges. The challenges are complex and multifold in the presence of the inter-linkages between water, environment, and socio-economic well being. These challenges primarily include:

Intricacy of aquifers and human behaviour

Intricacy of aquifers and human behaviour cannot be unravelled with a broad-brush approach. It is a big challenge to the policy makers to find feasible solutions to the groundwater problems that are spread over large landscapes.

Excessive irrigation

Lack of a comprehensive and integrated land-use planning framework caused declining water tables. Change to water intensive crops for more income has resulted in overuse of the irrigation water.

Unregulated extraction

Availability of cheap electricity shifted farmers' dependency from surface water to groundwater. The continued groundwater extraction has induced low recharge rate evidenced by the rise in the stage of extraction of groundwater. The rising groundwater depletion within aquifers was mainly attributed to an increase in the number of wells, its increasing abstraction capabilities which resulted in well failure, increased well investment, and reduced farmer income.

Groundwater Legislation

Regulating the utilization of groundwater to manage the growing demand is a huge challenge for India as the rights of groundwater and its extraction is owned by the landowners. It is a politically sensitive subject and hence any change of legislation in this context restricting the use of groundwater can be a big challenge for the governments.

Poor knowledge of groundwater management systems

Lack of clarity on the aquifer boundaries and weak local institutions due to the technical incompetence and insufficient financial support affect the local bodies' ability to manage groundwater.

Groundwater pollution

Groundwater quality has also deteriorated from the chemical contamination of fertilizers and pesticides. Groundwater in India has arsenic contamination on account of anthropogenic activities. Falling water tables have further increased the pollution risk, as water becomes more concentrated with pollutants.

Subsidies on electricity

Access to free or cheaper electricity and increased level of subsidies have contributed significantly to the overall carbon emissions from the agriculture sector.

Poor Technical Efficiency in Artificial Recharge

The soils in the hard rock regions have very poor infiltration capacity as they are associated with hard rock, crystalline rocks and sedimentary sandstone and limestone aquifers. Therefore the water harvesting structures such as tanks, ponds, and check dams in these locations suffer due to poor infiltration. On the other hand, evaporation losses from these structures are very high since these structures have significant surface area in relation to the total amount of water they store.

Groundwater governance

Empowering the community-based systems of decision-making is not forthcoming even though efforts are being made by government departments/NGOs. Modifying the existing legal framework and groundwater management institutions to support and foster community action is also not successful due to inherent difficulties to mobilize community efforts.

Informal groundwater market

The existence of informal groundwater market has contributed to the problem of overexploitation of groundwater aquifers. In case of Eastern Gangetic Plains, groundwater availability is not a constraint, but its access and efficient use for irrigation in an energy-efficient manner is critical. Small and marginal farmers lack of investment capacity and mostly depend on informal rental market with price and timely supply issues. The cumulative effect of all the above-mentioned challenges is intensified by the climatic adverse effect experienced by the country.

4.8 Groundwater Development and Management—The Way Forward

The National Water Policy (2012) asserts that the groundwater should be managed as a community resource under public trust to achieve food security, livelihood security, and equitable and sustainable development. Focus of the future policy on groundwater must rely on regulating groundwater development, increasing the area under surface irrigation, appropriate rationing of electricity with tariff rate, community participation in groundwater management, promoting micro irrigation technologies, supply augmentation strategies, and conjunctive use of water, etc., apart from introducing an effective Groundwater Regulation Act. Adaptive management of groundwater is essential to accomplish the long-term goals such as securing ecological integrity, availability of quality water, etc.

Key policy interventions listed below could improve the development and management of groundwater in the country.

Currently, wind and solar energy has surpassed geothermal electricity. With costs per unit going down and addressing the perceived risks of the conventional generation, and their shorter payback periods, renewable energy sources can potentially replace fossil fuels in the irrigation sector. Development and promotion and use of wind and solar energy must be promoted at national level with cooperation from states. India's power sector is a major source of carbon emissions and represents about half of the country's total energy-related CO_2 emissions which can be reduced significantly by increasing the use of renewable energy.

There is also a need to promote energy-efficient solar pumpsets at scale level. Solar pumpsets of 2–3 horsepower can pump water easily where the water table is high, which can be supplied to smallholders. Government of India has plans to deploy about 3 million solar pumps, and approximately 180,000 had been deployed till 2018. Though solar pumpsets may not directly address the problem of declining groundwater resources, it will definitely contribute to the agriculture–water–energy nexus, and climate mitigation.

Drip subsidy is hindering the adoption and expansion of micro irrigation in several states. Currently, only about 12.5 m.ha of irrigated area has been covered under micro irrigation in the country compared to the potential of about 72 m.ha. To boost up the adoption, subsidy norms need to be re-examined and simplified and made available to all farmers eligible for micro irrigation. Micro irrigation subsidies to producers could also boost the production.

Indeed, there is a close nexus between development of the surface irrigation and groundwater. Further, the conjunctive use of groundwater and surface water is also one of the efficient practices for optimal allocation of available water specifically in the water scarcity regions. The stabilization value of groundwater when used along with the surface supplies is very high thus justifying the conjunctive use. Conjunctive use can be very effective in tank irrigation systems of south India where tanks have inadequate supply for most of the years and groundwater recharge is very effective under tank irrigation systems.

To control the overuse of energy for groundwater extraction, suitable interventions focused on regulatory measures such as groundwater acts, rules, and well spacing norms should be introduced and practised. Rethinking on ways and means of reviving the community management of groundwater in rural areas, restrictions on electricity connection for new tubewells in over-exploited areas, metering the tubewells for energy auditing are important interventions to be considered in future planning and implementation programmes on groundwater.

Further, suitable actions for artificial recharge to stabilize the groundwater reservoir and effective utilization policies focusing on regulation and maintenance of the groundwater systems are highly warranted for sustainable management of groundwater.

As the groundwater gets depleted predominantly by irrigating high-value and water intensive crops, climate change in the future may have substantial impacts on groundwater supply due to possible increase in irrigation water demands. Further, increasing frequency and intensity of extreme rainfall events interspersed with severe and prolonged droughts could cause changes in the recharge and discharge of groundwater systems along with irrigation demands. Hence, climate adaptation strategies like artificial recharge structures like irrigation tanks, check dams, and percolation tanks and climate-smart crop and water management practices need to be prioritized in the investment planning.

It is important to devise proper planning and management of transboundary aquifers through legal provisions, protection measures, and pilot projects. This will help to minimize the aquifer pollution through anthropogenic activities, recharge mechanisms, and hydrogeological characteristics of the groundwater system.

The application of remote sensing and geographical information system needs to be intensified as it has been highly effective in identifying the potential recharge zones over different regions. This will act as cost effective intervention in exploiting the groundwater potential. Application of GRACE approach in a larger scale can also be explored.

Development of aquifer management framework is important. The state and groundwater legislations need to act as the facilitator of partnership between government departments, local government, people's institutions, and the civil society organizations to promote scaled approaches to groundwater management. Participatory groundwater management will comprise of the required scientific knowledge, and community level engagement will have the capacity to influence rural government systems like Panchayats. Re-engineering the existing legal framework and groundwater management institutions is also important.

A systematic review of the existing methodologies for assessing the groundwater resources, and bringing in an aquifer basis to the assessment is warranted. Further, to make groundwater development and management more meaningful, the following interventions are needed: (a) comprehensive mapping of aquifers and monitoring groundwater quality at appropriate scales in the various hydrogeological settings, (b) developing typologies of groundwater resources as the basis for a national-level groundwater management programme, (c) setting up a robust and transparent groundwater data collection and sharing mechanisms, mainly as decision support tools in programme implementation and monitoring.

References

Rajan A, Ghosh K (2019) Energy productivity of Indian agriculture: Are energy guzzling districts generating higher agricultural value? In: 3rd World Irrigation Forum (WIF3), Bali, Indonesia ST-1.1 W.1.1.071. https://www.icid.org/wif3_bali_2019/wif3_1-1_7-min.pdf

AditiMukherji (2022) Solarirrigation in South Asia: Emerging evidence of impacts on mitigation, adaptation, equity and groundwater. https://solar.iwmi.org/2022/02/15/solar-irrigation-in-south-asia-emerging-evidence-of-impacts-on-mitigation-adaptation-equity-and-groundwater/

Ministry of Agriculture & Farmers Welfare (2021) Agricultural Statistics at Glance, 2021. Directorate of Economics & Statistics. Government of India

Amami HE (2018) How does groundwater depletion contribute to unequal access to the resource among classes of farmers? evidences from case studies in the center of Tunisia. Int J Hydro 2(4):445–451. https://doi.org/10.15406/ijh.2018.02.00108

Bekele Shiferaw (2021) Addressing groundwater depletion: Lessons from India, the world's largest user of groundwater. IEG, World bank group. https://ieg.worldbankgroup.org/blog/addressing-groundwater-depletion-lessons-india-worlds-largest-user-groundwater

Central Ground Water Board (2022) http://cgwb.gov.in/documents/2022-11-11-GWRA%202022.pdf

Dierauer JR, Whitfield PH, Allen DM (2018) Climate controls on runoff and low flows in mountain catchments of Western North America. Water Resour Res 54(10):7495–7510. doi.org/https://doi.org/10.1029/2018WR023087.

Dubash NK (2007) The electricity-groundwater conundrum: case for a political solution to a political problem. Econ Pol Wkly 42:45–55

Energy Statistics (2020) 27th Issue. National Statistical Office Ministry of Statistics and Programme Implementation Government of India. www.mospi.gov.in/sites/default/files/publication_reports/ES_2020_240420m.pdf

Gerland P, Raftery AE, Ševčíková H, Li N, Gu D, Spoorenberg T, Alkema L, Fosdick BK, Chunn J, Lalic N, Bay G, Buettner T, Heilig GK, Wilmoth J (2014) World population stabilization unlikely this century. Science 346(6206):234–237. https://doi.org/10.1126/science.1257469

Harpold AA, Kohler M (2017) Potential for changing extreme snowmelt and rainfall events in the mountains of western United States. J Geophys Res: Atmos 122(24):13219–13228. doi.org/https://doi.org/10.1002/2017JD027704

India Energy Outlook (2021) OECD Publishing. Paris. https://doi.org/10.1787/ec2fd78d-en

IPCC (2001) Climate change: impacts, adaptation and vulnerability third assessment report of intergovernmental panel on climate change. Cambridge University Press, Cambridge

Roome (2022) India seeks to arrest its alarming decline in groundwater. End Poverty in South Asia. https://blogs.worldbank.org/endpovertyinsouthasia/india-seeks-arrest-its-alarming-decline-groundwater

Joshi PK, Jha AK, WaniSuhas P, Sreedevi TK, Shaheen FA (2008) Impact of watershed program and conditions for success: A meta-analysis approach. Global Theme on Agroecosystems Report no. 46. International Crops Research Institute for the Semi-Arid Tropics, Patancheru 502 324, Andhra Pradesh, India, p 24. https://www.un-igrac.org/areas-expertise/groundwater-climate-change#:~:text=Climate%20change%20does%20not%20only,difficult%20to%20reverse%20the%20process.

Long D, Chen X, Scanlon BR, Wada Y, Hong Y, Singh VP, Chen Y, Wang C, Han Z, Yang W (2016) Have GRACE satellites overestimated groundwater depletion in the Northwest India aquifer? Sci Rep 6:24398

National Water Policy (2012) Ministry of Water Resources.Government of India. http://jalshakti-dowr.gov.in/sites/default/files/NWP2012Eng6495132651_1.pdf.

Narayanamoorthy A (1997) Impact of Tariff Policies on the use of electricity and Groundwater: Arguments and Facts. ArthaVijnana 34(3):323–340

NITI Aayog (2017) Micro-irrigation through public private partnership: "From source to roots". NITI Aayog, New Delhi

Palanisami K, Suresh Kumar D (2003) Power pricing, groundwater extraction, use and management: comparison of Andhra Pradesh and Tamil Nadu states, Water, Livelihoods and Environment in India: Frontline Issues in Water and Land Management and Policy, January 27–29, 2003, Institute of Rural Management Ahmedabad, Gujarat

Palanisami K, Raviraj A, Ian Gale (2008) Augumenting groundwater resources by artificial recharge in Kodangipalayam Watershed, Coimbatore District, Tamil Nadu—A case study. In: Palanisami K, Ramasamy C, Thangarajan M (eds) Ground water Management and Policies, pp 114–126. Macmillan India Ltd, Delhi,

Palanisami K, Suresh Kumar D, Wani SP, Mark Giordano (2009) Evaluation of watershed development programmes in India using economic surplus method. Agric Econ Res Rev 22(July-December):197–207

Pascolini-Campbell M, Reager JT, Chandanpurkar HA, Rodell M (2021) A 10% increase in global land evapotranspiration from 2003 to 2019. Nature 593:543–547. doi.org/https://doi.org/10.1038/s41586-021-03503-5

PMKSY (2019) Pradhan Mantri Krishi Sinchayee Yojana. Ministry of Agriculture and Farmers Welfare. Government of India. https://pmksy.gov.in/AboutPMKSY.aspx

Sahil Ali (2018) The future of Indian electricity demand, How much, by whom, and under what conditions?. https://www.brookings.edu/wp-content/uploads/2018/10/The-future-of-Indian-electricity-demand.pdf

Sarkar A (2020) Groundwater irrigation and farm power policies in Punjab and West Bengal: challenges and opportunities. Energy Policy 140:111437. https://doi.org/10.1016/j.enpol.2020.111437

Shah T, Deb Roy A, Qureshi AS, Wang J (2003) Sustaining Asia's groundwater boom: an overview of issues and evidence. Nat Res Forum 27:130–141. https://doi.org/10.1111/1477-8947.00048

Shah (2009) Climate change and groundwater: India's opportunities for mitigation and adaptation. Environ Res Lett 4(035005):(13pp). https://doi.org/10.1088/1748-9326/4/3/035005

Shah T, Giordano M, Mukherji A (2012) Political economy of energy-groundwater nexus in India: Exploring issues and assessing policy options. Hydrogeol J 20(5):995–1006

Shah T (2014) Groundwater Governance and Irrigated Agriculture. Global Water Partnership Technical Committee (TEC). Report 19. Global Water Partnership (GWP). Sweden

SoniA and Syed TH (2015) Diagnosing land water storage variations in major Indian river basins using GRACE observations. Glob Planet Ch 133:263–271

Suresh Kumar D, Palanisami K (2019) Managing the water–energy nexus in agriculture–adoption of water management technologies. Econ Polit Wkly 54(14).https://www.epw.in/journal/2019/14/special-articles/managing-water%E2%80%93energy-nexus-agriculture.html

UNESCO (2022) Groundwater—Making the invisible visible. The United Nations Educational, Scientific and Cultural Organization, 7, place de Fontenoy, 75352 Paris 07 SP, France. ISBN978–92–3–100507–7. http://www.indiaenvironmentportal.org.in/files/file/world%20water%20development%20report%202022.pdf

WHO (2022). https://www.who.int/news-room/fact-sheets/detail/drinking-water

World Bank (2012) India groundwater: a valuable but diminishing resource. https://www.worldbank.org/en/news/feature/2012/03/06/india-groundwater-critical-diminishing

World bank (2020) Unshackling India's Energy-Water-Agriculture Nexus: Is Solar Power the Panacea it Promises to be?. https://www.worldbank.org/en/news/feature/2020/06/05/unshackling-india-energy-water-agriculture-nexus

Chapter 5
Reviving Tank Irrigation

Abstract There are about 0.25 million tanks in India and the tank irrigated area has declined from 3.3 m·ha in 1953 to 1.67 m·ha in 2018. Encroachment of the tank catchment, weak tank-based institutions, erratic rainfall and poor tank filling, inadequate budget for maintenance, improper land use in the upstream, and siltation of the tank water spread over years had resulted in poor performance of the tanks. Tank rehabilitation programmes by various agencies have started around four decades ago and the performance was not satisfactory due to poor implementation protocols followed. Looking at the past rehabilitation performance and need of the future, six time-tested options for tank modernization are suggested: Option 1: Desilting only in selected locations in the tank water spread area. Option 2: Desilting the entire tank water spread area to a given depth. Option 3: Converting the tanks into percolation ponds. Option 4: Providing 1–2 fillings to the tanks during the tank season. Option 5: Full (optimum) groundwater development in the command area. Option 6: Adopting sluice rotation (opening and closing the sluices in alternate weeks) so that groundwater and tank water will be simultaneously (conjunctively) used throughout the crop season. At national level, a tank authority can be established mainly to coordinate and mobilize funds for tank investment and management by closely working with the states.

Keywords Tank irrigation · Encroachment · Tank siltation · Groundwater supplementation · Stabilization value · Rehabilitation and modernization · Time-tested measures · Social forestry · Multiple uses · Tank WUAs

5.1 Introduction

A tank is a low earthen bund formed across the shallow valley to hold the rainfall runoff from the catchment area above. Tanks may be either isolated or in a chain or cascade. Tank irrigation contributes significantly to agricultural production in large parts of South and Southeast Asia. Tanks are rainfall dependent, and majority of the tank systems are located in arid and semi-arid regions. There are about 0.25

K. Palanisami and U. S. Nagothu, *India's Water Future in a Changing Climate*, Advances in Geographical and Environmental Sciences,
https://doi.org/10.1007/978-981-97-1785-9_5

million tanks in India irrigating about 1.7 million ha. The southern states account for about 60% of the tank irrigated area. Together, the tank irrigated areas of these states produce about 4.5 million tonnes of rice per year, approximately 25% of rice production in these states.

Years of disrepair, neglect, and lack of proactive management have adversely affected storage, productivity, and efficiency of tanks. Most tanks across the country have suffered from siltation in the water spread area or tank bed, resulting in the decline of their storage capacity. Siltation has occurred partly because of natural processes and partly because of changes in the land use in the upstream location. Originally, the bulk of the tank catchment area was uninhabited and water demand in the catchment area was negligible thus facilitating free flow to the downstream tanks. With changes in land use and consequent siltation of the supply channels, the flow capacity of supply channels gradually reduced. Siltation also reduces the permeability of topsoil and impedes groundwater recharge.

Over the years, an increasing number of water harvesting and storage structures due to changes in land-use upstream of the tanks have further reduced the inflows downstream. With this, the free catchment available to the tanks has declined and so has rainwater runoff. Further, beneficiary participation in all aspects of tank management starting from catchment management to field water distribution has reduced over the years. Lack of maintenance of tanks including desiltation works have also led to their decline. An even more important reason is the dysfunctional social/village dynamics around the tanks (Raju and Shah 2000, Palanisami 2000). In sum, the potential of tanks seems to be sinking where the productive capacity of most tanks has been reduced to a fraction of the potential because of long neglect and disrepair.

5.2 Why Tank Systems Should Be Revived?

Given the constraints in future expansion of irrigation under canals and groundwater development, tanks are considered as neglected opportunities, as irrigation tanks have now re-emerged as some of the finest examples one can find of extensive irrigation that spreads a given quantum of available water over as large an area as possible. This maximizes the productivity per cubic foot of available water rather than per acre of land.

Resource poor farmers (owning less than 2 ha) still account for the major share of tank irrigated area in India. Marginal (less than 1 ha) and small farmers (1–2 ha) together account for about 55% and the large farmers account for 6%. This is also true across different states where tank irrigation has considerable presence even today (Narayanamoorthy 2007, Narayanamoorthy and Suresh 2016).

Expected climate change impacts in Asia in general and South Asia in particular will have more intensive rains/floods in shorter periods and prolonged drought spells for a longer period. During floods, tank systems offer more scope to store the excess water and allow for both irrigation and groundwater recharge during times of water shortage. Because these tanks are distributed across the landscape, they can capture

Fig. 5.1 Tanks with long bund, Tamil Nadu

more water, which otherwise would be wasted as runoff. There is considerable interest in tank irrigation systems as one of the best adaptation strategies to climate change. Moreover, most of the river basins are approaching their irrigation potential and any further expansion in area under irrigation has to come from rainwater harvesting (Asian Development Bank 2006).

Current cost of development of new major and medium (M&M) irrigation projects will be Rs 8–12 lakh/ha compared to Rs 3–4 lakh/ha for rehabilitation. Most of the potential sites for construction of M&M projects have already been developed. In the case of minor irrigation tanks the cost of rehabilitation was around Rs 60–70,000/ha (as per 2007 prices). If the hidden costs behind M&M projects are included, the M&M projects would prove even more costly. Hence, investment cost-wise, tanks offer more scope for future expansion in irrigation. Hence, the tanks need urgent revival considering their technical and managerial issues. A typical tank with a long bund is shown in Fig. 5.1.

5.3 Performance of the Tanks

In the absence of accurate data across all the states, the number of tanks and ponds in India are reported to vary from 0.2 to 0.35 million (Asian Development Bank 2006). The number of tanks in use assessed by the MI census in 2000–01 was 0.23 million in India. Besides, an estimated 42,955 tanks were not in use in 2000–01, which has gone up to 85,807 in 2010–1[1]. Together, this provides an estimate of number of tanks in the range of 0.3 million in India. The Southern region accounts for 35% of the total tanks in use in India (2000–01), followed by West (20%), East (16%), and North (14%). East has the highest proportion of tanks not in use (21,471 out of 42,740 in 2010–11) followed by south (35,742 out of 1,02,991 in 2010–11), the West, and North regions (Reddy et al. 2018). The high density of tanks in use as well as decayed tanks in South could be the reason for the focus the region received in terms of tank renovation programmes.

In terms of irrigated area, tank irrigated area in India has declined from 3.3 m·ha in 1953 to 1.67 m·ha in 2018 (almost about 50% reduction in tank irrigated area during this period)[2]. The share of tank irrigated area to total irrigated area thus reduced from 15.6% in 1953 to 2.3% in 2018, whereas the share of groundwater irrigation has increased from 30.8 to 63.9% during this period. Because of the decline in tank irrigation, the land productivity is also observed to be low under tanks compared to other sources of irrigation. For example, in Tamil Nadu, compared to tanks, land productivity per ha was 13% higher under canals and 183% higher under wells (Palanisami 2009). The yield gap under the tank irrigation has increased over years due to poor tank management and water control. This calls for revival and improvement of tanks so that tank irrigation will remain as a viable and sustainable irrigation source in the future.

5.4 Status of Tank Irrigation Across Regions in India

The variations in the density of tanks across states are basically due to geo-hydrological, topography, and rainfall patterns. Among the regions, southern India is noted for the intensity of tanks. Unlike the northern region, the rivers in the south are mostly seasonal and the plains are not very extensive. The local topographic variations have been effectively exploited to impound rainfall in tanks which are used to raise irrigated rice crop and simultaneously serve as means of improving groundwater recharge in the command areas.

To get an insight into the tank performance across region, all the major States and Union Territories (UTs) in India are grouped under five regions (as per the Planning Commission) viz., South, North, East, and West.[3] Irrigated area under tanks has been declining since 1950–51, while all other sources have recorded an increase.

[1] These figures are provided by indiastat.com.

[2] The number of tanks and the area irrigated by tanks are varying over years due to the addition of small water bodies like ponds etc. Hence, actual area irrigated by tanks may vary between states and the total area under tanks in the country may be around 2 m·ha. However, this report used the available data on records to analyse mainly the trend in tank irrigated area among the regions.

[3] Northern region covers Jammu and Kashmir, Punjab, Haryana, Himachal Pradesh, Delhi, Uttar Pradesh, Uttaranchal, Madhya Pradesh, Chhattisgarh. Eastern region covers Bihar, Jharkhand, Orissa, West Bengal. Western region covers Maharashtra, Goa, Rajasthan, Gujarat. Southern region covers Andhra Pradesh, Tamil Nadu, Karnataka, Kerala, Pondicherry. North-eastern states are not

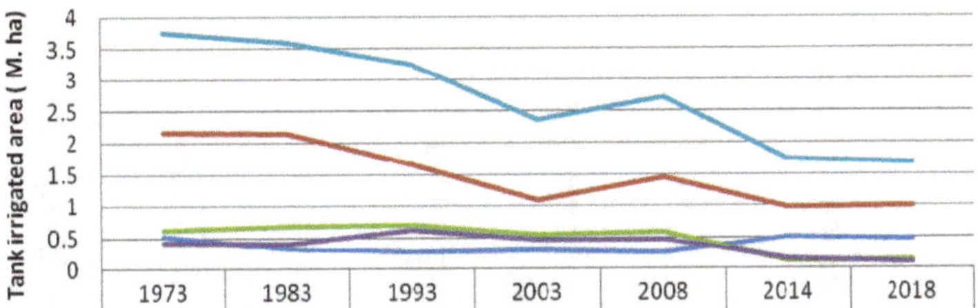

	1973	1983	1993	2003	2008	2014	2018
North	0.52	0.33	0.27	0.28	0.24	0.51	0.46
South	2.17	2.16	1.66	1.07	1.43	0.96	0.98
East	0.62	0.68	0.69	0.53	0.57	0.11	0.13
West	0.43	0.41	0.62	0.47	0.46	0.14	0.09
All India	3.74	3.58	3.24	2.35	2.7	1.72	1.67

Fig. 5.2 Area under tank irrigation over the years across regions (m·ha).

Area under tank irrigation between 1973 and 2018 has declined steadily in north and south regions as well as at the country level, though there was some revival between 2003 and 2018 (Fig. 5.2). This could be due to tank rehabilitation programmes in some of the southern and eastern states. Over the last 3 decades, area under tank irrigation has declined by 1 million hectares at the country level. Of this, more than 0.7 million hectares of area was from southern region consisting of erstwhile Andhra Pradesh, Tamil Nadu, Karnataka, and Kerala. The share of different regions in the total area under tanks in India has changed between 1973 and 2018. In the South, it has marginally increased from 58% to 59%, and in the North from 14 to 28%. In the East and West, it had reduced from 17 to 8% and 12 to 5% respectively (Fig. 5.3). On the other hand, relative importance of tank irrigation to overall (total) irrigated area has come down substantially over years.

At the country level the share of tank irrigation has declined from 12% in 1973 to 2% in 2018 (Fig. 5.4). Within the respective regions the importance of tank irrigation declined substantially over these years. Tank irrigation, however, holds relatively greater importance in South (8%) and East (2%) to total irrigated area in the country. This could be one of the reasons for the attention these regions get in terms of tank rehabilitation initiatives in the last few decades. In the case of other regions, tank irrigation accounts for 0.5% of total irrigated area in West and 1% in the Northern regions of India.

At the national level, a decline was observed from 12 to 2% between 1973 and 2018, indicating an urgent need for relooking at the tank irrigation investment and rehabilitation. This could be relevant especially in the South, East, and West regions where higher tank density coupled with increasing numbers of non-functional tanks exists.

---

included here as tank irrigation accounts for a negligible per cent of the total tank irrigated area in India, and also not much information available on tank-related aspects.

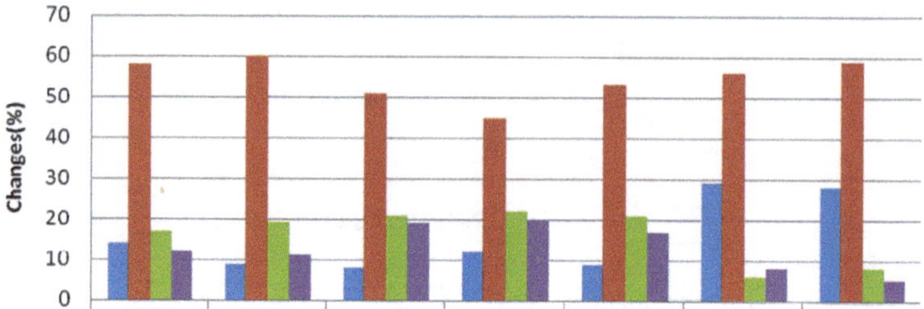

	1973	1983	1993	2003	2008	2014	2018
North	14	9	8	12	9	29	28
South	58	60	51	45	53	56	59
East	17	19	21	22	21	6	8
West	12	11	19	20	17	8	5

Fig. 5.3 Changes in relative contribution of tanks in each region to all India level (%). *Source* GoI (different years), compiled from: Directorate of Economics and Statistics, Ministry of Agriculture & Farmers Welfare

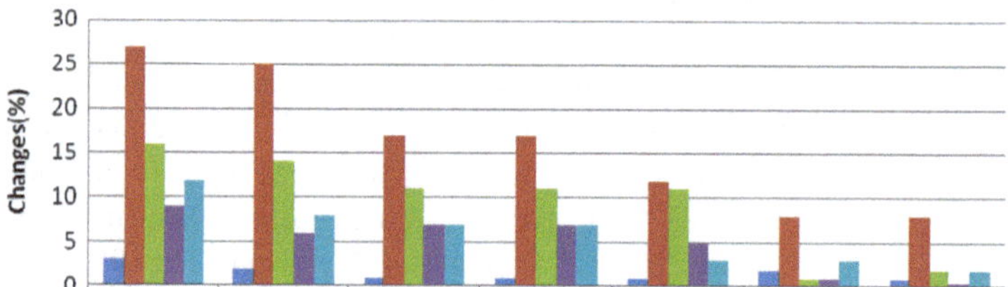

	1973	1983	1993	2003	2008	2014	2018
North	3	2	1	1	1	2	1
South	27	25	17	17	12	8	8
East	16	14	11	11	11	1	2
West	9	6	7	7	5	1	0.5
All India	12	8	7	7	3	3	2

Fig. 5.4 Changes in relative importance of tank irrigation in each region and India (%)

5.5 Declining Tank Performance: Causes and Concerns

In most of the regions of India, especially in the drought-prone semi-arid tropics, declining tank irrigation and expansion of groundwater irrigation was observed over years due to decline in benefits from community-based tanks. Many studies clearly indicated that socio-economic changes, weakening of the tank management institutions and poor governance contributed significantly for the decline of tank irrigation (Von Oppen and Rao 1980a, 1980b; Reddy 1995; Van Koppen et al. 2009; Reddy 2015, Mosse 1999, Shankari 1991; Sivasubramanian 2006, Janakarajan, 1993; Reddy et al, 1993; Reddy et al, 2018, Sakthivadivel et al. 2004, Palanisami 2006; Pant and Verma 2009; Palanisami et al. 2011a, 2011b, Nehlin 2016, Raju and Shah 2000, 2002). The decline of traditional systems, therefore, is a cumulative effect of mainly the policy and institutional neglect. As a consequence, other associated socio-environmental benefits such as groundwater percolation, retarding negative impacts of floods and droughts, community-led management and maintenance of tanks, and enriching biodiversity in the vicinity of the cascading tanks got collapsed, creating a considerable imbalance in the hydrological and ecological systems in most of the tank irrigated regions of India.

In the case of Eastern India, most of tanks are small and mostly used for irrigation, fishery, domestic use, and livestock management. About 90% tanks were used for irrigation while 68% were used for fishery. Only 59% were used for domestic use and 68% for livestock management. 62% tanks were used both for fishery and irrigation. None of the tanks are allowed for irrigation if the pond depth is less than 3 ft (0.9 m). Diesel pumps are not allowed in most cases to drain water. Netting of fish is usually done in lean season when there is no need for irrigation. Stocking of fish fingerlings in rainy season and harvesting in summer gives yield of 2–3 tonnes of fish/acre of water spread area. Supplementary irrigation provides scope for yield improvement and taking up a second crop after harvest of first crop. An optimal combination of irrigation and fishery will ensure the sustainability of tank irrigation. Not many studies have addressed these issues in detail in Eastern India along with needed interventions even though several studies have focused on these issues in South Indian tanks (Sakurai and Palanisami 2001; Kajisa et al. 2006, 2007; Ranganathan and Palanisami 2004).

In sum, the declining tank performance can be summarized as follows

Palanisami and Easter (1983, 1984, 1987, 1991, 2000) have mentioned a number of possibilities for the low level of tank irrigation, which are presented here: "(1) The procedures used to calculate the water yield to tank storage are no longer applicable to the current situation and can lead to an over estimation of water availability, (2) There has been a change in the rainfall patterns over time and changes in catchment characteristics due to abuses in the upper part of the watershed, (3) Encroachment in the tank foreshore and along the feeder channels, as well as siltation and tree planting in the tanks has reduced tank storage capacity, (4) In several cases, the construction of dams/anicuts in upper watersheds has prevented the water supplies from reaching downstream tanks, (5) Low level resource mobilization in the villages and low state budget allocations for O and M, (6) Community management of the tanks has declined due to conflicts between different groups in the villages, (7) Intensive groundwater development in tank command areas has caused tank management to decline due to declining collective action, and (8) The poor design of new tanks has resulted in low level of performance.

Since, the tank-related issues are seemingly complex to comprehend due to interplay of time and space, it is important to differentiate the issues relating to decline in tank performance in terms of spatial and temporal scale, as many issues that are significantly affecting tank performance have been observed in the recent 2–3 decades. Such an analysis will help isolate the factors influencing tank management and will highlight factors that may be important while dealing with tank improvement investment.

In spatial scale, the major factors contributing for the decline of tank irrigation in India and south India are the following:

(a) System level: erratic rainfall pattern and inadequate and unreliable water supply to the tank, encroachment of tank bed and supply channel by the government, public, and private people, silting of tank water spreads and supply channels, large-scale infestation of trees/weeds and loss of grazing land in the tank bed, damaged sluices and weirs, dilapidated and weak tank bunds, delinked tanks in the tank cascades, less investment on new tanks (Palanisami 1990, 1991, 1993, 2000, 2005, 2006, 2013).
(b) Command area level: unlined and damaged field channels, fast growing *proposphis* plants in the fields, inadequate tank water, declining groundwater levels in the wells, well abandonment, in appropriate crop pattern, poor crop yield, low income and lack of water management practices, pollution of urban tanks, functioning of tanks into percolation ponds (Palanisami 1993, Palanisami and Flinn 1988, 1989, Palanisami and Easter 2000).
(c) Tank community level: lack of community involvement in tank management and maintenance, non-functioning of tank-based local institutions, meagre resource allocations for maintenance, out-migration of the potential labourers and farm family members for other jobs due to fast development of roads and communication facilities, expansion of urbanization in the tank villages which are close to towns, non-profitability of tank irrigated agriculture due to increased labour wages and input costs (Palanisami et al 1997, 1998, 2008a, 2008b, 2010, 2011a, 2011b, Kajisa et al. 2006, 2007).

In temporal scale the declining tank performance can be summarized under different phases as follows:

In the first phase, the declining performance has been related to the past governance of the tanks. During this period (pre-independence to 1950), institutional arrangements (such as *Dasabandam* and *Kudimaramat*of South India, *Aher-Pyne* systems of South Bihar; *Chandeli* tanks of Bhundelkh; and *Johads and Pals* of Rajasthan), were in place to protect these systems from decay. These institutional arrangements nurtured by the benevolent local rulers who have been central to development and sustenance of the tank systems over centuries. However, the policy shift towards major and medium irrigation during the British period, coupled with the changes in policy perception of irrigation development, that is, treating irrigation as a productive (revenue-generating source) rather than a protective source, has resulted in the degeneration of these institutions.

The second phase (1950–1980) of the declining performance was attributed to the govt's investment priorities on medium and large-scale irrigation projects under five-year plans. Advent of the energization of groundwater lifting mechanisms also contributed for mass expansion in groundwater irrigation. The benefits of tanks as common property has been transferred as private prosperity and water markets also started functioning even though they couldn't sustain in the long run. The new technologies in pumping systems during the 1980s coupled with the benefits from green revolution technology have resulted in an unprecedented expansion of groundwater development. Further, poor farmers were not in a position to adopt these technologies because of their capital intensive nature, especially during the initial stages. Also, the socio-economic changes in the village dynamics made the traditional tank associations non-functional. Population pressure with expansion of village roads and housing areas resulted in encroachment of tanks.

The third phase (1980–2000) of declining tank performance has been related to the growth and spread of socio-political set-up in the villages where regional political parties and local caste groups started strengthening their hold by attracting youths and dominant caste groups in the villages. Increasing well failure (including bore wells) in the tank commands has become a common phenomenon in the recent years thus making both tanks and wells become inactive in most of the periods. Further, several governments initiated social security nets such as fair price (ration) shops, national rural employment guarantee schemes (MGREGA) provided some food security and employment for the rural poor and small-scale industries, expansion of village roads, etc., facilitated the easy movement of the rural labour classes towards higher income earning options.

The fourth phase (2000 onwards) of the declining performance has been due to climate change and the associated anthropogenic pressures related events where the intensive rains in few days and prolonged droughts for more number of days resulted in less water supplies to tanks in most of the years resulting in uncertainty in tank irrigation thus making the tank irrigated agriculture a non-profitable, risky and unsustainable livelihood option in the rural areas. Landowners slowly abandoned their agriculture and migrated to towns thus leaving their agriculture lands as fallow lands. Urban tanks witnessed a major fall during this phase where city expansion and increase in land values made urban tanks as storage ponds with polluted waters. The temporal changes will capture adequately the impact of several factors on tank management and its attributes (Palanisami et al. 2008a, 2010, 2011a, 2011b) (Table 5.1).

5.6 Tank Rehabilitation and Modernization

Given the existing as well as emerging challenges (hotspots) in tank irrigation, it is important to see how best the tanks should be restored to the extent possible as well as to sustain their function. Indeed, there are many suggestions for improving tank irrigation systems, through tank modernization incorporating both technical and management interventions. Broadly speaking, the tank modernization strategies can work to save water and extend the surface irrigation or increase recharge and conjunctive use of groundwater or both. Hence the tank modernization is considered as an appropriate measure to address the future of the tanks.

Earlier review of India‘s irrigation sector (World Bank 1998) begins with this observation, "— There is need for a major shift in India‘s irrigation strategy. There is a need for shift from the past near exclusive reliance on irrigation expansion to a strategy emphasizing improving the performance of irrigation and irrigated

Table 5.1 Temporal changes in tank management and their impact

No.	Phases	Factors influencing	Major impact on	Impact
1	Phase 1 (pre-independence to 1950)	Policy shift and governance on irrigation investment	Tank system*: (a) Above outlet (b) Below outlet (c) Tank institutions (d) Village: (I) Farmers • well owners • non-well owners (ii) Labourers (iii) Others	Low Low Low Low Low Low Low
2	Phase 2 (1950–1980)	Emphasis on major and medium surface irrigation, green revolution, emphasis on groundwater development, and energy use, crop production technologies	Tank system: (a) Above outlet, (b) Below outlet (c) Tank institutions (d) Village: (i) Farmers • well owners • non-well owners (ii) Labourers (iii) Others	Medium Low Medium Low Medium Low Low
3	Phase 3 (1980–2000)	Socio-economic development programmes such MGREGA, population expansion, peri-urban influence, development of easy access roads and communication facilities, labour migration. Village safety nets, Poor functioning of tank-based institutions	Tank system: (a) Above outlet, (b) Below outlet (c) Tank institutions (d) Village: (i) Farmers • well owners • non-well owners (ii) Labourers (iii) Others	High Medium High Low High Low Medium
4	Phase 4 (2000 +)	Climate change, visible changes in rainfall pattern with more drought years. Population pressure, Pollution of urban tanks, expanded well investment, new caste/political party-based local groups tanks into percolation ponds.	Tank system: (a) Above outlet, (b) Below outlet (c) Tank institutions (d) Village: (i) Farmers • well owners • non-well owners (ii) Labourers (iii) Others	Very high High High Medium High Low Medium

Note * above outlet refers the tank's hydrology components such as catchment area, foreshore, supply channel, water spread area and tank bunds, sluices, and surplus weirs. Below outlet refers mainly the command area, the field channels, and wells agriculture". There are compelling reasons for giving much greater attention and resources to small-scale surface irrigation schemes (Vaidyanathan 1999).

The rationale for tank modernization is valid not only from the equity and stability points but also from the economic angle where unit costs of modernization are marginal compared to investment costs in creating new irrigation systems. Further, irrigation tanks are more equitous. Research on costs and benefits of rehabilitating irrigation tanks made it clear that benefits outweigh costs in all situations irrespective of tank size though benefits are proportional to the size of the tanks (Palanisami 2005; Reddy and Bhagirath 2009a,2009b, Kumar et al. 2011).

Recharging of groundwater appears to be one of the most pressing reasons for tank restoration given the fact that groundwater is the single largest source of irrigation in most parts of India. It is observed that in the absence of replenishing mechanisms like tanks or canals water supply available from wells is much limited. In most regions open wells have dried up and water levels go down rapidly in the deep bore wells in the absence of well managed tanks in the vicinity, especially during the low rainfall years (Reddy et al. 2018). The stabilization value of groundwater in the tank systems is about 15% to supply value which further justified that wells are value added to tank ecosystem[4] (Palanisami et al. 2011a, 2011b). This not only emphasizes the rational for the revival of tanks but also points to the need for conjunctive use of surface and groundwater resources. In this context, a review of the past rehabilitation/modernization activities will help arrive an appropriate models for tank modernization.

The newly formed Telangana state has embarked on an ambitious tank rehabilitation programme, viz., "*Mission Kakatiya*". Under this programme it is planning to restore 9,306 tanks every year (20% of total tanks) with an eventual target of restoring all 46,531 tanks in 5 years, in a phased manner to bring 0.45 million ha into command. Main activities include desilting, repairing of sluices, weirs, etc., strengthening of tank bunds, repairing the feeder channels, and re-sectioning of irrigation channels.

5.7 Cost of Tank Rehabilitation and Modernization Programmes

A study by ADB has indicated that the present investment strategy is skewed towards physical rehabilitation and very little is provided towards institutional development, and maintenance and management (Asian Development Bank 2006). With the available average value of the catchment area, water spread area, and the command area of the 875 tanks studied, the tentative cost per unit is as follows:

(a) Catchment area = Rs 20,000 per ha

[4] The groundwater first augments the total water supply for irrigation and then stabilizes the fluctuations in the supply of irrigation water to the crops. The total value of groundwater is thus the sum of the above two groundwater generated benefits. The benefit generated by the variability reducing function of the groundwater is called the Stabilization value of groundwater

(b) Water spread area
(c) Improvements to tank bund, sluices, weirs, desilting = Rs 30,000 per sq km
(d) Command area development = Rs 5,000 per ha

This cost will constitute 75% of the total project cost. The remaining 25% of the cost towards institution and maintenance and management is to be arrived and added to get the total project cost.

The allocation in thc total project cost is:
Institutional development 10%
Physical rehabilitation[5] 75%
Maintenance and management 15%

Using this allocation for typical tanks, the cost per ha of the command area works out to Rs 34,000 (as per 2005 prices). This cost at current (2021) price will be about Rs. 102,000.

The focus of rehabilitation in the future is to be more on water acquisition. Hence, it should be based on rupees per sq km of catchment area for works in the catchment including feeder channel improvement; for tank storage improvements, it should be based on rupees per ha of water spread area; and for the command area, it should be based on rupees per ha of command area (Asian Development Bank 2006).

In the case of World Bank funded tank modernization programmes completed recently in 4 states, the cost structure has changed depending upon the projects (World Bank 2014, Annexure 1). The composition of the costs is as follows[6]:

Total number of water bodies/tanks covered: 12523.

Average total project cost/ha of command area= US$ 933. The cost (at 2021 price) will be about US$ 1330.

The composition of the project cost is:

Infrastructure = 66%

Agriculture= 16%

Institutional support= 12%

Project management= 6%

[5] Catchment area Rs 20,000 per ha; Water spread area (improvements to tank bund, sluices, weirs, desilting) Rs 30,000 per sq km; Command area development Rs 5,000 per ha

[6] Based on the averages of the 12423 tanks covered in 4 states under 4 major programmes viz., Karnataka Community-based Tank Management Project (KCBTMP), Andhra Pradesh Community-based Tank Management Project (APCBTMP), Orissa Community Tank Management Project (OCTMP), Tamil Nadu Irrigated Agriculture Modernisation and Water-bodies Restoration and Management (TN-IAMWARM) Project

5.8 Relooking Tank Modernization: Changing Dynamics

Until 1990s not much emphasis was given on alleviating poverty by generating employment opportunities, engaging local communities for institutional development, and considering tanks cascade system as a whole instead of an individual tank. Starting from early 2000s, focus shifted to alleviating poverty coupled with enhanced agricultural productivity. This shift occurred as a result of experiencing more advantages while augmenting structural treatments under semi-arid or dry conditions in cascading tanks against the individual tank (non-system tank). For instance, canal lining for provisioning of water supply to attain last mile connectivity in the cascading network has resulted in water conservation by about 21% (Asian Development Bank 2006).

The past experiences on tank modernization clearly indicate that a successful programme of tank rehabilitation has to work on the invidious tank village dynamics and create a socio-technical system that is able to align the interests of all users with the sustainability of the tank management. A robust, self-managing tank water users' organization will be a necessary condition for better tank management. However, the difficulty in designing an effective modernization project aiming at management regeneration is that there are no such working models of sustainable tank management as sustainability is key parameter of better tank management. However, the scope for deriving a working model for each region given the proactive stakeholder groups, resource availability, and livelihood enhancement of the local groups and stability of the tanks in getting adequate supplies still exists. Even though, NGOs are acting as catalysts in promoting local tank level organizations, their roles mainly dependent on continuous external financial support. Hence there is a win-win situation for both the NGOs and the WUA.

Once the modernization process is initiated, it needs to assess the wider impact of proposed rehabilitation work on the entire array of its stakeholders who include, the command area farmers, petta (foreshore) cultivators, groundwater users, livestock owners, fish culturists, pastoralists, and other non-agricultural users, including domestic users, particularly rural women. Similarly, investment in rehabilitation needs to be assessed and justified on grounds of multifarious economic, social, and environmental benefits and costs of rehabilitation. An ABC analysis of sorts needs to be done with farmer participation to evolve a medium or long-term programme of rehabilitation beginning with low-cost-high-pay-off repair works and building up to more expensive items of physical structures. The cost curves for each of the modernization options proposed to be drawn to get meaningful inferences on investments.

Contrasting the techno-economic perspective (structural components) with a socio-ecological perspective (structural and managerial aspects) leads to a different understanding of the role tanks play in the lives of the people around them. It may also result in a radically different notion of the kind of rehabilitation that will enhance the relevance and overall social value of tanks. It will also produce new insights into the strategy of tank management by the user groups that is likely to make sense given the diverse interests and varied stakeholder participation in the village. Already available evidence has shown that multiple uses of tanks result in different levels of tank performance than just merely looking only on irrigation benefits as is done in most of the impact evaluation studies (Palanisami et al. 2010). Hence, enhancing the benefits from multiple uses of tanks is highly warranted and future modernization strategies need to consider this option as well in order to sustain the benefits from tank systems.

5.9 Tank Modernization—Time-Tested Measures

Heavy rains in different regions of the country in the recent years had ensured that storage and management of the rainwater are more relevant for tanks. Once the tanks are full, assured groundwater recharge will also be possible and feasible number of wells in the tank command can be calculated based on the annual groundwater draft. This will act as an upper limit for well expansion. Solar pumps are getting popular in tank commands for supplemental irrigation (Fig. 5.5). Using Tamil Nadu state as a case, future tank modernization activities are suggested under the following six options (Table 5.2).

Fig. 5.5 Solar pumps in tank commands, Andhra Pradesh

Table 5.2 Aquifer-tank ecosystem development options

Indicators of change	Aquifer-tank ecosystem development options					
	S1: Tank partial desilting	S2: Tank full scale rehabilitation	S3: Tank into percolation ponds	S4: converting non-system into system tanks	S5: Tank with full groundwater development	S6: Conjunctive use
Water supply increase	5%	12%	6%	10%	15%	20%
Multiple use enhancement	30%	35%	15%	40%	25%	25%
Financial viability	Low (IRR<5%)	Low (IRR=5–7%)*	Medium (IRR=7–10%) (for limited tanks)	High (IRR=10-15%). (for limited tanks)	High (IRR=10–15%)	Very high (IRR=15–20%)
Unit cost	Moderate (Rs10000/ha)	High (Rs60000/ha)	Low (Rs7500/ha)	Medium (Rs.30,000/ha)	High (Rs.100,000 for wells & Rs. 2.5 lakhs with solar power)	Medium (Rs 20000/ha)
Sustainability of aquifer-tank ecosystem	Medium	High	Medium	High	High	Very high
Increase in number of wells	5%	10%	15%	10%	20%	20%
Groundwater externality	Positive	Huge positive	Negative	Huge positive	Negative	Positive
Tank ecosystem/ multiple use benefits	Medium	High	Low	Very high	Low	Medium

(continued)

Table 5.2 (continued)

Indicators of change	Aquifer-tank ecosystem development options					
	S1: Tank partial desilting	S2: Tank full scale rehabilitation	S3: Tank into percolation ponds	S4: converting non-system into system tanks	S5: Tank with full groundwater development	S6: Conjunctive use
Role of active WUAs & quantum increase in tank benefits	Medium & 20%	High & 25%	Low & 5%	High & 30%	Medium & 35%	High & 35%
Social benefits	Medium	High	Medium	High	Low	High

Note * IRRs are low because costs are additive and benefits are non-additive. Furthermore, tank storage-based benefits are not uniform in all years

Enhancement of multiple uses refers the quantum of benefits from multiple uses expected when the particular modernization option is implemented. This is more or less correlated with tank ecosystem benefits.

Quantum increase in tank benefits due to active WUAs refers the role of WUAs in managing the water allocation during scarcity periods and also managing the tank structures with minimal water losses.

Option 1: Desilting only in selected locations in the tank water spread area. Normally it covers 20 to 30% of the water spread area.

Option 2: Desilting the entire tank water spread area to a given depth.

Option 3: Converting the tanks into percolation ponds by closing the sluices when the tank water storage is less than 40% in the last 5 years.

Option 4: Providing 1–2 fillings to the tanks during the tank season from the nearby canal/anicut/river systems.

Option 5: Development of full groundwater in the command area by working out the feasible (optimal) number of wells in the tank command area using annual groundwater draft. It is expected that in the tank commands, about 20% increases in the number of wells will be possible.

Option 6: Adopting sluice rotation (opening and closing the sluices in alternate weeks) so that groundwater and tank water will be simultaneously (conjunctively) used throughout the crop season.

Social benefits refer the overall benefits to the village community in terms of on-farm and off-farm employment and other livelihood options to the tank-based households and institutions.

It could be seen that most of the options are having possibilities to implement depending upon the tanks and the budget availability. These options are not mutually exclusive. Option 3 (converting into percolation ponds) and Option 4 (converting into system tanks) will have lesser applicability compared to other options as these options can fit into only to those tanks with lesser storages and those close to/within the canal/anicut systems. Once the groundwater development in the tank commands is completed, it is possible to manage the conjunctive use through sluice rotations as a possible strategy to sustain the tank irrigation in the long run. Also among the system and non-system tanks, the rate of return to modernization is higher with system tanks due to their adequate water supply 3 out of 5 years compared to non-system tanks which have adequate supplies mostly 1–2 years out of 5 years.

Tank rehabilitation had undergone many notable changes over the years in terms of their objectives and focus of rehabilitation, funding pattern, and funding sources, physical components prioritized for rehabilitation, implementation pattern, and institutional changes. Looking at the past rehabilitation performance and need of the future, it is important that given the increased benefits and equity considerations, improving the livelihoods of the rural community through increasing the total benefits from tank multiple uses should be the objective of future tank rehabilitation programmes. In this context, it is important to examine, how future of tank irrigation could respond to the emerging hotspots.

5.10 Emerging Hotspots—Key Questions and Policy Suggestions

Given the above issues and challenges, it is important to examine what questions need to be answered and what policy suggestions could address these questions adequately. The list of key questions and the policy suggestions are summarized (Table 5.3).

Once the hotspots are appropriately addressed through policy suggestions, it is important to bring in the bright spots of tank irrigation to make sure that tanks will be sustainable irrigation structures in the near future. The following big ticket questions will help locate the bright spots in tank modernization.

5.11 Future of Tank Modernization—Ways and Means

What exactly will be able to address the future tank modernization? What will be key components of tank modernization? How the modernization options could be prioritized in the tanks? What are the factors that will make tank modernization sustainable? How the tank modernization could be done in a cost effective manner? What kind of modernization cycle could be followed in the long run to cover all the tanks in the region/state? What kind of institutional mechanisms will safeguard the tanks and efficient irrigation management?

The facts that may help to address the above questions are discussed:

(a) What exactly will be able to address the future tank modernization?

Climate change will affect the water supplies in south Asia, where floods and droughts with high intensity are expected in the future. Increasing water storage is a key adaption response. The experience of irrigations tanks, water harvesting structures dating back centuries across much of peninsular India, illustrates both the potentials and challenges of this adaptation response. Analysis of climate and hydrological data at an appropriate scale should be given high precedence. Further, instead of focusing purely on financial gains, providing livelihood to all including the landless should be the key focus of future modernization.

(b) What will be key components of tank modernization?

Tank modernization should primarily comprise catchment treatment, foreshore plantations, and creation of dead storage for community and livestock use, improvement to supply channels, improvement to tank structure, on-farm development works, crop and water management, and provision of community wells wherever feasible. Rehabilitation focus should be on distribution of water based on crop demand and adoption of crop and water management practices through on-farm development.

The rehabilitation budget should be allocated in the ratio of 10, 75, and 15% to institutional development, physical works, and maintenance, respectively. A one-time investment of 15% for maintenance and management activity will be allocated

Table 5.3 Hotspots—key questions and the policy suggestions

No.	Hotspots	Key questions	Suggestions
1	Rainfall pattern and tank filling	The weak relationship that exists between the rainfall and area under tank irrigation leads to a genuine question: Do other factors influence area under tank irrigation more than the level of rainfall?	(a) What appears from this is that any improvement in the level of rainfall is not going to make a perceptible increase in area under tank irrigation, unless adequate arrangements are made to improve the storage capacity as well as the flow of water supply into the tanks. (b) Even in normal or good rainfall years, several tanks couldn't get the adequate tank filling and many of the tank farmers reported that there is no adequate rain in the tank catchments and officials rely on nearby station rainfall data for quantifying the runoff into the tanks. Currently the rain-gauges are available in the block offices only and periodical measurement of rainfall intensity is not followed. Hence, installation of rain-gauge stations at different locations of the tank cascade to measure rainfall intensity and estimate the exact relationship between tank storage and rainfall is suggested.
2	Tanks and groundwater development	Which type of tanks (system versus non-systems, PU vs. PWD) has more potential for expansion of well irrigation? What will be the optimum number of wells in the tank command area?	(a) Tank management will be complementing the well development and hence strategies that help improve the tank performance need to be identified and implemented. Water markets at tank command area can help supplement the tank irrigation benefitting non-well owners. (b) Using the annual groundwater draft, feasible number of wells in each tank command needs to be worked out for optimal use of tank and groundwater. (c) Conjunctive use of tank and groundwater will enhance the water use efficiency of both tank and groundwater.

(continued)

Table 5.3 (continued)

No.	Hotspots	Key questions	Suggestions
3	Change in village socio-economic dynamics	How do the changes in socio-economics profile of the households influence tank management? What measures will ensure sustainable tank management?	Increasing socio-political changes among the village households hinder effective tank management. Hence encouraging self help groups with micro finance options will ensure better livelihood options and out migration of the labour households can be prevented. Further, village support programmes like livestock and poultry development programmes, social forestry protection, and fishery development programmes can be encouraged in the tanks for better multiple use benefits among the stakeholders.
4	Climate change and impacts	What exactly the impact of climate change on tank water storage and getting supplemental irrigation for crops? What kind of climate-smart practices will support tank-based agriculture? What kind and type of tanks are most affected by climate change?	(a) Tank modernization could help improve supply channels leading from catchment, increase/stabilize storage capacity of the tanks thereby providing scope for storing more rainwater during heavy runoff. It is important to restore the original storage capacity of tanks by desilting. In some cases, as a least cost option, increasing the bund height may help to store more flood water provided no submergence in the catchment. Tanks with short bunds may be good to implement this. Also this will help recharge groundwater in the tank commands. Since more dry spells will be common due to climate change, groundwater supplementation with solar pumps for groundwater pumping, Climate-smart agricultural and water management practices, changes towards less water consuming crops, could help address the emerging climate change influence in agriculture. (b) A tank manual should be prepared for all the tanks in the cascades.

from the rehabilitation funds and this amount will be deposited in a bank. The interest accrued out of this fund can be used only for maintenance and management. Capacity building of SHGs and vulnerable groups is a built-in component of the tank rehabilitation project. There should be a subcommittee in the tank user groups to do social audit of the water distribution to ensure equity among the users.

(c) How the modernization options could be prioritized in the tanks?

Even though tank modernization is seemingly a good option for reviving the tank irrigation as a whole, a major criticism observed in the past rehabilitation works is its focus as a single-time activity. Also most of the modernization programmes followed a standard ***package*** (blue print) approach where same package is applied for all tanks irrespective of their major attributes such as location in the cascade, command area, water storage, and physical structures. This approach resulted in poor post rehabilitation activities with more number of rehabilitated tanks becoming degraded. In several tanks, impact of tank modernization is not yielding benefits as expected (Palanisami et al. 2008a). Also, lack of effective involvement of those stakeholders, for whom the tank is still the primary livelihood, especially in the process of planning, implementation, cost sharing, operation, and maintenance of the rehabilitated tanks, is still a major problem. Therefore, there is a need to relook the tank rehabilitation packages in terms of modernization package to sustain the benefits from tanks. For greater cost effectiveness, it is important to identify ***selective*** (customized) modernization strategies to suit different tank typologies. For example, in the Eastern and West regions the key attention of tank rehabilitation is still on increasing agricultural productivity, where the benefits remain concentrated to the landowners (landlords) while the landless remains merely as labourers. In the case of Southern regions, in most cases focus is given for improving livelihood opportunities wherever household in the village receives benefits in one way or another from tanks multiple uses thus creating a confidence in tank management.

In the case of construction of new tanks and rehabilitation of the existing tanks, it is always important to see the suitability of the sites for new tank construction along with hydraulic particulars such as rainfall runoff, terrain characteristics, etc. Based on the experiences, it is suggested that in the case of Eastern region, where small tanks/ponds are common, construction of new tanks/ponds can be an attractive investment option compared to other regions where already large number of tanks are existing and where watershed programmes are competing for rainfall storages, and it is always economical to invest in rehabilitation of the existing tanks in these regions.

(d) What are the factors that will make tank modernization sustainable?

The focus of rehabilitation in the future will be more on water acquisition, storage, and distribution. The cascade approach should be followed in restoring tanks if the full benefits of harvesting the runoff from a micro watershed and effective groundwater recharge are to be realized. All tanks in the cascade, small and large and irrespective of the size of their command area, have to be renovated by restoring the link canals between them on top priority.

The foremost driver behind modern-day rehabilitation programmes observed was the engagement with stakeholders under a decentralized governance system. The MNREGA manpower can be better used in tank improvement works so that the impact of the MNREGA labour could be measurable.

Integration of tank and groundwater system: It is important to manage the conjunctive use of tank and groundwater through appropriate water management strategies. For example, sluice rotations are considered as a possible strategy to sustain the tank irrigation in the long run wherein the stabilization value of groundwater will be high compared to current practice of using tank and groundwater separately, i.e., tank water in the first few months and then later months groundwater (Palanisami and Easter 2000).

Integration of tanks and social forestry also needs attention. A study (Palanisami 2006) on the water consumption by the tree has shown a linear increase in uptake and utilization with the age of the trees with a corresponding increase in the biomass production. But there is no significant loss in tank water compared to tanks without the social forestry plantations as the tree cover prevented the evaporation of water from the tank. However, only the tank foreshore area can be permitted to go for social forestry as this will prevent the foreshore encroachment and also the extensive growth of *prosophis* tress which have less economics values. The revenue should go to the tank maintenance funds.

(e) How the tank modernization could be done in a cost effective manner? How to make the funding mode more effective?

Regarding investment and cost sharing in tank modernization, a one-time investment of a maximum of 15% for maintenance and management will be allocated from the total modernization budget and the bank interest accrued out of this fund can only be used for tank maintenance and management on regular basis. The modalities to draw the interest from bank for tank-related works need to be worked out in consultation with the stakeholders where the provision for a matching grant from WUAs will be assured. A higher level of contribution from the WUAs will be an added advantage to help mobilize more funds from government and other funding agencies. In the case of water acquisition, the cost calculations should be based on rupees per square kilometre (sq km) of catchment area for works in the catchment including feeder channel improvement.

Given fiscal constraints, the issue concerns the balance between creating new works and making better use of the already existing tanks. If the existing thousands of tanks and ponds are rehabilitated in a cyclic manner, it will contribute significantly to not only increasing food production but also providing a variety of livelihood options to the rural poor, especially women and also these rehabilitated tanks will act as buffer to store more flood water due to climate change induced heavy rains. This appears to be the best cost effective option than creating new irrigation works.

Since budget constraints may surface out when going for large-scale rehabilitation, it is important to examine the relevant co-financing options in tank modernization programmes. As such several funding agencies mostly in a single agency mode are funding the tank modernization. For example, EEC, NABARD, World Bank funded programmes, etc., are mostly done based on their concept note and work plan already agreed with the implementing government departments such as irrigation or agriculture departments. Some agencies might be having most expertise to implement both structural and non-structural investment options and many may not have the technical expertise resulting in weakening of the modernization. It is observed that in most of the tank modernization programmes, technology adoption and capacity building aspects are weak resulting in poor performance of the modernized tanks over years (Palanisami and Mohanasundari 2020). Hence, co-financing with national or international agencies with different expertise in modernization will enhance the benefits from modernization. For example, NABARD can consider the co-financing options with ADB or WB for utilizing their specific expertise in the field. So there is a win-win situation for both the funding agencies. Co-financing will also provide better opportunities for cross learning and main streaming the investment programmes in a better manner.

(f) What kind of institutional set is needed for the future of the tanks? How the tank farmers associations and other tank-based institutions could be revived?

Tank water users association should be made mandatory for all the tanks. One single agency such as PWD or Rural Development Department must be made the nodal agency to coordinate with the tank farmers associations. This will facilitate efficient fund flow from government departments to the tank management. Also, the revenue generated from tank multiple uses should be shared with tank farmers associations through appropriate government norms or Acts. At national level, a tank authority can be established mainly to coordinate and mobilize funds for tank investment and management by closely working with the states. Tank authority should ensure better convergence of different programmes by government departments, external funding agencies, and corporate initiatives on tank improvements, and cross learning tank management issues across states, prioritizing of investment options and enhanced livelihood options in the rural sector. The role of government (funding agencies), NGOs, and tank user groups at tank, cascade, and federation levels are increasingly important in achieving these enhanced livelihood options through effective modernization.

In sum, given the climate change regime and the constraints in further expanding canal and well irrigation sources, irrigation tanks (both new and existing) offer the best investment strategy for stabilizing the irrigation potential of the country besides benefitting mostly small and marginal farmers and other rural households in the country.

5.12 Reviving Tank Irrigation—The Way Forward

Given the constraints in future expansion of canal irrigation and groundwater development in the country, irrigation tanks have now re-emerged as the best example of providing extensive (protective) irrigation in the semi-arid tropics. Further, tanks will also serve as the best climate changes adaptation strategy by storing excess inflows during heavy rains. Looking at the past rehabilitation performance and need of the future, improving the livelihood of the rural community through increasing the total benefits from tank multiple uses should be the objective of future tank rehabilitation programmes. Accordingly, six time-tested options for future modernization are suggested: 1. Desilting only in selected locations in the tank water spread area. Normally it covers from 20 to 30% of the water spread area, 2. Desilting the entire tank water spread area to a given depth, 3. Converting the tanks into percolation ponds by closing the sluices when the tank water storage is less than 40% in the last

5 years, 4. Providing 1–2 fillings to the tanks from the nearby canal/rivers during excess flows, 5. Groundwater development in the command area by working out the feasible (optimal) number of wells, and 6. Adopting sluice rotation (opening and closing the sluices in alternate weeks) so that groundwater and tank water will be simultaneously (conjunctively) used throughout the crop season.

References

Asian Development Bank (2006) Rehabilitation and management of tanks in India: a study of select states. Publication Stock, No. 122605, Philippines

Janakarajan S (1993) In search of tanks: some hidden facts. Econ Political Wkly 28(26):A53–A60

Kajisa K, Palanisami K, Sakurai T (2006) Effects on poverty and equity of the decline in collective tank irrigation management in Tamil Nadu. November, Un Published Paper, IRRI, Manila, Philippines, India

Kajisa K, Palanisami K, Sakurai T (2007) Effects on poverty and equity of the decline in collective tank irrigation management in Tamil Nadu, India. Agric Econ 36:347–362

Van Koppen B, Smits S, Moriarty P, Penning de Vries F, Mikhail M, Boelee E (2009) Climbing the water ladder: multiple-use water services for poverty reduction. Technical Paper Series 52, The Hague, The Netherlands, IRC international water and sanitation centre and international water management institute and challenge program on water and food

Kumar M, Dinesh, Panda R, Vedantam N, Bassi N, Mukherjee S (2011) Tank ecology and multiple uses of water in villages of western Odisha: technology choices, Economics, institutions and livelihoods. Report submitted to IWMI-Tata Water Policy Research Program, IWMI South Asia Regional Office, Patancheru, Andhra Pradesh, Institute for Resource Analysis and Policy, Hyderabad & Manav Adhikar Seva Samiti (MASS), Jabalpur, Odisha

Mosse D (1999) Colonial and contemporary ideologies of community management: the case of tank irrigation development in South India. Mod Asian Stud 33(2):303–338

Narayanamoorthy A (2007) Tank irrigation in India: a time series analysis. Water Policy 9(2):193–216

Narayanamoorthy A, Suresh R (2016) Does urbanisation affect tank irrigation development in Tamil Nadu?, a macro-level analysis. Rev Dev Change 21(2)

Nehlin M (2016) Management of local irrigation systems and stakeholder perceptions in southern Tamil Nadu, India, Master's Thesis. Environmental Management and Physical Planning, Department of Physical Geography, Stockholm University

Palanisami K (2000) Tank irrigation—revival for prosperity. Asian Publication Services, New Delhi

Palanisami K, Easter WK (1987) Small scale surface (tank) irrigation in Asia. Water Resour Res 23(5):774–780

Palanisami K, Easter WK (2000) Tank irrigation in the 21st century—what next? Discovery Publishing House, New Delhi

Palanisami K, Balasubramanian R, Ali MA (1997) Present status and future strategies of tank irrigation in Tamil Nadu. TNAU Press, Coimbatore

Palanisami K, Balasubramanian R, Ali MA (1998) Analysis of tank irrigation performance in Tamil Nadu. J Indian Water Resour Soci 18(4):3

Palanisami K, Balasubramanian R (1998) Common property and Private Prosperity: Tanks versus. Private Wells in Tamil Nadu. Indian J Agric Econ 53(4):Oct-Dec

Palanisami K, Easter WK (1983) The tanks of South India- a potential for future expansion in irrigation. Economic Report, ER83-4. Department of agricultural and applied economics, University of Minnesota, St Paul. Minnesota, USA

Palanisami K, Easter WK (1984) Tank irrigation in India and Thailand: problems and prospects. Irrigation Management Network, Agricultural Administration Unit, London, Network Paper 10E, pp 1-12

Palanisami K, Easter WK (1991) Hydro-economic interaction in tank irrigation systems. Indian J Agri Econ 47(2)

Palanisami K Flinn JC (1988) Evaluating the performance of tank irrigation systems. Agri Syst 28(3)

Palanisami K Flinn JC (1989) The Impact of inadequate water supplies on input use and yield of rice crops. Agric Water Manag 15

Palanisami K, Mohanasundari T (2020) Sustaining tank irrigation in South India through time-tested measures. Chapter 4. Kochi Fujita et al (ed) Toward sustainable development of India and South Asia. Springer, Singapore

Palanisami K, Gemma M, Ranganathan CR (2008a) Stabilization value of groundwater in tank irrigation systems. Indian J Agric Econ 63(1)

Palanisami K, Jegadeesan M, Fujita K, YasuyukiKono (2008b) Impacts of tank modernisation programme on tank performance in Tamil Nadu State, India. Kyoto Working Papers on Area Studies, No. 5 (G-COE Series 3), Centre for Southeast Asian Studies, Kyoto University, Kyoto, Japan

Palanisami K, Meinzen-Dick R, Giordano M (2010) Climate Change and water supplies: options for sustaining the tank irrigation potential in India. Econ Political Wkly 45(26&27)

Palanisami K, Mark G, Krishna K, Ranganathan CR (2011a) The stabilization value of groundwater: evidence from Indian tank irrigation systems. Hydrogeol J. https://doi.org/10.1007/s10040-011-0793-3.Nov.4

Palanisami K, Ranganathan CR, Umetsu C (2011b) Groundwater over-exploitation and efficiency in crop production: application of data envelopment analysis. J Oper Appl Res 3(1)

Palanisami K (1990) Tank irrigation—what next ? IIMI/ODI Irrigation Management Network, London, July 1990 and also in Water Resour ESCAP/United Nations. ST/ESCAP/SER.C/167. pp 50–54

Palanisami K (1991) Conjunctive use of tank and well water in tank irrigation systems in future directions for Indian irrigation. In: Svendsen M (ed) Research Policy Issues, Dick RM. IFPRI, Washington DC

Palanisami K (1993) Karnataka tank rehabilitation project-evaluation report. Consultancy Report. World Bank, New Delhi

Palanisami K (2005) Sustainable management of tank irrigation systems in South India. Working Paper Series No.2. Afrasian centre for peace and development studies, Kyoto, Japan

Palanisami K (2006) Sustainable management of tank irrigation systems in India. J Dev Sustain Agri 1(1)

Palanisami K (2009) Irrigation tanks—a new way forward. In. Duraisamy P, Srinivasan R (eds) Indian rural economy, chapter 4. University of Madras, Chennai. pp 83–112

Palanisami K (2013) Scoping study to assess constraints and opportunities for sustaining the tank irrigation ecosystems and small holder livelihoods in East India (Bihar, Odisha and West Bengal) and Nepal. ACIAR Scoping Report, Canberra and IWMI, Hyderabad

Pant N, Verma RK (2009) Tanks in Eastern India: a study in exploration, Hyderabad, India: International Water Management Institute, IWMI-TATA Water policy research program. Centre for development studies, Lucknow, Uttar Pradesh, India, p 232. https://doi.org/10.5337/2010.228

Raju KV, Shah T (2000) Revitalisation of irrigation tanks in Rajasthan. Econ Political Wkly 35(23):1930–1936

Raju T, Shah KV (2002) Rethinking rehabilitation: socio-ecology of tanks in Rajasthan, north-west India. Water Policy 3:521–536

Ranganathan CR, Palanisami K (2004) Modelling economics of conjunctive surface and groundwater irrigation systems. Irrig Drain Syst 18:127–143

Reddy PR (2015) An over view of irrigation tanks rehabilitation in semi-arid hard rock terrain. J Indian Geophy Uni 19(4):481–487

Reddy VR, Bhagirath B (2009a) The economic and ecological impacts of tank restoration in South India. Eur J Dev Res 21(1):112–136

Reddy VR, Bhagirath B (2009b) Economic viability of rehabilitation of traditional water harvesting system: empirical evidence from South India. Q J Int Agric 48(4):321–336

Reddy DN, Barah BC, Sudhakar T (1993) Traditional water harvesting system: tank in the drought prone areas of Andhra Pradesh. Indian J Agric Econ 48(1):76–87

Reddy R, Reddy VMS, Palanisami K (2018) Tank rehabilitation in India: review of experiences and strategies. Agric Water Manag 209:32–43

Reddy MA (1995) Irrigation tanks in Rayalaseema. In: Barah, BC (ed) Development of water resources and its ecological implications: the role of minor irrigation. Delta Publishing House, New Delhi

Sakthivadivel R, Gomathinayagam P, Shah T (2004) Rejuvenating irrigation tanks through local institutions. Econ Political Wkly 39(31):3521–3526

Sakurai T, Palanisami K (2001) Tank irrigation management as a local common property: the case of Tamil Nadu, India. Agric Econ 25:273–283

Shankari U (1991) Tanks: major problems in minor irrigation. Econ Political Wkly 26(39): A115-A119, A122-A125

Sivasubramaniyan K (2006) Sustainable development of small water bodies in Tamil Nadu. Econ Political Wkly 41(26):2854–2863

Vaidyanathan A (1999) Water resources management: institutions and irrigation development in India. Oxford University Press, New Delhi

Von Oppen M, Rao SKV (1980a) Tank irrigation in semi-arid tropical India, Part I: Historical development and spatial distribution, ICRISAT, Economics Program progress Report 5. Hyderabad, India

Von Oppen M, Rao SKV (1980b). Tank irrigation in semi-arid tropical India. Part II: Technical features and economic performance. ICRISAT Economics Program Progress Report 8, Hyderabad, India

World Bank (1998) India-water resources management sector review: groundwater regulation and management, The World Bank, New Delhi

World Bank (2014) Review of tank projects in India. FAO—World Bank cooperative programme. Draft Report—24 (unpublished), New Delhi

Chapter 6
Maximizing Wastewater Reuse

Abstract The core drivers to increase the reuse of wastewater include frequent shortage of water sources, and technology and cost viability/efficiency of treating wastewater resources. The current installed sewage treatment capacity is 31,841 MLD, but the operational capacity is 26,869 MLD (26%). The amount of water needed by the manufacturing sector will increase by 2.2 times between 2010 and 2050 and the energy sector will increase by 3.7 times. Wastewater reuse decreases the demand for freshwater by 20–50% depending on the extent of the treatment and reuse possibilities. Cost of freshwater per m^3 is less under recharge borewell followed by percolation ponds and if treated wastewater is considered, it will be much cheaper. Using Life Cycle Cost Assessment (LCCA) analyses four wastewater treatment technologies were compared for cost efficiency. Challenges encountered in reclaimed water reuse include technical challenges, institutional challenges, economic challenges, and social challenges. Investments in wastewater treatment projects are also a social investment like investment in major and medium irrigation projects and hence future government priorities in water sector investment should also include wastewater treatment projects. Policies to augment and efficient wastewater reuse are discussed.

Keywords Wastewater recycling · Wastewater reuse · Treatment capacity · O&M costs · Price of water · PPP models · Success cases · Life cycle costs · Zero discharge options

6.1 Introduction

The worldwide total water consumption for agriculture alone accounts for 92%. Freshwater mostly from rivers and underground aquifers accounts for about 70% of the irrigation requirements (WRI 2020). Rapidly depleting and increasing demand for freshwater resources, make wastewater reuse as one of the most important strategies of the future.

K. Palanisami and U. S. Nagothu, *India's Water Future in a Changing Climate*, Advances in Geographical and Environmental Sciences,
https://doi.org/10.1007/978-981-97-1785-9_6

Table 6.1 Water demand for industries and power sectors in India

Year	Water demand by industries and power sector (BLD)	Percentage in total water demand
2010	41	5.9
2020	65	7.0
2025	80	8.6
2050	144	10.2

Source Authors' compilation. *Note* BLD—Billion litres per day

Currently, efforts to treat and reuse the wastewater for irrigation are gaining recognition. It is also becoming popular to sue treated wastewater for non-potable purposes, like groundwater recharge, washing vehicles, flushing toilets, fire fighting, and cooling thermal power utilities. Among these uses, treated wastewater use in irrigated agriculture supports the livelihoods of millions of smallholder farmers. At global level, reuse of treated wastewater in irrigated agriculture ranges from 1.5 to 6.6% where more than 10% of the global population uses agriculture-based products harvested from treated wastewater for irrigation (WHO 2006).

It is reported that at global level, about 380 billion m^3 of wastewater is generated annually. Among the regions, Asia accounts for the largest volumes of wastewater (42% of the global total) and among the South Asian countries, India has the highest potential for wastewater generation (GoI(a) 2022). Hence, wastewater reuse in Indian agriculture will be highly complementary to the use of freshwater water resources. Industries primarily obtain their water from surface water (41%), followed by groundwater (35%) and municipal water (24%). Urban and peri-urban areas have a limited supply of it (FICCI 2011). The amount of water used by the industrial sector is not precisely known. However, the expected industrial demand shows an increasing trend over years where the percentage of industrial demand is projected to go up to 10% of the total water demand in 2050 (Table 6.1). This chapter outlines how the wastewater reuse can be made a reality in India's water futures.

6.2 Importance of Wastewater Reuse in India

Two key factors are propelling the business case for recycle and reuse of treated wastewater in India. First is addressing water security, as several cities in India are facing water crisis even in normal monsoon periods and during periods of droughts, as both industry and agriculture sector suffer in terms of loss in production. Second is increasing government support for reuse of treated wastewater through several programmes such as Clean Ganga Mission. Accordingly, government policies are gradually aligning to support efforts to treat wastewater and promote reuse. This chapter outlines the potential for wastewater reuse, current efforts ongoing in India, and the ways to improve in the future.

The core drivers to increase the reuse of wastewater are: (i) frequent shortage of water sources due to climate extremes and competing uses of freshwater resources, (ii) availability of good quality treated water as a result of compliance with more stringent wastewater standards, and (iii) technology and cost viability/efficiency of treating wastewater resources. Several industries and bulk water users will see wastewater as an economically viable option to meet their future water requirements.

Instituting wastewater recycling reduces the demand for conventional water and sewerage infrastructure. It is important to note here that these costs may not be higher, and in fact it would be less than what would be required to build, extend, and operate a conventional sewerage system. As a result, costs can be cut down and instead can redirected for other purposes—for expanding access to piped water and sanitation. Further, wastewater recycling and water conservation are already a must in many situations.

Wastewater reuse decreases the demand for freshwater by 20–50% (Government of Andhra Pradesh 2019), depending on the extent of the treatment and reuse possibilities. In the long run, the costs for wastewater treatment, infact, will not be prohibitive and are actually lower than those associated with developing, extending, and running a traditional sewage system. Further, wastewater recycling reduces pollution and other related health hazards as vast majority of wastewater is directed straight to open drains contaminating both the surface and groundwater resources.

Use of treated wastewater in agriculture sector accounts for roughly 83% of all applications, followed by industrial use (12%) and other uses (5%) (IIT 2022). Wastewater reuse for different purposes is broadly shown in Fig. 6.1.

The wastewater reuse for different purposes in different states and the process of wastewater generation, treatment, and reuse are explained in Figs. 6.2 and 6.3 respectively.

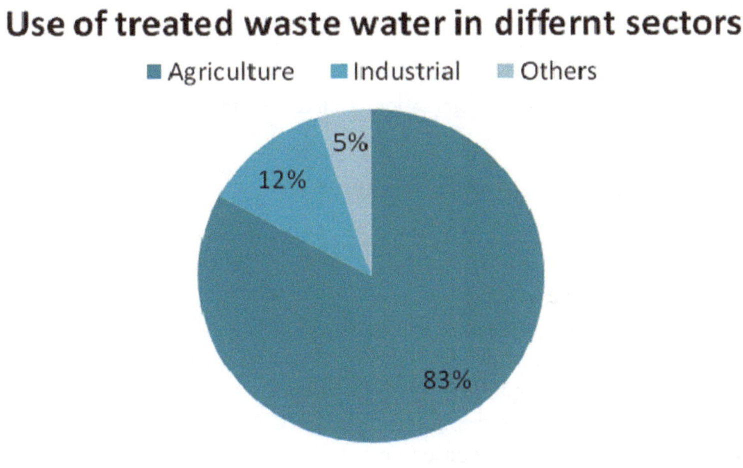

Fig. 6.1 Use of treated wastewater in different sectors, India

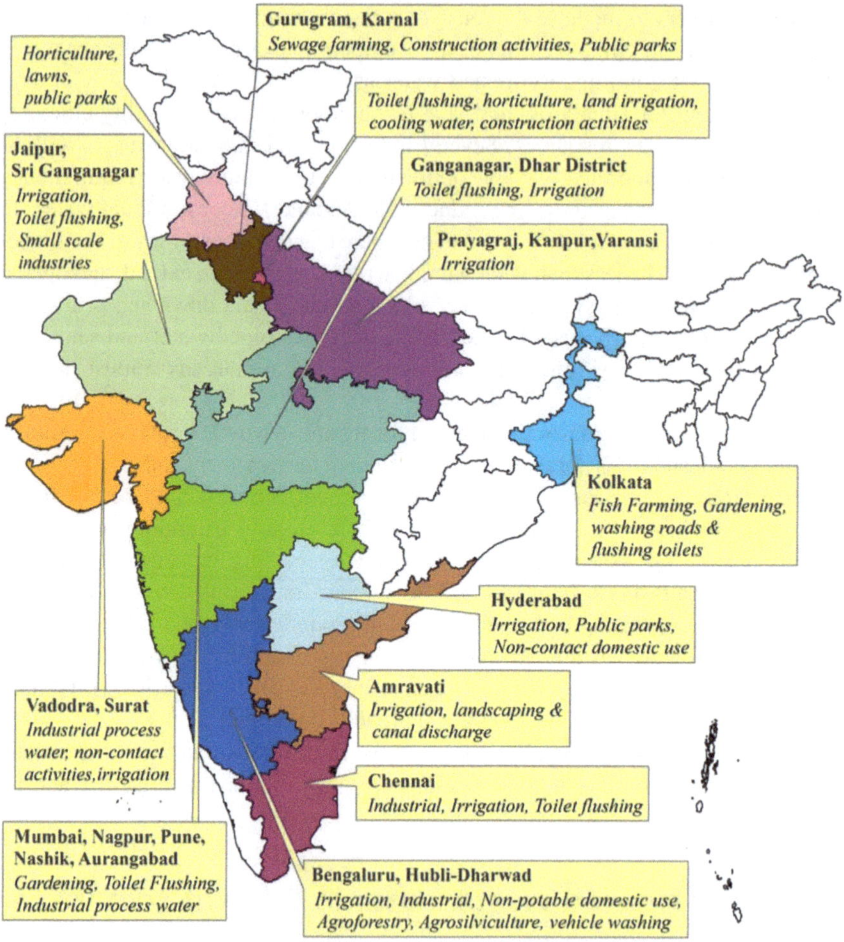

Fig. 6.2 The wastewater reuse for different purposes, India. *Source* IIT, Roorkee, 2022

6.3 Benefits of Wastewater Reuse

The major benefit of wastewater treatment will be the reduction in value of the freshwater used for urban or industrial use. This would minimize the cost of getting the water supplies through more expensive sources such as water tankers. In addition, reuse prevents untreated wastewater discharge to coastal and groundwater systems with ecosystem and health benefits. Depending on the local situation, it could benefit the farmers through reduction in pumping groundwater, while the nutrient present in the wastewater could save expenses on fertilizers (Table 6.2). It could benefit to the local environment from reduced flows of untreated wastewater (FAO 2011; Toze 2006).

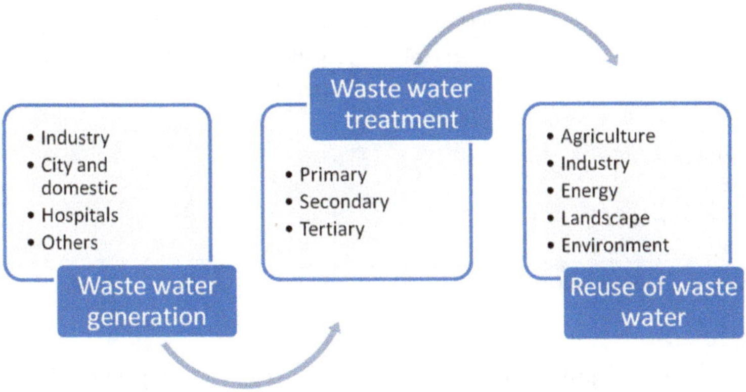

Fig. 6.3 Process of wastewater generation to reuse

Table 6.2 Benefits of wastewater reuse for irrigation in select cities in India

City	Crops cultivated	Increase in yield (%)	Decrease in fertilizer use (%)	Average annual incremental benefit (Rs/ha)
Indore	Wheat (Rabi)/ vegetable (summer)	30–40	50	36,752
Nagpur	Wheat (Rabi)/ vegetable (summer)	30–40	33	26,951
Jaipur	What (Rabi)/ vegetable (summer)	30–40	50	37,799
Bangalore	Rice (Rabi), Sapota, flowers (Summer)	30–40	100	33,849
Ahmedabad	Rice and wheat (Rabi)	–	–	–14,640
Delhi	Okra	67	60	8500
Kanpur	Paddy and wheat	Decrease in yield	–	6166 (paddy) 954 (wheat)

Source Ravikumar et al. (2014)

A comparative analysis of the effects of treated effluent on the chemical characteristics of the soil, crop development, yield, and irrigation water use efficiency (IWUE) has indicated that drip irrigation with treated effluent had the highest maize grain production of 7.1 t/ha and the highest irrigation water use efficiency (IWUE) of 1.08 kg/M^3 when compared to flood irrigation (Table 6.3) (Suganthi et al. 2019).

Table 6.3 Irrigation water use efficiency of maize under surface and drip irrigation

Irrigation method	Grain yield (t/ha)	Applied water (M^3/ha)	IWUE (kg/M^3)
Treated effluent + surface irrigation	6.5	7120	0.91
Treated effluent + drip irrigation	7.1	6580	1.08
Well water + surface irrigation	6.0	7240	0.83
Well water + drip irrigation	6.3	6580	0.96

In the course of producing paper and pulp, paper and pulp mills discharge a significant amount of salts into the effluent. Salts build up in the soil surface and rhizosphere (zone of chemical, biological, and physical influence generated by root growth and activity) when paper and pulp mill effluent is used as a source of irrigation in the long run. In this work, the phyto-desalination capability of halophytes (group of plants which are accumulating salt) was analysed and results showed that the use of suitable halophytes successfully phyto-desalinated (uptake of salt from soil and accumulating in their leaves) the saline soils and ensured that crops grown with paper and pulp mill treated effluent are sustainable (John et al. 2022).

The combination of treated paperboard mill effluent, well water, and solid wastes such as biomanure (BM), vermicompost (VC), farmyard manure, and fly ash (FA) has been tried to grow the vegetable crops chillies and brinjal while simultaneously examining the effects of these wastes on the quality of the soil and crops. The soil organic carbon, accessible nutrients (N, P, K), and exchangeable cations (Ca, Mg, K, and Na) had increased due to the application of FA, BM, and VC in conjunction with effluent irrigation. Increased soil health and crop growth are a result of the soil's microbial population and enzyme activity. The treated effluent irrigated plot with solid waste application (BM + VC + FA) had a high plant height and branch count. According to the experimental findings, irrigation with treated effluent increased the production of chillies and brinjal compared to irrigation with well water. As comparison to the control (100% NPK alone), the yield rose by 20.90% for chillies and 17.28% for brinjal (Udayasoorian and Ponmani 2009).

Thus, the use of treated wastewater in agriculture can help farmers increase their income. When considering the average landholding size of 1 ha in the country and channelling the entire amount of treated wastewater towards agriculture, it is possible to support 2 million farmers and increase their annual farm earnings by Rs. 17,000 per ha or about 30% over the baseline income levels, generate about 2.8 million person-days of employment and reduce greenhouse gas (GHG) emission by 73.7 million Mg CO_2–e per ha (Minhas et al. 2022). Therefore, the potential for improvement is significant (Ravikumar et al. 2014).

A suitable wastewater treatment infrastructure is however, a prerequisite for effective wastewater management, which is currently lacking in India. India's present capacity for wastewater treatment is only 26%, according to a recent assessment by the Central Pollution Control Board (March 2021). In practice, most of the wastewater treatment facilities in the country do not operate at a full capacity.

6.4 Current Stage of Wastewater Generation and Treatment in India

About 35% of India's total population is concentrated in urban centres and the estimated wastewater generation is about 72,368 MLD (million litres per day) compared to 39,604 MLD in the rural sector for the year 2020–21 (CPCB 2022). The current installed sewage treatment capacity is 31,841 MLD, but the operational capacity is 26,869 MLD (Table 6.4), which is only about 26% of the actual quantity of wastewater generated and about 74% of wastewater is left untreated and allowed in rivers/lakes/groundwater leading to contamination and health concerns, particularly where dilution levels of the receiving water body are low (CPCB 2022). If we consider the wastewater generation and treatment capacity at urban level alone, the gap is about 67% for Class I cities (having more than 10 lac population) and 95% for class II towns (having less than 1 lac population) indicating that the amount of wastewater generation is influenced by the quantum of installed treatment capacity (Fig. 6.4). The number of treatment plants till 2020 was only 1469 and it is planned to add another 162 units in the coming years with a capacity of about 5000 MLD (CPCB 2022).

It is estimated that the projected wastewater from urban centres may cross 120,000 mld by 2051. Rural sector in addition, will generate not less than 50,000 mld (Kamyotra and Bhardwaj 2011). Further, manufacturing clusters can generate around 13,500

Table 6.4 Sewage generation, installed treatment capacity, and actual treatment in states with treatment of more than 1000 MLD

States/UTs	Current sewage generation (in MLD)	Installed capacity (in MLD)	Number of STPs installed	Operational treatment capacity (in MLD)
Maharashtra	9107	6890	154	6366
Gujarat	5013	3378	70	3358
Uttar Pradesh	8263	3374	107	3224
National Capital Territory of Delhi	3330	2896	38	2715
Karnataka	4458	2712	140	1922
Haryana	1816	1880	153	1880
Punjab	1889	1781	119	1601
Tamil Nadu	6421	1492	63	1492
All India Total	72,368	31,841	1469	26,869

Source (CPCB Annual Report 2022)

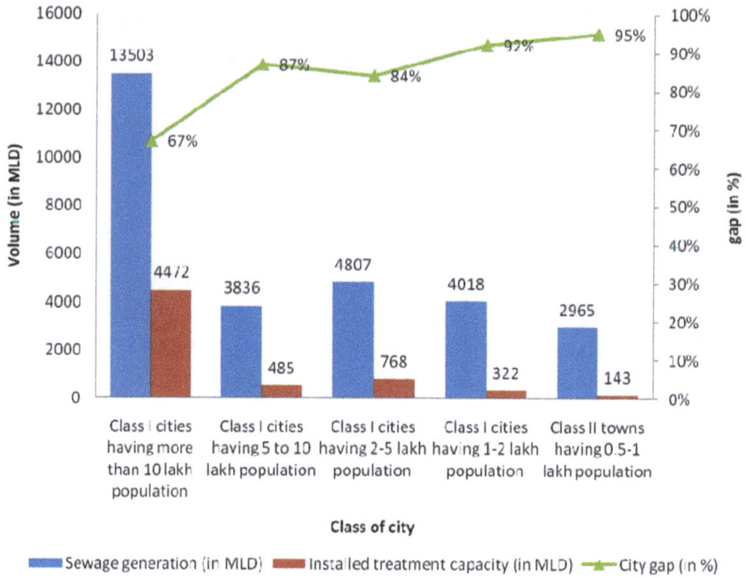

Fig. 6.4 Wastewater generation and treatment capacity gap in India (2001 population basis). *Source* (CPCB 2022)

mld of industrial wastewater, 60% of which is processed at the country's 193 common effluent treatment plants (CETPs). In general, the STPs are concentrated in larger cities and CETPs are unevenly distributed across states and hence the shortages in treatment capacity are magnified at the local level.

More than 39% of treatment plants could not meet the general norms under the Environmental (Protection) Rules for discharge wastewater into streams. The operation and maintenance of existing plants and sewage pumping stations are inadequate and inefficient. Further, the existing treatment capacities are underutilized in most cities. Hence, the intermediate pumping stations (IPS) and main pumping stations (MPS) of all STPs need to be provided with power backups to improve the operation of the plants (CPCB 2016).

6.5 Costs of Wastewater Treatment

The benefits to urban and industrial users will be relatively high and will justify the investment. In the case of agriculture, the benefits will depend on local situations, quantum of water reuse compared with their rainfed cultivation or alternative sources such as surface or groundwater. However, in this case, their net benefits will not meet the full costs of the investment where cost also includes the additional investment for reservoirs, pipelines, pumping stations, etc., and operation and maintenance (O&M) costs. Further, the net impact of the wastewater treatment project on the local and downstream environment will be location specific and in most cases it will be financially feasible.

The reuse of wastewater involves three treatment processes. The primary technique of treatment is the physical sedimentation/coarse screening method for eliminating suspended particles. Bar screens, fine screens, grid chamber, oil skimmer, sedimentation tank, and settling tank are the examples of primary treatment. Secondary treatment reduces the BOD level in treated wastewater by removing organic particles. Several technical methods that are eligible for secondary treatment are implementable in accordance with regional needs. Some of them are: Activated sludge processes (ASP), waste stabilization ponds (WSP), upflow anaerobic sludge blankets (USAB-anaerobic digesters), extended aeration processes (EA), moving bed bio-film reactors (MBBR), sequential batch reactors (SBR), membrane bioreactors (MBR), submerged aerobic fixed film (SAFF), biological filtration and oxygenated reactors (BIOFOR-secondary and tertiary). Tertiary treatment is an expansion of secondary treatment methods such as carbon absorption, flocculation/precipitation, membranes or advanced filtering, ion exchange, dechlorination, pressure sand filter, activated carbon filter, membrane filter, and reverse osmosis are used to remove pathogenic bacteria, e coli, viruses, giardia cysts, protozoa, etc. An alternative method for wastewater reuse is bio-remediation. It is the application of biological process to the treatment of wastewater with hazardous chemicals. Bioremediation is a natural process and less expensive. It could be suitable for rural areas, possible at different scales, and environment friendly. Through this method it generates less sludge and it transformed hazardous chemicals into harmless products. The low-cost treatment can also be an option when the stabilization ponds and constructed wetlands are used. Further, the net cost of treatment can also be minimized through the use of biogas for energy and power (FAO 2011).

Recycling wastewater offers some financial benefits as a source of income for ULBs and can satisfy various water needs. The initiatives taken by several ULBs in India to sell treated water to industries are listed (Table 6.5). In Hyderabad, it was planned to provide treated wastewater to industries at a charge of Rs 1 per kilolitre, however in Surat, it was sold at a higher rate of Rs. 22 per kl to industries. The cost can vary according to locations and type of treatment process.

Life cycle cost for wastewater treatment

Life Cycle Cost Assessment (LCCA) compares different treatment methods with their costs which normally include the project capital expenditures, operations, maintenance, repair, replacement, and disposal costs. For example, in an LCCA study, four wastewater treatment technologies were compared for cost efficiency utilizing secondary data and literature (Table 6.6). The UASB (Upflow Anaerobic Sludge Blanket) is a financially viable option if there is no land limit and effluent quality is not the primary treatment concern. If land is not a major consideration, ASP (Activated Sludge Process) is a viable option. It can be configured to provide centralized treatment or to handle more sewage at once. In terms of elimination, UASB performance is much less than ASP. Due to reduced footprint, BIOFOR (Biological Filtration and Oxygenated Reactor) might be a sensible choice for decentralized treatment.

Table 6.5 O&M cost of wastewater treated in India

Place	Capacity/STPs	O&M cost (Rs/Kl)
Delhi	20 MLD	4
Bangalore	4 STPs	12–15
Bangalore	135 MLD (Tertiary)	12
Surat	40 MLD (Tertiary)	22
Hyderabad	3 STPs	1
Gurgaon	2 STPs (15 MLD; 25 MLD)	4
Jaipur	62.5 MLD	–
Chennai	740 MLD (Secondary)	4.08
Nagpur	110 MLD	3.4

Source (Desai 2017)

In comparison to MBBR (Moving Bed Bio Reactor), BIOFOR eliminates a greater percentage of COD and TSS. The remarkable treatment efficiency of MBBR allows for its use in circumstances when cost is not a factor. Decentralized treatment is available with MBBR installation. The treatment medium, or filter, is what results in MBBR having the highest upfront and recurring costs. MBBR is used to clean wastewater from both homes and businesses. Changes to any of the cost components have an impact on the LCCA and are dependent on the actual site conditions where it is adopted (Rathore et al. 2022).

The present value costs of each cost component were considered by the Life Cycle Cost (LCC). The LCC computations are shown in Table 6.9, where the costs are first converted to present costs by multiplying them by the appropriate discount factors, and then the total LCC is calculated. The last row shows how much cost (in terms of LCC) is roughly needed to treat 1 MLD of wastewater. Assessing the benefits of a wastewater reuse project is generally difficult due to the difficulties encountered in quantifying those benefits. However, valuing the benefits using opportunity cost of water will provide some insights in the estimation of the benefits.

6.6 Waste Water Reuse—Case Studies Across India

(a) Agricultural water reuse in Kanpur, Uttar Pradesh

The city of Kanpur, which has a population of over 3 million in the state of Uttar Pradesh (UP) produces roughly 340 MLD of sewage, and its cluster of tanneries produces 26 MLD of effluent. Three wastewater treatment facilities were put into operation at Jajmau as part of the Ganga Action Plan—Phase I (GAP-I) in 1986: a 130 MLD STP based on ASP, a 36 MLD CETP based on UASB, and a 5 MLD pilot plant based on UASB. About 2,500 acres of peri-urban farms are served by a 4 km concrete irrigation channel that transports the treated sewage from ASP along with tannery effluent from the UASB (166 MLD). GAP-I was described as a programme to enhance

Table 6.6 Life cycle cost calculation for wastewater treatment

Costs (Rs in Million/MLD)	Wastewater treatment technologies			
	ASP	UASB	BIOFOR	MBBR
Capital base cost	5.10	3.20	1.14	11.47
Capital present value cost	4.85	3.00	1.08	10.90
Annual O&M base cost	0.90	1.25	1.20	11.00
Annual O&M present value cost	11.21	15.58	14.95	880.20
Annual energy base cost	0.66	0.44	0.98	0.82
Annual energy present value cost	8.19	5.46	12.19	10.19
Replacement base cost	0.50	0.30	0.11	1.14
Replacement present value cost	0.31	0.18	0.07	0.70
Salvage base cost	0.30	0.17	0.06	0.46
Salvage present value cost	0.11	0.06	0.02	0.17
Life cycle cost	24.44	24.19	28.27	158.67

Note ASP—This process uses aeration and floc to separate sludge from water. UASB—methane-producing digester, which uses an anaerobic process and forming a blanket of granular sludge and is processed by the anaerobic microorganisms. BIOFOR—aerobic or anoxic process biological reactors that use attached growth technology for application in municipal or industrial wastewater treatment and it is used primarily for the removal of BOD, TSS, and ammonia pollution in secondary and tertiary treatment. MBBR—uses plastic carriers covered in biofilm to decompose waste. In addition to being an effective means of removing organic substances, it is also an innovative method for nitrification and de-nitrification. *Source* (Rathore et al. 2022)

water quality by intercepting, rerouting, and treating home and industrial pollutants. Nevertheless, the technological set-up of the CETP in Kanpur was insufficient to handle the specific tannery pollutants or the increased volume of tannery effluent (from 9 MLD in 1993 to 36 MLD in 2015). The plants were also poorly run and frequently had electricity outages. Mixed treated wastewater from the STP and CETP produced poor irrigation water that exceeded requirements for suspended particles, BOD, COD, and heavy metals like chromium. As a result, the peri-urban agriculture suffered from significant soil and water contamination, decreased milk and crop yield, and increased human and livestock health problems.

(b) Agricultural water reuse in Hyderabad, Telangana

The Greater Hyderabad area with a population of about 10 million requires freshwater 2,350 MLD for domestic uses. Only 43% of the 1,800 MLD of wastewater that is created, a mixture of sewage and industrial effluents, can be treated. In addition to a few smaller, decentralized STPs, there are two major, centralized plants (Amberpet, 339 MLD, and Nagole, 172 MLD). Out of 23 STPs in the Greater Hyderabad area, many are not operational or undergoing maintenance. Most of the treated and untreated wastewater is dumped into the river Musi, a tributary of the river Krishna that is permanently filled with effluent outside of the monsoon season. At various downstream locations, water from the Musi river is channelled into irrigation channels to irrigate 10,000–12,000 acres of agriculture.

The economics of wastewater use in irrigation was also worked out, where the capital cost was Rs. 93.51 crores and the operating cost was Rs. 11 crore per year for 339 MLD (in 2008 prices). Assuming a 30 year life and 10% interest rate, the annualized cost of capital was Rs 0.8/m^3 and annual O&M cost was Rs 0.9/m^3, thus making the annual cost of treated wastewater as Rs 1.7/m^3. The benefits were calculated using different crop combinations such as rice, vegetables, and fodder crops. Except for rice, cultivation of other crops using treated waste was found more profitable (Hopper 2015).

Even though several government departments are responsible for construction, maintenance, and checking the quality of wastewater generated, in practice there are no measures in place to keep an eye on the quality of wastewater used for irrigation by farmers downstream. There is no well-developed infrastructure for delivering treated water to farmers' crops. Only farmers who are close to the STPs or who need a small amount of water for high-value crops will therefore find it practical to carry treated water to their fields using tankers. Other reasons for the discrepancy between farmers' formal use of treated wastewater and informal use of partially treated wastewater include the lack of institutional capacity to implement and monitor rules, the inadequate freshwater supply and sewage pricing system, and insufficient attention to environmental issues.

(c) Industrial water reuse in Nagpur, Maharashtra

The Bhandewadi STP is in operation and MahaGenCo pays Rs 15 crores per year for 110 MLD treated sewage and Rs 2.3 per m^3 of additionally treated sewage compared to buying freshwater from other sources at a cost of Rs 9.6 per m^3. The PPP contract and revenues from MahaGenCo helped to cover the operation and maintenance of their wastewater treatment plants (World Bank Group 2019). The Nagpur case is a successful business model for industrial reuse.

(d) Industrial water reuse in Chennai, Tamil Nadu

The total wastewater generation in the city is about 1,100 MLD, and the current installed capacity is around 70%. Overall, 49% of the treated wastewater is reused, which can meet 15% of the city's water demand (IWA 2018). A total of 33 MLD of secondary treated wastewater is supplied to industries, including tanneries, fertilizer, and petro chemical units (Natarajan 2020). Price of treated wastewater is very low (Rs 18.4 per m^3) compared to the price of freshwater supply (Rs 73–145 per m^3) and still the small-scale industries show less interest as substantial capital investment will be required to further treat the supplied wastewater to meet with the industrial requirements (Anumol et al. 2016; Lena et al. 2022). There are also successful "zero discharge" cases which are explained in Box 6.1

6.7 Government Initiatives on Wastewater Reuse

The Indian government has made significant efforts in recent years to encourage wastewater reuse through policy regulations and investments. It started with limiting industrial water use and requiring enterprises to meet statutory water reuse standards. The state governments have also followed up enacting policies to reuse treated wastewater, which are showing positive impacts. Some of the initiatives are discussed below.

The Gujarat government has launched "Reuse of Treated Waste Water Policy" in 2018. The policy will support setting up of sewage treatment plants (STP) in all major towns and cities across the Gujarat state. It will encourage the use of treated wastewater by the industries. It is also indicated that after meeting the industry demands, the treated wastewater will be supplied to agriculture/irrigation purposes to meet the demand (Government of Gujarat 2018).

The Punjab government has initiated "The State Treated Wastewater Policy 2017" to promote recycling and reuse of treated sewerage for non-potable applications. Till date, 47 projects have been completed successfully by the Department of Soil and Water Conservation for using 243.3 MLD treated wastewater of STPs. These projects have been implemented by laying underground pipeline system for irrigation water conveyance covering an area of 7652 ha (DECC 2020).

The Government of Karnataka has issued an official directive to take all necessary steps to ensure that only tertiary treated sewage is used for non-potable purposes, like all gardening, including parks, resorts, and golf courses (GoI(b) 2022).

The state of Delhi has set ambitious reuse targets to treat and recycle 25% of its total sewage output by 80% by 2027 (GoI 2012).

The following are some of the treatment plants set up across the country in recent years:

- Surat Municipal Corporation (SMC) developed a 40 MLD reuse plant in 2014 in order to deliver reclaimed water to Pandesara Industrial Estate (GoI 2019).
- Nagpur Municipal Corporation (NMC) and Generating Company (MAHAGENCO) jointly funded a reuse project where treated water from an STP is further processed and used as cooling water.
- Chennai Metro Water Supply and Sanitation Board (CMWSSB) in 2016 awarded a PPP-based reuse project contract where the design, build, and operate (DBO) approach is used to create a 45 MLD reuse capacity to deliver non-potable water to industries.
- A tertiary treatment facility with a 10 MLD capacity has been constructed near Yellahanka by Bengaluru's water utility to provide reclaimed water to the airport (Menezes 2016).
- The secondary treated sewage water is purchased and treated for use in cooling water makeup in major industries like Madras Refineries, Madras Fertilizers, GMR Vasavi Power plant in Chennai as also in Rashtriya Chemicals and Fertilizers in Maharashtra and most recently in the Indira Gandhi International Airport in Delhi and Mumbai International Airport (GoI(b) 2022).

6.8 Implementable PPP Models for Wastewater Reuse in India

Though several models are suggested by being researchers to suit different locations, the review of these models suggests that the following three models can give an idea about the future of the implementation models for wastewater reuse (Table 6.7) (Chauhan et al. 2016).

Table 6.7 Comparison of the implementable models

PPP model	Players	Advantages	Possible risks
Three party fixed price (TPFP) model	Water utility/ ULB, Private sector developer, Industrial bulk consumer	Utilizes existing infrastructure and raises treatment quality by including the private sector throughout the entire treatment process With a fixed pricing approach, the private developer has incentives to increase effectiveness	Customer's bulk water purchase guarantees and a minimum guaranteed revenue from the utility The STP's supply quality and variability risk is moderate. The project is still viable even without significant subsidies, therefore the private player's financial risk is just moderate
Reuse utility buy-back model	Water utility/ ULB, Private sector developer	The utility's buy-back guarantee lowers the private developer's revenue risks. Although the utility is the final beneficiary, it can subsidize the tariff by offering capital subsidies	The ability of the utility to make timely payments for the treated water poses the biggest risk. The utility's ability to effectively collect and transport sewage from homes to the STP is a significant concern
End user reuse PPP (EURP) model	Water utility/ ULB, Private sector developer	This approach is totally driven by end users, so mobilizing resources for the project will go much more quickly. It offers more financial benefit because there is no involvement from an intermediary treatment company. Due to this, reuse operations of this nature are competitive	The risk is that the supply will be highly variable, because the utility will be in charge of the secondary treatment most of the time The end user's responsibility for technical design, including selecting the best technology option also a risk

6.9 Comparison of Cost of Water and Investment Options

Different investment options indicate that cost of freshwater per m^3 is less under borewells followed by percolation ponds (Table 6.8) (Palanisami and Suresh Kumar, 2017). If treated wastewater is considered, it will be much cheaper.

Box 6.1 Closed-Loop or "Zero Discharge" Cases

There is increasing demand for closed-loop or zero discharge projects that can be easily planned and managed by institutions or industries. Two such cases are worth mentioning: These models transform consumers into producer–consumers and this has interesting implications for governance over wastewater management.

One is at the IIT in Chennai. The IIT invested in a state-of-the-art system to recycle wastewater and re-engineer the entire campus for dual plumbing. Their STP provides recycled water for toilet flushing in the dormitories, campus gardening, air conditioning, and other building and cleaning needs. The IIT-M STP treats close to 3 million litres per day, including 8 lakh litres [800,000 L per day] of wastewater generated by the IIT-M research park. After meeting the requirements, it is diverted into ponds linking with the groundwater recharge wells. Overall, no water is wasted.

The second closed-loop system is in the Renaissance Hotel and Conference Center in Mumbai. They created two STPs with capacities to treat 750 and 650 kld underground at the back side of the complex. This provided more than enough water for their AC cooling towers, for flushing toilets, and for watering the gardens. The Management was able to install dual or double plumbing in the entire hotel to enable them to use recycled water for flushing toilets and city piped (pure) water for the kitchen, faucets, and showers in the rooms. *Source:* Krishna Chaitanya and Krishna 2017; Kelly et al. 2018.

Table 6.8 Cost of water under different investment measures

Investment measures	Cost (Rs/m^3)
Bore wells	1.73
Community level percolation ponds	2.82
Storage farm ponds in the fields	5.6–8.6
Water purchase from tankers in urban areas	125–140
Wastewater reuse	1.35–2.36*

* The cost will be high when secondary and tertiary treatments are considered with different treatment technologies

6.10 Challenges and Potentials for Wastewater Reuse in India

Challenges encountered in reclaimed water reuse

India is still in the earliest phases of wastewater treatment and reuse. The development of wastewater treatment and reuse in the country is being constrained by the numerous obstacles associated with the implementation of treated wastewater reuse projects. The removal, handling, and treatment of sludge have been found to be the most neglected aspects of running sewage treatment plants (STPs) in India. The facilities built to treat wastewater do not work well and are frequently closed due to inappropriate design, poor maintenance, frequent energy outages, and a lack of technical manpower. In the majority of cases, the use of biogas produced by UASB reactors or sludge digesters is likewise insufficient. The fact that none of the currently available options provides a direct economic return is one of the main issues with wastewater treatment techniques. Local authorities typically have no interest in starting a wastewater treatment programme since there is no financial reward. According to a CPCB performance study of STPs conducted in a few cities, 26 of the 92 STPs examined, did not meet the required BOD requirements, rendering the waters unfit for domestic use. Due to this, even though the nation's wastewater treatment capacity has expanded over years, only 26% of the sewage generated is adequately treated, with the remainder making its way into natural ecosystems and causing significant contamination (Kaur et al. 2012).

Technical challenges: Only about 30% of the municipal wastewater and 60% of the industrial wastewater is now treated due to low sewerage network coverage. Additionally, there are issues with the availability of qualified staff as well as the lack of effective treatment technologies that are user friendly and cost effective.

Institutional challenges: Data on water supply and demand, gaps, wastewater volumes, water consumption, etc., are not as readily available due to the fragmentation of the water sector with different departments controlling them, which have an impact on the planning for treated wastewater reuse.

Economic challenges: Initial expenditures are very exorbitant, and there is very little access to funding and technology for the ventures. Also the viability of reuse projects must be taken into account in order to meet the growing water demand using treated water. The increase in capital and operations and maintenance (O&M) costs of reuse projects will have an impact on the water tariff. Further, the availability of treated water close to industrial areas, access to alternative water sources, competitive water tariffs, and conveyance distance also influence the future of the wastewater treatment and reuse.

Social challenges: The public's reluctance to use treated wastewater in other applications is further fuelled by inadequate treatment technology, hazards to public health, and associated risks. More than the reaction itself, a barrier is the public utilities' views of how consumers will respond to the reuse of recycled water as the quality of water is not ensured always. Awareness has to be increased through education, providing better information to the users and transparency. We also need political will and prioritization by the state governments in the future.

The development of treated wastewater reuse is hampered by these issues. Nonetheless, the environment is rapidly shifting as a result of technological advancement and economic expansion. Treated water reuse in the country has a significant potential as we enter into an era where climate extremes and their increased frequency is becoming a new normal. Wastewater should be treated as a resource for development and economic growth.

6.11 Wastewater Reuse in India—The Way Forward

The recycled wastewater is crucial for ensuring a steady supply of water for various uses in India. In addition to offering a chance for environmental restoration by keeping wastewater discharge out of water bodies and supplying an assured supply of water to meet the increased water demands, treatment and reusing treated wastewater is highly warranted and quite advantageous because it lessens the strain on freshwater resources.

Further, future water supplies available for the industrial and energy sectors will decrease from 491 billion cubic metres to 135 billion cubic metres in 2050 if the annual demand for freshwater for domestic uses increases (Kumar 2020). The amount of water needed by the manufacturing sector is predicted to increase by 2.2 times between 2010 and 2050, while the amount needed by the energy sector will increase by 3.7 times. Hence, this outlines the scope for wastewater use in industry and energy sectors in the future.

In this context, appropriate policy interventions are needed exclusively for the wastewater reuse. Important policy recommendations include the following:

It is important to see how the agricultural-to-urban water transfers raise the economic productivity of agriculture. The magnitude of this productivity increase depends on local agricultural practices and demand for agricultural products. Since most of the urban water transfers are from projects upstream (river basin level), the possible expansion of the area under wastewater irrigation (mostly downstream) needs to be planned carefully for achieving the higher economic water productivity. To address the future wastewater reuse in an effective manner, a national wastewater reuse policy with targets, regulatory and financial measures should be formulated and implemented in a phased manner.

The wastewater sector in India is currently driven by government initiatives and incentives. It may be useful to improve public–private partnership models, in the future. Private investments in wastewater treatment and reuse should be encouraged to reduce the burden on governments. Hence, sound policy and effective regulatory mechanism by the central and state governments are important for planning and launching of innovative reuse projects through different models or partnerships. As the affordable cost of groundwater exploitation makes wastewater reuse unattractive, one important regulatory intervention will be to prevent industries from groundwater abstraction at a level that leads to overexploitation which is happening in several regions of the country.

Detailed guidance documents on sewage treatment and reuse technologies, and effective financing mechanisms for sufficient cost recovery for long-term operation and maintenance of sewage treatment infrastructure are needed with strong Institutional and monitoring support.

Public–private partnerships models with cases of success stories should be encouraged with the inclusion of financial (as well as non-financial) incentives for different users mainly the industry and agriculture sectors as government incentives for the PPP models should be ensured for private sector participation. At the city level, a first step would be to prepare an urban wastewater reuse plan, showing the supply and demand followed by selecting an implementable model with detailed feasibility studies. Corporate bodies and various institutions in the country should be encouraged to adopt zero discharge models as a mandatory requirement of efficient water management. The state industrial development company, the urban local bodies, and State Pollution Control Board should collaborate to establish a project based on industrial water demand assessment. This would help to determine the level of treatment required and other design aspects involved. In order to cover ongoing O&M costs, utilities/ULB should also be needed to develop a finance strategy. It is important to ensure that treated wastewater should be cost competitive compared to alternative options available to industries.

Furthermore, the Ministry of Environment, Ministry of Agriculture, and Ministry of Water Resources should jointly provide quality standards for different grades of industrial and agriculture water, which will help standardize design of reuse systems across states.

Investments in wastewater treatment projects are also a social investment like investment in major and medium irrigation projects and hence future government priorities in water sector investment should also include wastewater treatment projects. Workshops and trainings at the state level might be held to educate utility management on water reuse and cover the available technologies, new standards, legislative incentives, implementation difficulties, and procurement best practices.

References

Anumol T, Vijayanandan A, Park M, Philip L, Snyder SA (2016) Occurrence and fate of emerging trace organic chemicals in wastewater plants in Chennai. Environ Int 92–93:33–42. https://doi.org/10.1016/j.envint.2016.03.022

Ashar S (2017) Maharashtra Govt makes reuse of toilet water mandatory for cities. The Ind Express. https://indianexpress.com/article/cities/mumbai/maharashtragovt-makes-reuse-of-toilet-water-mandatory-for-cities-4918336/. Accessed 15 April 2020

Bhardwaj (2005) Water quality monitoring in India—achievements and constraints, International work session on water statistics, Vienna

Chauhan S, Jensen O, Sengaiah G, Screenivas N (2016) Closing the water loop: reuse of treated wastewater in urban India (September). www.PwC.in

CPCB (2016) Evaluation of operation and maintenance of sewage treatment plants in India 2016, Control of Urban Pollution Series: CUPS/68/2007. Central Pollution Control Board, India

CPCB (2022) Status of water supply, wastewater generation and treatment in class I cities and class II towns of India. Series: CUPS/70/2009–10. Central Pollution Control Board, India

DECC (2020). Action plan for reuse of treated wastewater. Directorate of Environment and Climate Change, Department of Science, Technology and Environment. Government of Punjab, India

Desai SS (2017) Recycling, reuse and economics of treated wastewater in urban India. Inter J Eng Res Mech Civil Eng 2(12)

FAO (2011) Wealth from waste: the economics of waste water use in agriculture. Wat Report 35, UN. ISBN 978–92–5–106578–5

Federation of Indian Chambers of Commerce and Industry—FICCI (2011)Water use in Indian industry survey, FICCI Water Mission, New Delhi

GoI(b) (2022) Urban waste water scenario in India. NITI Aayog, Government of India

GoI (2012) Water policy for Delhi. Government of India

GoI (2019) Action plan for reuse and recycle of treated wastewater. Government of India

GoI (a) (2022) National framework on safe reuse of treated water. Government of India

Government of Andhra Pradesh (2019) Action plan for usage of treated waste water from sewage treatment plants. Municipal administration and urban development department, Andhra Pradesh

Government of Gujarat (2018) Policy for reuse of treated waste water. Gujarat Pollution Control Board, Ghandinagar

Hooper VL (2015) The importance of the urban in agricultural-to-urban water transfers Insights from comparative research in India and China. Thesis submitted to the University of East Anglia, UK for the degree of Doctor of Philosophy. University of East Anglia School of International Development. https://ueaeprints.uea.ac.uk/id/eprint/58493

IIT (2022) Water reuse-A vital step towards water value. Roorkee

IWA (2018) The reuse opportunity. The International Water Association, London, UK

John JE, Thangavel P, Maheswari M, Balasubramanian G, Kalaiselvi T, Kokiladevi E (2022) Bioprospecting halophytes for reclamation of paper and pulp mill effluent irrigated soil. J Curr Crop Sci Technol 109(1–3). http://masujournal.org/109/qD6BUHhcF6hDil9a6tLIZl8kLkBdyh.pdf

Kamyotra JS Bhardwaj RM (2011) Municipal wastewater management in India. India Infrastructure Report, pp 299–311

Kaur R, Wani SP, Singh AK, Lal K (2012) Wastewater production, treatment and use in India. In: Presented at the Second Regional Workshop of the Project Safe Use of Wastewater in Agriculture, 16–18 May 2012, New Delhi. http://www.ais.unwater.org/ais/mod/page/view.php?id=61

Kelly DA, Maurya N, Das S (2018) Parameters of successful wastewater reuse in urban India. Ind Poli and Policy 1(2). https://doi.org/10.18278/inpp.1.2.4

Krishna CSV, Krishna SV(2017) What drought? check out IIT Madras. It's an Oasis. Indian Express May 11, 2017. Accessed 14 July 2018. http://www.newindianexpress.com/cities/chennai/2017/may/11/what-drought-check-out-iit-madrasits-an-oasis-1603523--1.html.

Kumar G (2020) Water reuse in India: current perspective and future potential. Wastewater Treat Reuse—Lessons Learn Technol Dev Manag 6:33–63. https://doi.org/10.1016/bs.apmp.2020.07.011

Lena B, Gabriela CQ, Anshuman N, Nitin B, Nathaniel BD, Mayuri P, Saurabh K, Andraju NB, Anjin K, Paul C, Christine MH (2022) Perceived drivers and barriers in the governance of wastewater treatment and reuse in India: Insights from a two-round Delphi study. Resour., Conserv. & Recycl., 182. https://doi.org/10.1016/j.resconrec.2022.106285

Menezes N (2016) 4-MLD water treatment plant to come up in Cubbon Park, Bengaluru, DHNS. https://www.deccanherald.com/content/532359/4-mld-water-treatment-plant.html. March 03, 2016

Minhas PS, Jayanta KS, DotaniyaML AS, Madhumonti S (2022) Wastewater irrigation in India: current status, impacts and response options. Sci Total Environ 808:152001

Nataranjan KA (2020) Wastewater reuse in Chennai, Tamil Nadu. Smart Water Waste World, Maharashtra, India 182:106285

Palanisami K, Suresh Kumar D (2017) Drought proofing strategies by farmers in Southern India. Water Policy Research Highlights 04–2017. ITP, International Water Management Institute, Anand

Rathore P, Killedar DJ, Divyesh P, Akansha S (2022) IOP Conf. Series: Earth and Environmental Science, vol 1032, p 012006 IOP Publishing. https://doi.org/10.1088/1755-1315/1032/1/012006

Ravikumar J, Rajiv R, Shubhra J (2014) Wastewater recycle and reuse: an economical and sustainable option. Water and Sanitation Program, March

Suganthi R, Sebastian SP, Indhuparameshwari C, Avudainayagam S, l. Meena, Sadish O (2019) Potential use of reclaimed water from recycled paper board mill through micro irrigation system for maize crop production. Pollut Res 38 (3):716-722

Toze S (2006) Reuse of effluent water—benefits and risks. Agric Water Manag 80:147–159

Udayasoorian C, Ponmani S (2009) Effect of treated paperboard mill Effluent irrigation on soil health and yield of vegetable crops. J Environ Res Dev 3(3):879–889

WHO-World Health Organization (2006) Guidelines for the safe use of wastewater, excreta and grey water. In: Wastewater use in agriculture, 2nd ed. WHO-World Health Organisation, ISBN: 92 4 154685 9

World Bank Group (2019) Wastewater: from waste to resource the case of Nagpur. The World Bank, India

World Resources Institute-WRI (2020) Aqueduct country rankings. Available online https://www.wri.org/applications/aqueduct/country-rankings/. Accessed 3 Sept 2020

Chapter 7
Enhancing Water Productivity and Climate-Smart Technologies

Abstract The India water vision on enhancing water productivity aims for a technology-based improved water use by providing affordable, appropriate, and accessible water management technologies at all levels. About 22% of the available technologies/practices are currently adopted and based on the financial viability, only 8–10% of the adopted technologies are successful. Among the regions, southern and western regions account for comparatively a higher percentage of adoption (23–26%) and north-east region for a lower percentage (11%). The region-wise technology gap in water saving is comparatively lower (32%) in the southern region whereas the per cent gap in yield is lower in the northern region (48%). Two possible ways to increase the productivity per unit of water include bridging the technology gaps (i) by adopting the appropriate technologies/practices for different crops in those areas where traditional practices are followed (technology gap1), and (ii) by effectively adopting the technologies for different crops and soils in farmers' fields as per research station guidelines (technology gap2). Measures like digital agriculture, climate-smart agriculture (CSA), and transformative investment in CSA are also discussed. Refocusing water management research, strengthening the Pradhan Mantri Krishi Sinchayee Yojana (PMKSY) implementation, initiating capacity building programmes for farmers, and developing public–private partnership are suggested to improve technology up-scaling.

Keywords Climate-smart technologies · Climate-smart agriculture · Technology adoption · Technology gap · More crop per drop · Micro irrigation · Transformative policy · Water management research

7.1 Introduction

The United Nations (UN) 2030 Agenda for Sustainable Development and their supporting Sustainable Development Goals (SDGs) indicate that "by 2030, we need to substantially increase water use efficiency across all sectors and ensure sustainable withdrawals and supply of freshwater to address water scarcity, and substantially

K. Palanisami and U. S. Nagothu, *India's Water Future in a Changing Climate*, Advances in Geographical and Environmental Sciences,
https://doi.org/10.1007/978-981-97-1785-9_7

reduce the number of people suffering from water scarcity" (Meredith et al. 2017). This is because the water sector is facing increasingly difficult problems due to climate and weather extremes, growing demands for, and competition over scarce water resources, as well as the physical, economic, and environmental constraints for developing additional supplies. With agriculture being the largest user of water resources worldwide, the need to identify ways to achieve real efficiency gains and real water savings has been emphasized since the 90s, to provide "opportunities for increasing the productivity of water" (Seckler 1996). The water crisis in most situations is likely to worsen in the coming years as demand from different sectors continues to grow along with the effects of climate change. Given the large mandate of the agricultural sector in meeting the increasing food demand, providing adequate irrigation facilities through efficient use is highly warranted.

Technology adoption in the country is expected to increase from the current level, which will result in crop yield increase and water saving resulting in more income per unit of water used. Hence, expanding the water management research against a backdrop of rapid climate change and growing complexity is important to develop and deliver technical and policy-related solutions that are suitable for different agro-climatic regions. The chapter focuses on improving the productivity per unit of water under different technologies and the ways to achieve it in the short- and long-run perspective.

7.2 Current Status

The term productivity of water often means the ratio of physical production (in terms of crop yield) or value of production (in terms of gross value of product) relative to water used and expressed in kg/m^3 or Rs/m^3. Alternatively, the water footprint which is reciprocal of the water productivity can also be used where the aim will be to minimize the water footprint of the products and expressed in terms of m$^{3/}$kg of output.

Thus, in many instances, water productivity (WP) interventions have embraced more than one development objective. The WP ratios can be further used for making comparisons across users, sectors, and over time (Meredith et al. 2017). Consequently, at macro level, estimates of changes in WP will be useful to assess policy interventions (Barker et al. 2003; Kijne 2003; Wichelns 2014). Further, in the "widest possible sense" WP can also incorporate other aspects such as fisheries and ecosystem services (Rijsberman 2006; Molden et al. 2010).

The IMPACT-WATER (combining the International Model for Policy Analysis of Agricultural Commodities and Trade [IMPACT] model with a water simulation model) was used for the analysis of various WP scenarios for irrigated rice globally and regionally incorporating technology and management improvements, investments in agricultural infrastructure and research, and increased environmental flow requirements. It was observed that developed countries will have higher WP values than developing countries. However, the values could narrow down in the long run due to increase in irrigated rice yield and water use efficiency (Cai and Rosegrant 2003).

7.3 Climate-Smart Technologies

Most of the crop and water management practices or technologies, which have been practised by farmers traditionally are climate smart and have the capability to increase resilience against climate and weather extremes. Climate-smart agriculture (CSA) and water management practices ideally should encompass three main dimensions of sustainable development: economic, social, and environmental, i.e., sustainably increasing agricultural productivity and incomes; adapting and building resilience to climate change; and reducing and/or removing greenhouse gas emissions. It is thus important that while identifying CSA technologies the location and cultural and social sensitivities must be taken into account. In this chapter, CSA technologies that are relevant for improving WP are discussed.

7.4 Technology Adoption and Crop Yield

In-depth findings on the impact of technology on WP can be visibly seen from the water management research work in India. Notably, the research works carried out by various institutions under the Indian Council of Agricultural Research (ICAR since the 80s through various All India Coordinated Research Project (AICRP) on Water Management and Groundwater Utilization schemes are important. Also state Agricultural Universities periodically undertake water management research programmes/projects. These schemes/projects aimed to promote WP, i.e., more crop and income per drop of water. The concept "more crop per drop" gained importance in India with the implementation of the Farmers Participatory Action Research Program (FPARP) by the Ministry of Water Resources, GOI (2006), MoWR (2007). The FPARP was initiated by the Ministry of Water Resources during 2008–09 covering a sample of 5000 action research trials on farmers' fields covering 2001 villages in 23 states of India with the active participation of the State Agricultural Universities, ICAR Institutes, International institutes, water & land management institutes (WALMIS), and local NGOs. Each farmer trial was done on one hectare land. A total of about 402 crop, land, and water management technologies/practices were identified and used.

7.5 Performance of the Technologies

In this section, the performance of CSA technologies for up-scaling was assessed at three levels:

(i) Research station level: The performance of the various CSA technologies developed and tested in various research stations across the county under controlled conditions showed promising results in terms of increased crop yield, water saving, and financial viability (internal rate of return (IRR)) (Table 7.1). There are variations between states and this can be attributed to the type of technology developed/used, soils, crops, and other agro-climatic parameters (Palanisami et al. 2011, 2012).

(ii) Participatory research in farmers' field: detailed state-wise analysis was done on the performance of the technologies adopted by the farmers under the FPARP. The performance of the selected technologies in each state is given in Table 7.2. An increase in water saving has ranged from 8% in mixed cropping in vegetables in Orissa to 27.53% for INM in rice in Assam. In Jammu & Kashmir, water savings in Apple was 18.72%. Overall, reduction in number of irrigations has ranged from 0 to 13.5%, due to the awareness created by the FPARP on effective irrigation practices. The average water saving has ranged from 10 to 15% which is lower than the saving recorded in experimental research stations. The average yield increased as a whole under field conditions has ranged from 2.5 to 16.4%, which is about 38% of the yield increase reported in the research stations (Palanisami et al. 2015). The financial rate of return of the technologies/practices was comparatively low and was in the range of 5–18% with an average IRR of 10.8%. Private irrigation sources (such as wells) provided better water control and performed better than under public (surface) irrigation systems (Palanisami et al. 2015).

(iii) Promising/selected technologies at farmers' fields: Analysis was done by comparing the crop-wise cost and productivity of the important technologies. The cost of the technologies varied based on the type of technology where the transaction cost is considered to be important while promoting the technology spread and adoption. Some practices/technologies have no additional cost and some technologies are capital intensive where the cost per ha is comparatively high (such as precision farming, drip irrigation, etc.) but will be compensated by the additional yield. Among the crops and technologies, the physical productivity of water varied from 0.3 kg/m^3 for chickpea to 18.57 kg/m^3 for onion; the economic productivity varied from Rs 6/m^3 for rice to 520/m^3 for onion. All these values reflect the magnitude of crop yield and water saving and the type of technology used[1] (Table 7.3).

[1] Studies also express the physical productivity interms of kg/mm of water (i.e., 10 m^3 = 1 mm of water) applied. In the case of economic productivity, the issue is the price, where the physical water productivity may be lower but due to higher price of the products, the economic productivity may look higher. Hence generalizations of the water productivity should be carefully made taking into account the units used, type of technologies and location specific factors.

Table 7.1 Overall performance of water management technologies from research stations across States, India

State	Year of start	Adoption level (%)	Success rate (%)	Yield increase (%)	Water saving (%)	IRR (%)
Karnataka	1974	2–75	10–65	15–71	15–66	16–33
Kerala	1974	25–89	23–56	20–70	20–40	17–37
Tamil Nadu	1969	45–65	34–60	12–53	14–49	12–28
Gujarat	1970	30–56	30–55	10–51	6–57	16–36
Madhya Pradesh	1980	10–60	40–100	0–70	10–85	8–23
Chhattisgarh	1981	3–50	8–90	5–74	13–45	10–19
Maharashtra	1977	2–30	60–95	35–40	35–45	16–29
Rajasthan	1982	20–30	45–50	5–67	10–50	9–34
Haryana	1967	20–90	50–65	5–60	0–50	16–35
Punjab	1969	10–70	80–90	6–10	25–45	18–41
Uttar Pradesh	1980	8–20	25–35	5–20	15–40	6–18
Himachal Pradesh	1972	12–35	15–60	18–45	25–80	17–39
Jammu & Kashmir	1984	40–50	75–90	13–16	25–33	13–45
Uttarakhand	1975	20–34	25–30	11–40	36–87	10–25
Bihar	1980	20–50	40–90	12–40	30–40	12–23
Orissa	1969	20–52	55–85	10–30	10–30	16–31
West Bengal	1982	15–70	30–70	8–48	15–37	19–34
Assam	1971	2–44	70–90	5–200	12–60	18–41
Meghalaya	1975	14–40	80–95	40–200	10–45	11–45

Note The research stations mostly represent the All India Coordinated Research Project (AICRP) on Water Management and Groundwater Utilization schemes by the ICAR. The year of start refers to the start of the research works in the respective stations in the state. Data analysis was done for all the practices/technologies developed by the centres

The WP at farm level can be increased with training of farmers and supporting them with financial investments in technologies. Considering all the practices/ technologies adopted by the farmers, the region-wise performance gap was calculated where the per cent gap in water saving was comparatively lower (32%) in the southern region whereas the per cent gap in yield was lower in the northern region (48%) (Figs. 7.1, 7.2 and 7.3, Table 7.4). There are two possible ways to increase the WP and reduce the gap: (i) by bridging the Technology gap1 by adopting the appropriate technologies/practices for different crops in those areas where mostly the traditional practices are followed, and (ii) by bridging the Technology gap2 by effectively adopting the technologies in farmers' fields as per research station recommendations (Fig. 7.4). The drip irrigation in rice is shown in Box 7.1.

Table 7.2 Performance of the selected technologies at farmer field level across states, India

State	Crop/ practices	Irrigation depth (cm)		% water saving	% reduction in no. of irrigations	% increase in yield	IRR (%)
		Conventional	FPARP				
Tamil Nadu	Paddy/SRI	3.63	3.1	17.37	9.21	5.21	8.1
	Sugarcane/ drip and INM	5.91	5.21	11.85	8.41	11.51	16.1
	Ground nut/ Sprinkler	7.38	5.95	19.25	0.1	8.21	7.2
	Okra/ Sprinkler	6.42	5.03	21.55	0.1	10.31	14.1
	Floriculture/ Sprinkler	7.11	5.76	19.14	0.1	12.41	8.2
Gujarat	Wheat/INM	6.1	5.1	16.66	12.51	11.61	9.1
	Vegetables/ INM	6.2	5.56	8.34	0.1	8.91	9.1
	Onion/INM	2.1	2.1	0.01	0.1	9.21	8.3
	Rice/INM	8.1	7.1	12.51	7.41	6.72	6.4
Karnataka	Rice/SRI	7.2	6.1	14.29		6.12	10.1
Assam	Rice/INM	6.89	5.1	27.54	12.41	12.81	14.1
Haryana	Wheat/INM	7.55	6.2	18.68	13.51	9.22	10.2
Punjab	Rice/PLL	7.65	6.11	18.68	0.1	5.92	6.4
	Rice/PQW	7.51	7.55	0.01	0.1	7.81	7.2
	Rice/INM	7.53	7.55	0.01	0.1	9.21	8.2
	Wheat/INM	7.51	7.56	0.02	7.21	13.42	15.1
	Wheat/PLL	7.55	6.33	16.01	10.21	9.53	16.2
Himachal Pradesh	Vegetables/ INM & Farming system	5.1	4.22	16.02	9.71	15.41	18.1
Orissa	Vegetables/ MC	5.1	4.65	8.01	0.1	2.51	8.4
Madhya Pradesh	Wheat/ Sprinkler	16.72	13.8	17.38	–33.33#	9.41	14
Jammu & Kashmir	Apple/ (Water harvesting tank & pipe line	74.32	60.45	18.73	1.09	18.32	16.1
Andhra Pradesh	Rice/SRI	7.21	5.87	19.45	5.66	5.72	9.3
	Rice/SRI	6.92	5.56	20.21	14.91	5.82	9.3

(continued)

Table 7.2 (continued)

State	Crop/ practices	Irrigation depth (cm)		% water saving	% reduction in no. of irrigations	% increase in yield	IRR (%)
		Conventional	FPARP				
	Cotton/INM	5.1	4.1	20.01	18.21	7.12	13.4
Uttar Pradesh	Rice/SD & HYV	7.15	5.51	21.98	11.41	7.82	7.2
Rajasthan	Ground nut/ CSI	4.74	4.2	14.81	8.13	15.42	17.1
	Pearl millet/ CSI	3.53	3.1	14.21	6.11	14.32	13.3
Jharkhand	Wheat/INM	2.1	1.61	20.01	0.1	7.32	5.1
West Bengal	Rice/SRI	5.1	4.11	18.01	11.81	8.22	9.1
	Wheat/Zero tillage	5.55	4.73	15.17	6.12	16.42	15.1

PLL: Precision land levelling, PQW: Poor quality water, INM: Integrated nutrient management, SD & HYV: Short duration & High yielding Variety, SRI: System of rice intensification, MC: Mixed cropping, CSI: Critical stage irrigation, RI: Rotational irrigation. # more frequency of irrigations
Source FPARP data analysis

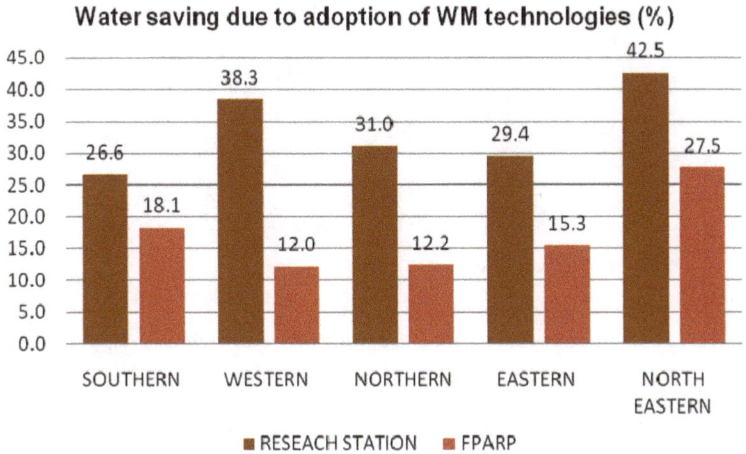

Fig. 7.1 Water saving at research station and farm level (FPARP) due to water management technologies

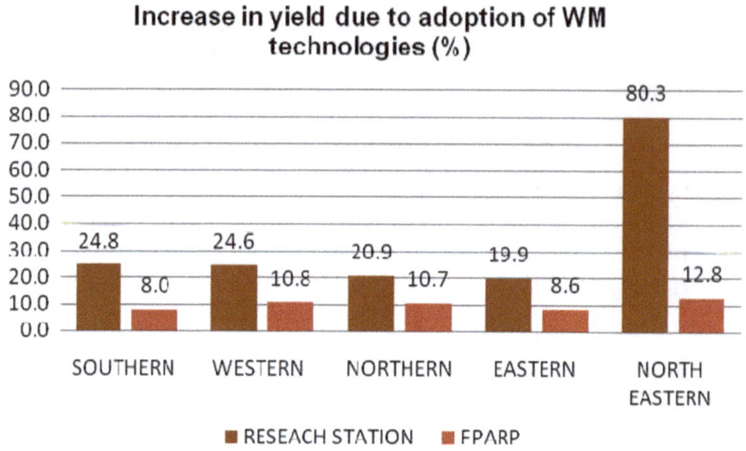

Fig. 7.2 Yield increase at research station and farm level (FPARP) due to water management technologies

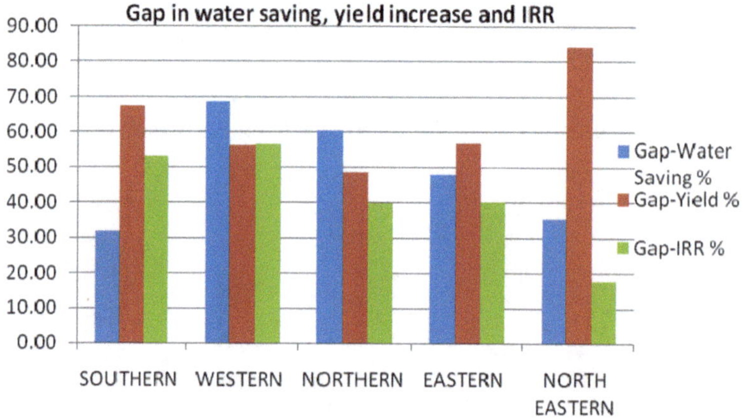

Fig. 7.3 Gap in water saving, yield increase, and IRR (%) between research station and farm level

7.6 Adoption Level of Technologies—Main Streaming and Up-Scaling

Even though improved water management technologies per se result in higher yield and increased WP, the success of the technology or the technology transfer depends upon how far the farmers are adopting them. The overall picture at country level shows that only 19.7% of the gross irrigated area (GIA) is under water management technologies. Among the regions, southern and western regions account for comparatively a higher percentage (23–26%) and north-east region for a lower

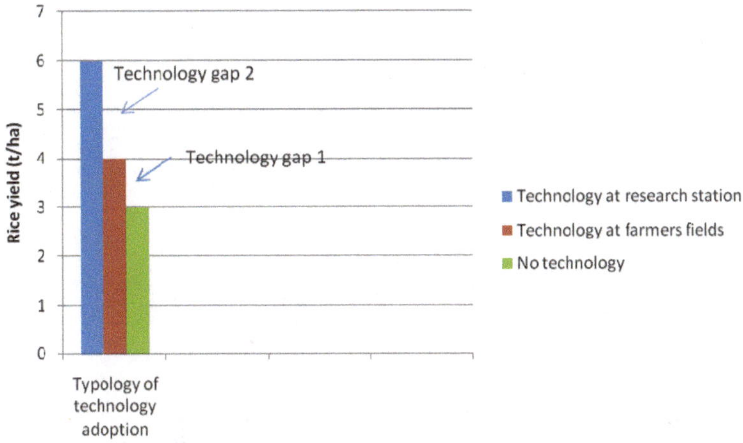

Fig. 7.4 Technology gaps in rice production

percentage (11%). Among the technologies, the surface/flood irrigation-based technologies account for about 59% of the area and the micro irrigation (MI) with drip and sprinkler accounts for the rest. However, this magnitude depends upon the potential area under each category. If we consider, the GIA of 95 m ha as the potential area for surface/flood irrigation-based technologies, then the area so far covered is about 11.6% and if the minimum MI potential of 45 m ha is considered, then the GIA potential for surface/flood irrigated-based technologies will be 50 m ha and the current technology coverage is about 22%.

In addition to the area-based technologies, further macro level analysis based on the number of technologies adopted by the farmers using the FPARP data was done. The rate of adoption and spread of these technologies is much lower where, for example, only 110 out of 402 technologies/practices (about 22%) have been adopted by the farmers. A sizeable portion of the practices currently practised by farmers is based on their traditional knowledge and practices. Hence, many of the scientifically developed technologies need to be up-scaled, which in turn will help save water to meet the future demand. Adoption of the MI technologies will help save water and energy. The current area under micro irrigation in the country is about 12.5 m ha and it signals the magnitude of water and energy saving in the agricultural sector. The energy used in conventional gravity irrigation and in micro irrigation showed that micro-irrigation resulted in significant energy saving (Table 7.5) (Raman and Tikadar 2011). The energy saving has varied from 100 kWh per ha in groundnut to 1,660 kWh per hectare in banana. The average energy saving per hectare is worked out to 410 kWh. For example, total energy saving at national level during the 2019–20 period works out to be 31891 lakh KWh where drip irrigation accounts for about 13900 lakh kwh and sprinkler irrigation accounts for about 17991 lakh kwh (Ministry of Agriculture and Farmers Welfare 2021). It is also expected that in the future micro irrigation can be practised for rice crop also (Box 7.1).

Table 7.3 Productivity per unit of water under different technologies, India

No.	Name of the technology	Name of the crop	Technology oriented cost		Technology-related attributes			
			Cost of technology (Rs/ha)	Difference in cost of cultivation (Rs/ha)	Water saving (M^3/ha)	Yield increase (Kg/ha)	Physical productivity (Kg/M^3)	Economic water productivity (Rs/M^3)
1	*Rice cultivation practices*							
	System of rice intensification (SRI)	Rice	918	−11500	3100	1172	0.61	8.19
	Direct sown rice (DSR)	Rice	221	−9660	3321	610	0.72	9.65
	Aerobic rice	Rice	2100	−7780	6020	577	0.98	13.06
2	Alternate furrow irrigation	Maize	600	−800	1000	1010	2.19	35.44
3	Border strip method of irrigation	Groundnut	800	−850	2622	402	0.75	39.23
4	Ridges and furrow method of irrigation	Sunflower	600	−3300	1100	214	0.46	15.29
5	*Irrigation scheduling IW/CPE ratio*							
	Maize IW/CPE = 0.6	Maize	600	−1900	190	425	12.95	209.77
	Groundnut IW/CPE = 0.4	Groundnut	600	−2150	85	144	3.27	171.29
6	Alternate wetting and drying	Rice	1600	−4100	1000	795	0.63	8.35
7	Improved water management practices (5 cm depth)	Rice	800	−600	1820	408	0.49	6.50
	5 days interval irrigation	Rice	450	−9430	4990	434	0.87	11.57

(continued)

Table 7.3 (continued)

No.	Name of the technology	Name of the crop	Technology oriented cost		Technology-related attributes			
			Cost of technology (Rs/ha)	Difference in cost of cultivation (Rs/ha)	Water saving (M^3/ha)	Yield increase (Kg/ha)	Physical productivity (Kg/M^3)	Economic water productivity (Rs/M^3)
	Spacing and irrigation:							
	a. 0.9 m lateral spacing & irrigation 0.8 PE	Turmeric	800	−1300	6060	4080	3.50	352.12
	b. 1.5 m lateral spacing & irrigation 1 PE	Turmeric	1000	−2190	4740	4310	2.84	285.96
	c. 1.5 m lateral spacing & irrigation 0.8 PE	Turmeric	800	−2100	5550	2890	3.18	320.63
8	Mulching (Plastic)	Bhendi	4662	−5838	3300	3890	2.48	49.62
9	Banana drip method	Banana	12644	−24281	2100	13750	5.98	143.80
	Drip and Fertigation	Banana	22397	−4553	5350	20000	7.08	170.10
10	*Sugarcane planting & irrigation*							
	Drip with conventional planting	Sugarcane	13547	−14343	2000	13000	6.47	18.45
	Drip with fertigation conventional planting	Sugarcane	14450	−14450	1800	15000	6.63	18.90
	Drip with improved planting method	Sugarcane	12644	−13356	4900	19000	7.71	21.96
	Drip & fertigation with improved planting (SSI)	Sugarcane	13547	−13343	3000	24500	7.79	22.21

(continued)

Table 7.3 (continued)

No.	Name of the technology	Name of the crop	Technology oriented cost		Technology-related attributes			
			Cost of technology (Rs/ha)	Difference in cost of cultivation (Rs/ha)	Water saving (M^3/ha)	Yield increase (Kg/ha)	Physical productivity (Kg/M^3)	Economic water productivity (Rs/M^3)
	Sub-surface drip 100%PE & 100% RDF	Sugarcane	12102	−16598	4390	10000	7.02	20.02
	Sub-surface drip 80 PE & 75% RDF	Sugarcane	12102	−14488	7620	16000	8.76	24.97
11	Sprinkler	Groundnut	6864	−7836	220	300	7.29	382.08
12	Sub-surface drainage	Sugarcane	2261	+2261	3000	43-,000	8.86	25.26
13	Precision land levelling	Onion	1450	+1450	400	800	18.57	520.00
		Turmeric	1600	+1600	7700	700	0.74	74.84
		Tomato	1450	+1450	3400	5000	6.15	80.00
14	Integrated nutrition management (INM)	Groundnut	680	+680	1500	1410	0.56	29.26
		Chickpea	680	+680	1400	220	0.30	17.46
15	Pond construction & giving 2–3 supplemental irrigation	Rice	7538	+2538	1000	926	0.48	6.33
16	Groundwater supplementation	Rice	800	900	950	1020	0.60	8.52
17	Deficit irrigation (tank irrigation)	Rice	860	490	5100	−500	0.43	5.7
18	High density plantation	Mango	15804	+309861	−28000	7500	0.54	21.43
19	Tissue culture	Banana	18062	+164937	7819	40000	10.18	244.57

Source Survey data

Table 7.4 Comparison of technologies at research station and farmers' fields

Regions	Increase in water saving (%)		Gap (%)	Increase in yield (%)		Gap (%)	Internal rate of return (%)		Gap (%)
	Research station	FPARP		Research station	FPARP		Research station	FPARP	
Southern	26.58	18.12	31.84	24.75	8.03	67.54	22.25	10.44	53.06
Western	38.34	11.98	68.75	24.63	10.79	56.21	25.00	10.86	56.57
Northern	30.95	12.23	60.49	20.86	10.72	48.60	19.08	11.44	40.01
Eastern	29.40	15.29	48.00	19.93	8.60	56.85	15.50	9.25	40.32
North-eastern	42.50	27.53	35.22	80.33	12.80	84.07	17.00	14.00	17.65
Average	32.73	14.85	54.61	26.21	9.72	62.92	20.96	10.8	48.48

Source Palanisami et al. (2015)

Table 7.5 Estimated crop-wise electricity energy saving with drip irrigation

Crops	Energy saving in kWh/ha	Area under drip in hectares
Potato	200	17841
Banana	1660	11360
Mango	371	8890
Papaya	200	4207
Vegetable	200	1248
Lemon	313	2821
Sapota	313	859
Groundnut	100	19742
Cotton	258	66968
Sugarcane	1250	7434
Castor	200	82980
Other Horti	313	10811
Other Agri	200	3169
Total		171946

Source Raman and Tikadar (2011) Estimated Energy Savings in Gujarat through micro irrigation. Paper presented in IGWC, Madurai

Box 7.1 Drip fertigation for rice

The experiments conducted in farmers' fields in Coimbatore district of Tamil Nadu by the Water Technology Centre, Tamil Nadu Agricultural University (TNAU) in association with Netafim India revealed the following:

Design:

Inline drippers; lateral spacing: 0.9 m; drippers spacing: 0.4 m; drippers discharge: 2 lph

The drip fertigated rice yield varied from 5.0 to 6.5 t/ha with a water requirement of 60–65 lakh litres. The irrigation water need was reduced from 1200 mm under conventional flooded conditions to 600 mm under drip irrigation method. The total cost incurred in installation of drip in paddy is about Rs. 65,000 per 0.4 ha

Recommendation:

1. Drip fertigation for rice could be recommended in non command areas where rice is grown utilizing groundwater resources
2. Drip fertigation in rice could be adopted in a cropping system approach

Source Water Technology Centre, TNAU, Coimbatore

The increasing demand for energy in all sectors is posing challenges to the power sector. The total subsidy on energy sales has been increasing over time, and it has become a financial burden on state governments. The water–energy nexus challenges had significantly impacted power supply, groundwater use, and agricultural production. Thus, adopting the micro irrigation at large scale would help to address these challenges as crop water productivity in groundwater irrigated farms can be between one and three times greater than in farms irrigated with surface sources of irrigation (DebRoy and Shah 2003).

7.7 Constraints in Adopting the Technologies

One of the major constraints in adopting the new technologies is the high capital cost of some of the technologies such as drip and sprinkler. More than 30 per cent of the sample farmers had reported the high initial cost as the major constraint. The other major constraints faced by the farmers are lack of labour supply followed by lack of technical skills and assured canal water release to adopt the recommended irrigation scheduling. Hence, relaxing these constraints will pave the way for better adoption of the technologies and up-scaling the same (Palanisami et al. 2015).

7.8 Future Scenarios of Technology Adoption

The National Water Policy suggests improving the present water use efficiency by 20% (GOI 2008). This in turn will improve WP if the policy is put into practice. In closed basins (i.e., no usable water leaves the basin), identifying opportunities to increase WP becomes increasingly important (Meredith et al. 2017). To start with, given the slow growth of technology adoption, it is important to define an achievable target for technology adoption, which in turn will help increase the productivity per unit of water used. Since technology adoption should consider first the priority/favourable areas for quick and effective spread of the technologies, it is important to have the potential for medium-term (5–15 years) planning. Looking at the spread of different land, crop, and water management technologies, a weightage is assigned to derive the medium-term potential. In the case of surface irrigation, about 30% is taken as the weight to arrive at the medium-term potential as this quantum will be easily achievable. Taking into account the magnitude of the surface irrigation-based improved practices/technologies that are practised already; it is assumed that 30% of the total surface irrigation potential will be achieved in the medium term. It is in fact a conservative estimate as the canal/surface-based gravity irrigations have poor water control to suit different seasons, soils, & crop patterns. Also, the growth rate in the adoption of the surface water management technologies is less than 1% per annum where the current adoption level accounts for only about 11% of the total potential (95 m ha) and 22% of the surface water potential (50 m ha). However, in the coming years, the technology adoption is expected to increase comparatively at a higher rate due to water scarcity, change in crop pattern, improved digitalization to enhance precision-based water management, government interventions, and better awareness among the implementing departments and farmers. Hence it is assumed to increase 1–3% per annum (i.e., about 1% low growth rate, about 2% medium growth rate, and about 3% high growth rate). Compared to the current level, the projected area increase under surface irrigation-based technologies in 2035 will account for 45, 55, & 68% of the surface irrigation technology potential under medium term (Table 7.6).

The potentiality of bringing more area under drip and sprinkler in the country varied from 42 M ha as estimated by Raman to about 72 M ha as estimated by (Raman (2010), Chand et al. (2020), Narayanamoorthy (2006), GOI (2004). Hence, it is important to have the MI potential for medium-term (5–15 years) planning. Giving weightage to the spread of MI (drip, sprinkler, and mini-sprinkler), the medium-term potential of 28 m ha can be arrived.[2]

Adoption of micro irrigation (both drip and sprinkler irrigation) in the past has been increasing at an average rate of 1.22% per annum till 2014 (Suresh Kumar and Palanisami 2019). However, in the coming years the growth in MI area is expected to increase comparatively at a higher rate due to improved government interventions and better awareness among the implementing departments and farmers. Hence it is assumed to increase by 3–5% per annum. This growth is possible in several states with (a) high-value crops, (b) canal commands (with water deficit or crop diversification) brought under MI, (c) increase in groundwater irrigation, & (d) increasing water scarcity. The future increase in MI area will account for 52, 63, and 77% of the MI

[2] The conservative estimate of 42–45 m ha MI potential is considered here. The medium term period potential has been worked out based on the adoption level of the sprinkler, drip, and mini-sprinkler in the last 2 decades, i.e., 55% of the total potential under sprinkler; 60% under mini-sprinkler and 80% of the total potential under drip will be covered in medium term.

Table 7.6 Future scenarios of technology adoption in India

No.	Water management technologies*	Potential area under water mgt technologies (m ha)		Future adoption level** (medium term) (m ha)		
		Long term (upto 2050)	Medium term (upto 2035)	Low growth rate	Medium growth rate	High growth rate
1	*Micro irrigation*					
	Growth rate/ annum (%)			3	4	5
	Area to be covered (m ha)	45.0	28.0	14.45 (52%)	17.70 (63%)	21.65 (77%)
2	*Surface irrigation*					
	Growth rate per annum (%)			1	2	3
	Area to be covered (m ha)	50.0	15.0	6.77 (45%)	8.33 (55%)	10.21 (68%)

Figures in brackets denote percentage to area to be covered in the medium term
* Even though the technologies here mean mostly water management related, they also include other component technologies such as land, crop, and soil management-related practices/technologies.
** The future adoption level for MI was calculated using the 2015 MI area as the base and the projections may be on the lower side as the growth is observed to be high only in the recent years. For surface irrigation, the technology/crop data handled in the All India FPARP project was used as the base

potential under medium term. Thus, the medium-term projection will make a true picture of the MI adoption and expansion in the country rather than just comparing the current adoption with the total potential as this will only show the poor adoption and spread which is not true always. Hence, the medium-term projections will pay the way for a quick technology spread and adoption at national level which in turn will encourage the faster adoption in the long run with appropriate interventions that are deemed to be important.

7.9 Strategies to Enhance Productivity Per Unit of Water

(a) Raising Farm-Level Water Productivity

The main emphasis in mitigating the emerging water crisis and establishing long-term sustainable water use for agriculture will be increasing WP (crop yield per unit of water consumptively used) (Kassam and Smith 2001). The vision, possible strategies, and the expected outcome are shown in Fig. 7.5.

India's food security will be dependent on the creation of additional irrigated lands, because of declining irrigation development investments, and growing competition for water from other sectors such as urban, industry, and environmental needs (Seckler

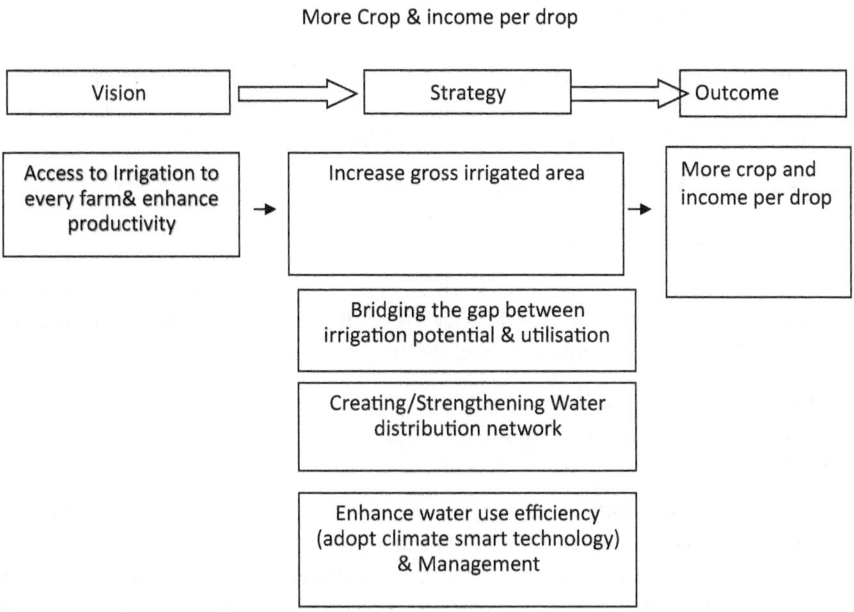

Fig. 7.5 Vision and strategies for achieving enhanced water productivity at national level

1996; Seckler et al. 1998). However, improving present WP by one per cent a year would assure full self-sufficiency in food-grains in India without additional supplies and increasing WP by 1.4% a year would meet all water demand in 2050 without increasing supplies (Upali et al. 2007). According to Ackermann (2012), improving India's WP will be able to free up water to meet urban and industrial demands (Ackerman 2012).

Improving water productivity can be possible by adopting the following three-pronged strategies:

1. Focus on crops in those areas where higher crop productivity is experienced
2. Diversify crops in those areas where lower crop productivity is experienced
3. Introduce improved land, crop, and water management practices/technologies in all the regions to achieve more crop per unit of water.

It is expected that strategies 1 and 2 can be possible in the long run and strategy 3 can be achieved in the medium term.

Overall, the above strategies will focus on:

- Adoption of the capital intensive water saving technologies like micro irrigation;
- Provision of less capital intensive practices/technologies. such as agronomic practices, laser land levelling, mulching, changing crop planting dates to match periods of less evaporative demand, deficit irrigation, etc. Several of these practices/technologies are shown in Table 7.3.

(b) **Increasing Farm-Level Income**

This option is possible by increasing crop production in a given cropping pattern or by changing the cropping pattern with higher-value crops with increased yield and reduced cost of technologies (Molden et al. 2003).

(c) **Refocusing WM Research and Technology Up-Scaling**

(i) **Water Management Research**

Till now, the water management research centres in India are giving more focus on continuing ongoing research/experiments with comparatively less focus on capacity building and outreach activities. Hence, the future focus should be on:

- Farmer participatory action research programmes
- Demand-driven research agenda
- Technology audit for all research centres.

Very recently, the Indian Institute of Water Management (2023) had demonstrated several success stories of irrigation water management. Similarly, the CGIAR Research Program on Climate Change, Agriculture and Food Security (CCAFS) (2020) demonstrated the benefits of solar powered irrigation systems in India (Mohanty et al. 2023; Shirsath et al. 2020). Solarification of agriculture in India is starting now, and is likely to expand in the future benefitting farmers practising irrigated agriculture.

(ii) **Micro Irrigation**

Given the poor adoption of MI, regional research institutes can design MI equipment that minimizes the cost and improve the fertigation schedules for different crops. The National Mission on Micro Irrigation (NMMI) has outlined the guidelines for better adoption of the MI systems in the country (GOI 2010; Global Agrisystem 2014).

In fact, drip irrigation subsidy disbursement is hindering the adoption and expansion of MI in several states of India. Farmers have reported about their losing interest in MI after waiting up to two years to get the eligible subsidy. Micro irrigation systems work effectively for the first two years then tend to create maintenance issues. Farmers need training in maintenance of the systems and in devising fertigation schedules. The experiences of the Gujarat Green Revolution Company (GGRC), which is successful in Gujarat, can be examined and replicated in other states (Palanisami et al. 2011).

Promoting micro irrigation in canal commands: Use of MI in canal command areas of the surface irrigation projects offers more scope for improving water use efficiency by up to 65–70%. Use of solar energy as a package with micro irrigation in canal commands will complement this initiative. Crop diversification, communication network, and produce marketing will become essential components of this intervention. A strong capacity building programme for farmers and irrigation service providers and public–private partnership models will strengthen this initiative. The

GOI's initiative, Pradhan Mantri Krishi Sinchayee Yojana (PMKSY) launched on 1st July 2015 with the objective to achieve convergence of investments in the irrigation sector at the field level aims at providing end-to-end solutions in irrigation supply chain, viz. water sources, distribution network, and farm-level applications. PMKSY adopts state level planning and projected execution that allows states to draw up their own irrigation development based on District irrigation Plans and State Irrigation Plan (GOI 2017).

(iii) **Capacity Building & Up-Scaling**

Capacity building programmes on these proven technologies can be built into the extension programmes of the research stations and state agriculture departments in each region. Details of the additional costs and benefits associated with new technology can be explained to the farmers. The transaction cost (hidden cost such as subsidy-oriented items) if any, should be explained to the farmers so that uptake becomes easier. The marginal cost curves drawn for these technologies will provide further insights into the feasibility of these technologies. Further, public–private partnership models in technology promotion and uptake such as drip and sprinkler irrigation can be explored by initiating local skill development programmes involving the drip manufacturers and suppliers and the potential local partners.

7.10 Digital Agriculture, Climate-Smart Agriculture (CSA), and Transformative Investment

Digital agriculture in the recent years is playing a major role in technology adoption where it either goes directly to the farmers or to expand products through suppliers and service providers. Most of the states in India are introducing the digital agri-platforms/portals to disseminate their innovations in technologies and products. The government's role is also seen visibly where several initiatives are taken to foster technology and innovation in the agricultural sector. For example, the government has been providing budget support to set up over 10,000 farmer–producer organizations (FPOs) in the next five years. The FPOs help grouping the farmers based on their crops and interest and can support agri-tech companies to access and scale up their business models.

It is important to see how best the transformative investment in digital agriculture as well as CSA is possible. The needed action steps to leverage technology and transformative investment options to accelerate and scale-up the adoption of climate-smart agriculture are:

1. While investments in CSA offer positive returns, many agriculture-based funding programmes have not identified climate-smart practices as an investment opportunity. Hence, awareness creation among the funding agencies and connect them to specific agriculture-based needs and opportunities.

2. Identify metrics, frameworks, and criteria. Database development is important on important technologies, which can be possible by providing needed support for the development of tools that collect on-farm data and share the results with the possible funding agencies who can see the range of benefits viz., social, environmental, and economic.
3. Integration of agricultural capital and outside investments can be better aligned to scale adoption of CSA practices where this integrated capital will be a key enabler.
4. Exploring private ecosystem services markets which can emerge as direct payments to farmers for the environmental benefits, like improved water retention, etc. This will help boost the technology adoption and up-keep.
5. Convincing private players like seed, food companies, and input suppliers who will be interested in pilot programmes to encourage and compensate farmers for adopting specific climate-smart practices.
6. Alignment with other policies and programmes. Although there are policies, programmes, and regulations intended to support resource conservation and management practices, policy can also be a barrier and hence better realignment among different policies will be useful.
7. Demonstrate clear risk/reward profiles of successful climate-smart farmers. Continue to share success stories of how different transformative investments are being made successfully to specific climate-smart agriculture opportunities to get a broader base of farmers.

7.11 Enhancing Productivity Per Unit of Water—The Way Forward

WP and use across scales—such as the crop, field, farm, irrigation system, or the basin level—need to be developed. In order to improve the precision in the WP estimates, hydrologic models coupled with crop models, remote sensing, and GIS database will facilitate the mapping across irrigation typologies. Promotion of digital agriculture and transformative investment options in CSA could be explored and practised to scale the adoption of CSA technologies. Agri-business models with private players like seed, food companies, micro irrigation system manufactures, and input suppliers who will be interested in pilot programmes to encourage and incentivize farmers for adopting specific climate-smart practices can be encouraged. Transformative policies both at state and central levels that will promote technology spread and adoption are also needed.

References

Ackermann R (2012) New directions for water management in Indian agriculture. Glob J Emerg Mark Econ 4(2):227–288

Barker R, Dawe D, Inocencio A (2003) Economics of water productivity in managing water for agriculture. In: Kijne JW, Barker R, Molden D, Wallingford UK (eds) Water productivity in agriculture: limits and opportunities for improvement. International Water Management Institute (IWMI), Sri Lanka, pp 19–35. (Comprehensive Assessment of Water Management in Agriculture Series 1)

Cai X, Rosegrant MW (2003) World water productivity: current situation and future options. In: Water productivity in agriculture: Limits and opportunities for improvement In: Kijne JW, Barker R, Molden D, Wallingford UK (eds) Water productivity in agriculture: limits and opportunities for improvement. International Water Management Institute (IWMI), Sri Lanka, pp 163–178. (Comprehensive Assessment of Water Management in Agriculture Series 1)

Chand S, Kishore P, Kumar S, Srivastava SK (2020) Potential, adoption and impact of micro irrigation in Indian agriculture. Policy Paper 36, ICAR-National Institute of Agricultural Economics and Policy Research (NIAP), New Delhi

DebRoy A, Shah T (2003) Socio-ecology of groundwater irrigation in India. In: Llamas R, Custodio E (eds) Intensive use of groundwater: challenges and opportunities. Swets & Zetlinger, The Netherlands, pp 307–336

Global Agrisystem (2014) Impact evaluation study on national mission on micro irrigation (NMMI), report submitted to Department of Agriculture and Cooperation, Ministry of Agriculture, Government of India

GOI (2004) Report of task force on micro irrigation, Ministry of Agriculture and Farmers Welfare, Government of India

Government of India (2006) Report of the sub-committee on more crop and income per drop of water. Advisory Council on Artificial Recharge of Ground Water. Ministry of Water Resources, New Delhi

Government of India (2008) Integrated water resources development and management, Central Water Commission, New Delhi

Government of India (2010) National mission on micro irrigation, operational guidelines, Department of Agriculture and Cooperation, Ministry of Agriculture, Government of India, New Delhi

Government of India (2017) Operational guidelines of per drop more crop (micro irrigation) component of PMKSY. Ministry of Agriculture & Farmers Welfare, New Delhi

Kassam A, Smith M (2001) FAO methodologies on crop water use and crop water productivity. Expert meeting on crop water productivity, Rome, Italy

Kijne JW (2003) Note on agronomic practices for increasing crop water productivity. In: Kijne, JW, Barker R, Molden D, Wallingford UK (eds) Water productivity in agriculture: limits and opportunities for improvement. International Water Management Institute (IWMI), Colombo, Sri Lanka, pp 319–321. (Comprehensive Assessment of Water Management in Agriculture Series 1)

Meredith G, Turral H, Susanne M, Scheierling, David O Tréguer, Peter G McCornick (2017) Beyond more crop per drop: evolving thinking on agricultural water productivity. IWMI research report 169. International Water Management Institute (IWMI) Colombo, Sri Lanka

Ministry of Agriculture & Farmers Welfare (2021) Agricultural Statistics at Glance, 2021. Directorate of Economics & Statistics. Government of India

Mohanty S, Sarangi A, Nanda P, Dasgupta P (2023) Success stories: AICRP on irrigation water management. Technical Bulletin No. 86, ICAR—Indian Institute of Water Management, Bhubaneswar, p 48

Molden D, Murray-Rust H, Sakthivadivel R, Makin I (2003) A water-productivity framework for understanding and action. In: Kijne, JW, Barker R, Molden D, Wallingford UK (eds) Water productivity in agriculture: limits and opportunities for improvement. International Water

Management Institute (IWMI), Colombo, Sri Lanka, pp 1–18. (Comprehensive Assessment of Water Management in Agriculture Series 1)

Molden D, Oweis T, Steduto P, Bindraban P, Hanjra MA, Kijne J (2010) Improving agricultural water productivity: between optimism and caution. Agric Water Manag 97(4):528–535

MoWR (2007) Water sector report. Government of India, New Delhi

Narayanamoorthy A (2006) Potential for drip and sprinkler irrigation in India. Gokhale Institute for Politics and Economics, Pune

Palanisami K, Mohan K, Kakumanu KK, Raman S (2011) Spread and economics of micro irrigation in India: evidence from nine states. Econ Polit Wkly 46(26 & 27)

Palanisami K, Kakumanu KR, Suresh Kumar D, Chellamuthu S, Chandrasekaran B, Ranganathan CR, Gioardano M (2012) Do investment in water management research pay?: an analysis of water management research in India. Water Policy 14:594–612

Palanisami K, Suresh Kumar D, Malik RPS, Raman S, Kar G, Mohan K (2015) Managing water management research: analysis of four decades of research and outreach programmes in India. Econ Polit Wkly 1(26 & 27)

Raman S (2010) State-wise micro irrigation potential in India: an assessment. Natural Resources Management Institute, Mumbai

Raman and Tikadar (2011) Estimated energy savings in Gujarat through micro irrigation. Paper presented in IGWC, Madurai

Rijsberman F (2006) More crop per drop: realigning a research paradigm. In: Giordano M, Rijsberman F, Saleth M (eds) More crop per drop: revisiting a research paradigm: results and synthesis of IWMI's research, 1996–2005. IWA Publishing, Colombo, Sri Lanka, International Water Management Institute (IWMI), London, UK, pp 8–21

Seckler D (1996) The new era of water resources management: from "dry" to "wet" water savings. International Irrigation Management Institute (IIMI), Colombo, Sri Lanka, p 20. (IIMI Research Report)

Seckler D, Amarasinghe U, Molden D, Silva R, Barker R (1998) World water demand and supply, 1990–2025: scenarios and issues. International Irrigation Management Institute (IIMI), Colombo, Sri Lanka, p 46. (IWMI Research Report 19)

Shirsath PB, Saini S, Durg N, Senoner D, Ghose N, Verma S, Sikka A (2020) Compendium on solar powered irrigation systems in India. CGIAR Research Program on Climate Change, Agriculture and Food Security (CCAFS)

Suresh Kumar D, Palanisami K (2019) Adoption of water management technologies: managing the water–energy nexus in agriculture. Econ Polit Wkly 54(14)

Upali A, Shah T, Singh OP (2007) Changing consumption patterns: implications on food and water demand in India. IWMI Research Report 119. International Water Management Institute, Colombo, Sri Lanka

Wichelns D (2014) Do estimates of water productivity enhance understanding of farm-level water management? Water 6(4):778–795

Chapter 8
Re-Estimating the Crop Water Requirement (CWR)

Abstract Crop water requirements (CWR) are changing over a period due to climate change. The CWR derived with limited field data in the past using broad assumptions couldn't represent the actual CWR to plan the water allocation in the present era of climate change. Reworking the crop water requirements (CWR), net irrigation requirements (NIR), and gross irrigation requirements (GIR) using scientific approach incorporating the climatic parameters has provided valuable insights into water allocation. CWR was re-estimated for different crops covering selected states in India. Comparing the NIR and GIR, at 60% efficiency level, only 66% more water is needed to meet the NIR, at 50% efficiency level, 100% more water is needed, at 40% efficiency level 150% more water is needed and at 30% efficiency level 233% more water is needed to meet the crop irrigation requirements indicating the importance of improving the project efficiency under growing water scarcity situations. Use of the re-estimated CWR is expected to alter the future irrigation demand scenarios in the country.

Keywords Crop water requirement (CWR) · Penman-monteith method · Climatological normals · Effective rainfall · Crop coefficient · Evapotranspiration · Delta · Net irrigation and gross irrigation requirement · Water use efficiency

8.1 Introduction

National and regional climate policy and planning aim to focus on an integrated approach to handle the impacts due to climate change. Increased water stress and future demands will therefore require strong decisions about how to allocate water resources between competing water uses, and at the same time there is an increasing need to relook at the allocation norms to address climate change mitigation and adaptation. Business as usual scenario is no longer an option and water requirement is to be viewed through a climate resilience lens (UN 2019).

As agriculture sector accounts for more than 70% of the total water use in the country, a better understanding of agriculture water use is necessary to manage

K. Palanisami and U. S. Nagothu, *India's Water Future in a Changing Climate*,
Advances in Geographical and Environmental Sciences,
https://doi.org/10.1007/978-981-97-1785-9_8

and utilize the irrigation more effectively. A good estimation of water demand is a prerequisite for preparing future action plans. This will help irrigation system authorities to understand spatial and temporal patterns of future water use, accordingly optimize system operations, and plan for future water investment in new systems. There are several approaches for estimating crop water requirement (CWR); these include simple rule of thumb based on depth of irrigation (delta), simple agronomic/field-based models with pan evapotranspiration data, and more detailed crop modelling incorporating different estimation methods. These models vary in complexity according to the number of parameters used and the extent to which, CWR is disaggregated by location, crops, and cropping systems and seasons.

8.2 Key Challenges and Options in Estimating Crop Water Requirement

Despite significant progress achieved during the last 4–5 decades, several challenges still have to be addressed while adopting the improved CWR estimation methodologies. The first challenge is how to integrate allied areas in the water requirement estimation such as water demand in other sectors that may have a bearing on system operation; the second challenge is to integrate the impact of climate change on crop water requirement; the third challenge is to incorporate the uncertainties relating to occurrence of rainfall while deriving the irrigation water demand.

8.3 Approaches in Crop Water Requirement Estimation

The National Commission on Agriculture (GOI 1976) assumed that an average delta of 0.65 ha m of groundwater is required to irrigate a cropped hectare and a delta of 0.9 ha m if the source is surface water, as conveyance losses are higher. The two together give an average delta of 0.8 ha m per cropped hectare. Further, with more scientific and economical use of water in the future, the average water depth for irrigation (delta) could be reduced to 0.7 ha m (GOI 1999). The total water requirement for irrigation in 2000 and 2025 were estimated at 672 km^3 and 770 km^3 respectively using different assumptions (GOI 1999).

Further, crop-wise data based on research trials are also used where depth of irrigation is multiplied by number of irrigations based on soil type and climate to get the total irrigation water requirement. One of the commonly used approaches in research stations is the ratio of irrigation depth (IW) to cumulative pan evaporation (CPE). This approach was largely adopted by the research workers for irrigation scheduling of crops (ICAR 2015). However, this ratio has restricted use outside research stations as CPE varies over locations and not readily available for non-researchers although the result of optimum IW/CPE is converted into frequency of irrigation (i.e., number of interval days for irrigating crops) for farmers' benefit. Yet, generalizations on CWRs have been made by researchers and water practitioners mainly to highlight the magnitude of CWRs among crops and used this as a standard reference in deriving rough crop water demand estimates and the water footprint for different crops.

The CWR thus derived with limited field data and broad assumptions couldn't represent the actual CWR. Hence, as an improvement over these estimates, consumptive CWRs have been worked out on climatological considerations for all the irrigation projects cleared for implementation since the 1970s (GOI 1999). However, such CWR calculations have been attempted only for those projects selected for implementation by the government where the estimation models and type of climate parameters used varied over time and between projects. However, researchers have started working out CWR using the crop models. For example, in the recent years, CROPWAT 8 model developed by FAO (1992) is being used by many researchers working on agriculture-related issues. Major weaknesses of such model-based CWR approach are: (a) most of them are micro level studies done focusing on 1–2 crops; (b) the parameters required for the CWR estimation in the inbuilt CROPWAT model will vary from location to location. (c) the weather data used also vary according to the researchers and their access to data; (d) many estimation procedures are difficult to handle such as effective rainfall calculation. Hence without using the location specific/project specific values, the CWR estimation may be biased (overestimated or underestimated the CWR).

A comparison of CWR derived from different approaches/methods is shown in Table 8.1 where high variability among the estimation methods is seen. As seen in Table 8.2, each method has its own strengths and weaknesses, and it is apparent that climatological approach with strong database with modern estimation procedures may provide meaningful estimates.

8.4 Need for a Scientific Approach to CWR Estimation

Further, there is a growing need to derive the irrigation water requirement of crops from the CWR. The CWR estimation depends upon only the weather parameters and crop growth stages (crop coefficient at different stages). Thus, in each location, the crop water requirement of a particular crop will not vary, whereas, depending upon the rainfall, method of irrigation, soil type, project efficiency, etc., the irrigation water requirement will vary. For example, more rainfall, higher project efficiency, and improved irrigation methods such as drip irrigation will reduce the overall irrigation water requirement of crop(s).

Even though estimation of crop water demand (CWD) is looking complicated as agriculture sector faces location specific issues in terms of varying weather parameters in a given time frame, it is possible to estimate and validate the CWR estimation procedure, as needed database for such model estimation is now available and further

Table 8.1 Comparison of different estimation methods of CWR

Crops/ methods	Estimation methods#					
	Method 1	Method 2		Method 3		Method 4
	Delta or depth of irrigation	Experimental stations trials with no. of irrigations	Experimental station trials with IW/CPE ratio	CWR data published in books	CWR data published in research reports	Improved methods with climate variables
Period of use	1960 onwards	1970 onwards	1980 onwards	1990 onwards	2000 Onwards	2010 onwards
Unit of CWR	ha m	mm	mm	mm	mm	mm
Surface water irrigated area	0.9					
Groundwater irrigated area	0.65					
Paddy				900–2500	1200–1600	729–1111
Sugarcane		1800–2000	1800	1500–2500	2800	1282–1368
Banana		1920	1900	1200–2200	2100	
Turmeric			695		1105	
Wheat				450–650		408–501
Maize			580	500–600	500–550	479–445
Sorghum				450–500	400	
Groundnut			617	500–600	450–500	536–591
Cotton				700–900	500	848–878
Chillies		960		500		
Green gram		400			250	476–660
Beans				300–500		
Brinjal		480			550	
Tomato				600–800	642	
Cabbage		480		380–500		
Onion				350–550		

Sources # different studies done in different time periods
(a) GOI (1999)
(b) Water Technology Centre (2016)
(c) ICAR (2015)
(d) SankaraReddi and Yellamanda Reddy (2002)
(e) Palanisami and Chandrasekaran (2001)

Table 8.2 Comparison of the methods on crop water requirement (CWR)

No.	Methods	Principle	Key aspects	Strength	Used for	Weakness
1	Delta	Total irrigation depth or gross irrigation requirement	Crop duration, duty and inflows used	Quick to quantify the gross water requirement of a surface as well as groundwater irrigated area	Macro level projection on irrigation water demand	Represents only gross irrigation requirement (GIR) of crops; Broad assumptions; too general data; climate parameters not considered
2	Water management research	Depth/ irrigation (IW/ CPE) to get crop water and irrigation water requirement	Farmers practice, research trials with selected weather parameters	Crop specific, soil specific irrigation water use; micro level estimation for irrigation scheduling	Irrigation scheduling of field crops	Location specific; no standardization of crop water requirements; only limited weather parameters considered
3	General CWR data from books and reports	Crop water requirement of crops	Average crop water requirement of crops	Easy for comparison of crop water requirement of crops	Crop water requirement of crops & water footprint calculation	Too general data; not based on strong scientific data/ analysis. Represents only average CWR of crops
4	Improved methods with climate variables	Crop water requirement, and irrigation water requirement	Climate data (long-term data), crop factor values, effective rainfall	Project/ system/basin level estimation possible using climate variables to represent present situation	Project level water demand; water sharing/ allocation	Mostly micro level studies for 1–2 crops; uses nearby weather stations; data for few parameters may be limited for few locations. Researchers often assume standard values when the actual data is not readily available Different authors use different periods and different equations in the estimation

the estimation methods are also scientific. Hence, estimation of CWD using scientific approach with strong database is crucial while analysing the performance of the various water supply systems.

8.5 Data and Methodology

Data: Calculation of CWR utilizes input data relating to climate, crop, soil, and rainfall for selected irrigation projects in Andhra Pradesh, Telangana, Karnataka, and Tamil Nadu states of India. The climatological normals (data) 1980–2010 from India Meteorological Department (IMD) were used. Few projects in Andhra Pradesh, Telangana, Tamil Nadu, and Karnataka with access to data have been considered for the CWR estimation. The details of the projects considered for CWR estimation are shown in Table 8.3.

Methodology: Crop water requirement is defined as "the depth of water applied to meet the water loss due to evapotranspiration (ET) of a crop which is actively growing in large fields under non restricting conditions of soil and water". The two separate processes where water is lost both from the evaporation from the soil surface and transpiration by the crop is generally referred to as evapotranspiration (ET) and is normally expressed in millimetres (mm) per unit time and per unit area. Generally, the ET rate expresses the quantum of water lost from a cropped area in units of water depth. Crop water requirement of a given area in a particular period/season mainly depends upon the crop pattern (i.e., area under different crops), growth stage of the crops, ET of each crop, special water needs of individual crop (such as percolation and seepage needs for crops like paddy), and the effective rainfall during that period.

In order to review and get an updated quantification methodology on crop water requirement estimation, the FAO has organized an expert consultation during 1990. The International Commission for Irrigation and Drainage and the World Meteorological Organization also joined the consultations. As a follow up of the consultation, the FAO Penman–Monteith method was suggested by which the ET of the reference surface (ETo) can be calculated. The Crop ET can be estimated using the weather-based reference ET values (ETo) and the crop coefficients derived from the crop coefficient curves (FAO 1998).

The Penman–Monteith equation is a close and simple representation of the physical and physiological factors governing the ET process and the mathematical relationship is given below:

$$\text{ETo} = \frac{0.408\Delta(Rn - G) + \gamma\frac{900}{T+273}U_2(es - ea)}{\Delta + \gamma(1 + 0.34U_2)}$$

where, ETo = Reference crop evapotranspiration (mm day^{-1}); Rn = Net radiation at the crop surface ($\text{MJm}^{-2}\ \text{day}^{-1}$); G = Soil heat flux density ($\text{MJm}^{-2}\ \text{day}^{-1}$); T = Air temperature at 2 m height (°C); U2 = Wind speed at 2 m height (ms^{-1}); es =

Table 8.3 Estimation of CWR based on climatological approach

State	Projects	Crop	CWR (mm) estimated by climatological approach	CWR (mm) from existing sources*	% deviation **
Andhra Pradesh	NSRC	Paddy in black soils	877	1000–1200	36.83
		Paddy in mixed soils	958	1000–1200	25.26
		Jowar	411	300–450	9.49
		Pulses	324	300–350	8.02
		Grams	383	300–400	4.44
		Groundnut	382	350–450	17.80
		Chillies	451	550–600	33.04
		Cotton	596	650–750	25.84
	NSLC	Paddy in mixed soils	928	1000–1240	33.62
		Groundnut	336	350–450	33.93
		Grams	284	300–400	40.85
	KDS	Paddy-black soils	787	1000–1240	57.56
		Pulses	243	300–350	44.03
		Chilies	445	550–650	46.07
Telangana	NSLC	Paddy in red soils	1058	1200–1400	32.33
		Paddy in mixed soils	928	1000–1240	33.62
		Paddy in black soils	848	1000–1240	46.23
		Ground nut	353	450–600	69.97
		Jowar	337	450–650	92.88
		Grams	285	400–450	57.89
		Cotton	583	650–850	45.80
		Chillies	384	600–800	108.33
	MUSI	Paddy-mixed soils	984	1000–1200	21.95
		Paddy-red soils	1063	1400–1600	50.52
		Jowar	385	400–600	55.84

(continued)

Table 8.3 (continued)

State	Projects	Crop	CWR (mm) estimated by climatological approach	CWR (mm) from existing sources*	% deviation **
Tamil Nadu	GA canal	Paddy-black soils	854	1000–1240	45.20
	Amaravathy	Paddy-mixed soils	994	1000–1240	24.75
Karnataka	KRS	Paddy-mixed soils	1112	1000–1240	11.51
	Harangi	Paddy-mixed soils	1165	1000–1240	6.44
Duration of crops: paddy 122–125 days, other crops 90–180 days					

* Existing sources are common sources without much authentication on how these CWR values were arrived, on what basis, and which time period. These sources mostly refer to reports, books, and other popular articles and publications. ** Deviation of the existing CWR data (i.e., upper limit) from the estimated CWR data using scientific approach

Saturation vapour pressure (kPa); ea = Actual vapour pressure (kPa)' (es − ea) = Saturation vapour pressure deficit (kPa); Δ = Slope vapour pressure curve (kPa°C^{-1})' γ = Psychrometric constant (kPa°C^{-1}). ETo can be calculated with different time steps (Ten day or monthly time step or daily or hourly also). The selection of the time step with which ETo is calculated depends on the purpose of the calculation, the accuracy required, and the time step of the climatic data available.

The performance of different ET estimation methods under different climatological conditions was studied by the Committee on Irrigation Water Requirements of the American Society of Civil Engineers (ASCE). The committee mentioned the FAO Penman–Monteith method as the standard method for computation of evapotranspiration. Several worldwide studies have also proved the merits of the P–M-based estimates over the other estimates under different climate conditions (Jensen et al. 1990; Kashyap and Panda 2001; Irmak et al. 2003). Hence it is important to apply this method to calculate the CWRs for different regions of India which can be used for future agriculture water demand estimations.

The Penman–Monteith equation incorporated in the CROPWAT model was applied in the CWR analysis. The model facilitated the calculation of reference ET, CWR, and irrigation requirements. After the ETo was estimated from CROPWAT model, the other related calculations for CWR estimation were done using the data viz., crop coefficients, effective rainfall, and project efficiencies relating to the selected projects.

The following are concepts and variables used in the estimation of CWR:

Reference Evapotranspiration (ETo) was calculated from actual monthly or daily or decade temperature (maximum and minimum in °C), humidity (%), sunshine (h day^{-1}), and wind-speed (km day^{-1}) according to the FAO Penman–Monteith method (FAO 1998) in addition to latitude, longitude, and altitude of the station by using CROPWAT model. The daily/monthly average of all these data were calculated from representative meteorological station of India Meteorological Department (IMD) stations in respective command area or the nearest and most representative meteorological station. The required climatic data was collected from IMD, Pune for the period from 1961 to 2003. Sunshine hours were obtained from CLIMWAT 2.0 model, wherever sunshine hours data was not available for the selected IMD station.

Crop Water Requirement (ETc) of the crops over the growing season was calculated by multiplying the ETo with the crop coefficients (Kc), which is the ratio of the crop evapotranspiration (ET_c) to the reference evapotranspiration (ETo). The Kc values normally vary with different crop growth stages (i.e., initial stage, development stage, mid-season stage, and late stage). FAO (1998) has presented updated values for crop coefficients for different stages of crops (Allen et al. 1998). For the present analysis, the crop coefficients for irrigated dry (ID) crops were adopted from FAO (1998) and for rice crop, the coefficients were taken from Ramanathan (2002). In all the crops, the crop stages were adjusted according to the duration of the crops as per the local conditions and using field data from the research stations. Time curves were drawn against Kc value and duration of crop growth stages and from these curves, 10 days interval Kc values were calculated and used in the estimation of ETc.

Soil data included mainly the types of soils that existed in the project command areas. Major soil type given in the respective detailed project reports (DPR) was used for calculations. The sand, silt, and clay present in the soils in various proportions determine the type of soil, its moisture holding capacity, etc. The percolation rates and puddling requirements under different soil types were arrived and used in the crop water requirement calculations of paddy. Percolation rates vary from 2 to 5 mm depending upon the soils. Percolation rate of 2 mm/day for alluvial and black cotton soils, 3 mm/day for clay loamy (mixed) soil, and 4 mm/day for red soils with long history of rice cultivation were used in the CWR calculations.

Water requirement for land preparation of paddy was taken as 150–200 mm depending upon soil type and duration of puddling and puddling requirements, 150 mm was used for main field puddling in the Krishna delta region and 200 mm was used for puddling as different soils require different puddling requirements. In the case of nursery 30 mm was used (assuming 300 mm for puddling 1 ha of nursery field which can be equal to 30 mm for 1/10 of the main field for nursery requirement). ETc was estimated by multiplying the ETo with the Kc. Given the cropping pattern for each project, planting date, crop duration, crop water requirement in 10 days (decade) interval was calculated and finally total requirement was arrived by summing the values.

Effective Rainfall (ER) is defined as that portion of rainfall falling during the crop growing period that infiltrates into the soil surface and is available for plant growth. In the CWR calculations, the effective rainfall was calculated for each crop separately using the methodology developed by Water Management Division, Ministry of Agriculture, GOI (1984). Monthly rainfall data was collected from IMD, Pune for the rain gauge stations located in the command areas of different projects for the period from 1951–52 to 2019–2010 and used in the estimation.

Irrigation Requirements refers to the quantity of irrigation water supplied when rainfall is insufficient to meet the evapotranspiration needs. Irrigation needs to be provided to the crops at the right time (period) and in the right quantity for maximum crop growth. The net irrigation water requirement (NIR) refers to the water required by irrigation to meet the crop evapotranspiration (ET_c) and auxiliary water needs that are not provided by water stored in the soil profile. The total amount of water inclusive of water loss in the field applied through irrigation is referred to as gross irrigation requirement (GIR), which is the net irrigation requirement plus application and other losses. Gross irrigation requirement at project level however is influenced by the level of project efficiency (PE) to deliver the required quantity of water from the canal head to field level.

$$\text{NIR} = \text{ETc} - \text{ER} + \text{percolation loss} + \text{puddling requirement (for rice crop)}$$
$$\text{NIR} = \text{ETc} - \text{ER (for ID crops)}$$
$$\text{GIR} = \text{NIR/PE}.$$

8.6 Major Observations

Comparison of the existing CWR estimates and the estimated CWR using scientific approach is given in Table 8.3. It is seen that invariably in all the crops the existing CWR data is higher than what is actually required under different soil types and this indicates the possible reduction in overall crop water demand even after incorporating climate parameters. When the estimated CWR values are used for future projection of water demand, there will be possible reduction in the overall irrigation water demand than being used now.

Crop Water Requirement and Irrigation Water Requirement. There is always confusion in using crop water requirement and irrigation water requirement for different purposes. While projecting future water demand and project operations, normally net irrigation requirement (NIR) and gross irrigation requirement (GIR) are important. NIR is arrived by subtracting the effective rainfall from the CWR whereas the gross irrigation requirement (GIR) (i.e., delta) refers to what will be irrigation water requirement at source which takes into account the project efficiency. Higher project efficiency means lesser the GIR. Table 8.4 and Fig. 8.1 show how the GIR increases at different project inefficiencies ranging from 30 to 60%. Hence improving project efficiency means less irrigation water demand at project level which is highly warranted under the present water scarcity situations.

Table 8.4 GIR for major crops with different project efficiencies

State	Projects	Crop	CWR (mm)	NIR (mm)	GIR (mm) at project efficiency (%) of			
					60	50	40	30
Andhra Pradesh	NSRC	Paddy-black soils	877	628	1047	1257	1571	2095
		Paddy-mixed soils	958	716	1193	1432	1789	2386
		Jowar	411	163	272	327	408	544
		Pulses	324	127	212	254	317	423
		Grams	383	152	254	305	381	508
		Groundnut	382	343	571	686	857	1143
		Chillies	451	180	300	360	450	600
		Cotton	596	284	474	569	711	948
	NSLC	Paddy-mixed soils	928	668	1114	1337	1671	2228
		Groundnut	336	269	448	537	672	896
		Grams	284	236	394	473	591	788
	KDS	Paddy-black soils	787	461	769	922	1153	1537
		Pulses	243	234	390	468	585	780
		Chilies	445	180	300	360	450	600
Telangana	NSLC	Paddy-red soils	1058	794	1323	1588	1985	2646
		Paddy-mixed soils	928	663	1105	1327	1658	2211
		Paddy-black soils	848	598	996	1195	1494	1992
		Ground nut	353	173	289	347	433	578
		Jowar	337	208	346	415	519	692
		Grams	285	240	401	481	601	802
		Cotton	583	273	455	545	682	909
		Chillies	384	243	406	487	609	812
	MUSI	Paddy-mixed soils	984	693	1155	1386	1733	2310
		Paddy-red soils	1063	819	1365	1638	2048	2730
		Jowar	385	378	630	756	945	1260
Tamil Nadu	GA canal	Paddy-black soils	854	690	1150	1380	1725	2300
	Amaravathy	Paddy-mixed soils	994	912	1433	1824	2280	3040

(continued)

Table 8.4 (continued)

State	Projects	Crop	CWR (mm)	NIR (mm)	GIR (mm) at project efficiency (%) of			
					60	50	40	30
Karnataka	KRS	Paddy-mixed soils	1112	847	1412	1694	2118	2823
	Harangi	Paddy-mixed soils	1165	933	1555	1866	2333	3110

Note Current project efficiency is 30–40%. In order to improve the performance of the irrigation projects, CWC has suggested an efficiency level of 60% (CWC 2016)

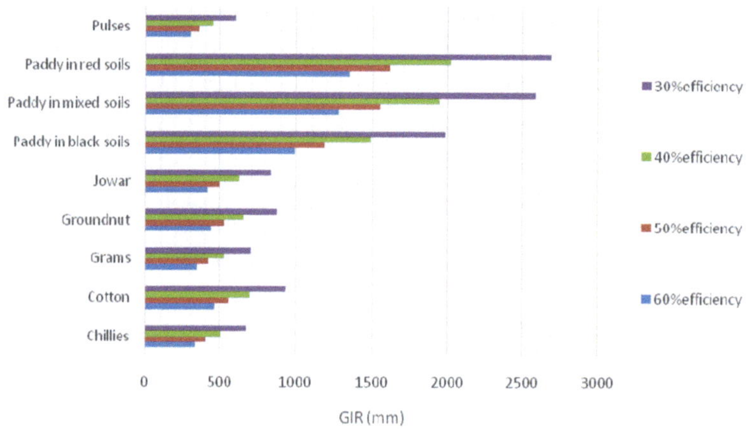

Fig. 8.1 GIR with different project efficiencies

A study by Central Water Commission (CWC) indicated that the project efficiency of the 35 projects studied across India ranged from 13 to 62% with a weighted average efficiency of 36% (CWC 2016). It is recommended that in the future the project efficiency should be increased to 60% level to save water and irrigate more area (CWC 2016). Lower project efficiency means more investment to get the same quantum of water for crops. It is seen from the table that compared to NIR, projects with 30% efficiency level, the GIR demand will be 3 times more which means only less area could be irrigated. Hence investment in improving project efficiencies is much more worth it than creating more irrigation potential at higher costs. Normally for improving efficiency, it costs Rs. 50000 to 100,000/ha compared to creating new potential which costs Rs. 350000 to 500000/ha depending upon the location, type of project, and implementation period.

8.7 Crop Water Requirements—The Way Forward

Projections of irrigation water demand are made mostly using the decades old crop and water requirement data, which ignored the climate change impact on crop water requirements. Reworking the CWRs and irrigation water requirements using scientific approach incorporating the climatic parameters has provided valuable insights in water allocation. Given the estimated net irrigation requirements (NIR) of different crops, at 60% project efficiency, only 66% more water is needed to meet the NIR, at 50% efficiency, 100% more water is needed, at 40% efficiency 150% more water is needed and at 30% efficiency level, 233% more water is needed to meet the NIR indicating the importance of improving the project efficiency under growing water scarcity situations.

The CWR derived from improved methodology will be helpful for both long-term forecasting (for system infrastructure and investment planning) and short-term forecasts (for setting up crop planning and scheduling of water allocation).

The results show that the scientific estimation of the CWR is more meaningful, and hence, CWR estimation is to be done on project basis covering all the crops and soil types in the command area of the projects. The empirical findings on water requirement estimation also show that water use can be altered by system level crop diversification due to profitability (e.g., rice to non-rice crops and other water conservation policies (e.g., canal-based micro irrigation), etc. This transition will help to improve the precision in estimation of the future irrigation water demand in the country. Further, improving efficiency of the existing irrigation projects will be more important and cost effective than investing in new projects at higher costs as new investments cost about 5 times more than the costs needed for improving the efficiency of existing projects. Adoption of improved soil and water management practices will save about 20% water, which can help improve the overall project efficiency. All the project engineers, planning dept officials, and policy makers can be familiarized with the estimation methods and the use of the revised estimates in their implementation programmes. The National and State Water Policies can incorporate the re-estimation of CWR as one of the policy agenda. Also the interstate water sharing protocols can also adopt these re-estimated CWR data.

References

Allen RG, Pereira LS, Raes D, Martin S (1998) Crop evapotranspiration. FAO Irrigation and Drainage Paper 56, Rome

Annual Report (2015) All India coordinated research project on irrigation water management. Tamilnadu Agricultural University, Coimbatore

CWC (2016) Summary report on water use efficiency studies for 35 irrigation projects. Central Water Commission New Delhi, India

Food and Agriculture Organization (FAO) (1992) CROPWAT: a computer program for irrigation planning and management, by M. Smith. FAO Irrigation and Drainage Paper No. 46, Rome

Food and Agriculture Organization (FAO) (1998) Guidelines for calculation of evapotranspiration. Irrigation and Drainage paper No: 56, Rome

GOI (1976) National Commission on Agriculture (NCA). Ministry of Agriculture, New Delhi

GOI (1984) A guide for estimating irrigation water requirements. Ministry of Irrigation. Water Management Division, New Delhi

GOI (1999) Report of the working group on perspective of water requirements. National Commission for Integrated Water Resources Development Plan, Ministry of Water Resources, New Delhi

GOI Indian Meteorology Department (IMD). Climatological normals 1980–2010, Pune

ICAR (2015) Annual Report all India coordinated research project on irrigation water management. Tamilnadu Agricultural University, Coimbatore

Irmak S, Allen RG, Whitty EB (2003) Daily grass and alfalfa-reference evapotranspiration estimates and alfalfa-to grass evapotranspiration ratios in Florida. J Irrg Drain Eng 129(5):360–370

Jensen ME, Burman RD, Allen RG (1990) Evapotranspiration and irrigation water requirements. ASCE manuals and reports on engineering practice (70). ASCE, New York

Kashyap PS, Panda RK (2001) Evaluation of evapotranspiration estimation methods and development of crop-coefficients for potato crop in sub-humid region. Agric Water Manag 50:9–25

Palanisami K, Chandrasekaran B (2001) Improving irrigation efficiency. Water Technology Centre, Tamilnadu Agricultural University, Coimbatore

Ramanathan SP (2002) Crop water requirement. Manual. Irrigation Management and Training Institute (IMTI), Trichy, India

SankaraReddi GH, Yellamanda RT (2002) Efficient use of irrigation water. Kalyani Publishers, New Delhi

United Nations (2019) Water policy brief on climate change and water

Water Technology Centre (2016). Report on CWR Estimation (unpublished). PJTSAU, Hyderabad

Chapter 9
Up-Scaling Micro Irrigation

Abstract Micro irrigation (MI) is one of the key interventions in water saving, energy saving, and improving the crop productivity and is gradually emerging as a demand management technology in India. The MI potential in the country varies from 42 to 72 M.ha by various estimates. At all India level, the MI area covered to the potential till 2019 is only about 17.4%. Among the drip and sprinkler coverage, the potential covered under drip is 28.2% and sprinkler is 12.9%. In terms of actual area coverage to the total MI area in the country, five states viz., Rajasthan, Andhra Pradesh, Maharashtra, Karnataka, and Gujarat have shared more than 10% each. Regarding MI implementation models, Gujarat Green Revolution Company (GGRC) model is comparatively performing better. The major suggestions to up-scale the MI adoption include provision of technical support for MI operation after installation, relaxation of farm size limitation in providing MI subsidies, supply of liquid fertilizers, improved marketing facilities and access to more credit, direct benefit transfer (DBT) scheme, MI in canal command area, and better implementation models like GGRC.

Keywords Micro irrigation · Jevons Paradox · MI subsidy · PMKSY · Gujarat Green Revolution Company (GGRC) · Drip fertigation · Direct benefit transfer · MI expansion

9.1 Introduction

The Micro irrigation (MI) technologies mostly drip and sprinkler are the key interventions in water saving, energy saving, and improving the crop productivity. Major focus on MI in the country was given during 1981 when the National Committee on Plasticulture in Agriculture (NCPA) initially approved the use of plastics in agriculture on a pilot basis. Subsequently, a centrally sponsored scheme on the use of plastics in agriculture was initiated in 1992. Now the government has merged all related schemes of irrigation under PMKSY. The main focus of PMKSY is to provide water to every field, improve on-farm water use efficiency, and adoption of

K. Palanisami and U. S. Nagothu, *India's Water Future in a Changing Climate*, Advances in Geographical and Environmental Sciences,
https://doi.org/10.1007/978-981-97-1785-9_9

water saving practices/technologies. It also facilitates the augmentation of recharge of aquifers and introduces sustainable water management practices. Further, the Direct Benefit Transfer (DBT) scheme was also launched by the Government of India in 2013 mainly to ensure accurate targeting of beneficiaries and faster flow of information on all aspects including allotted funds. A nodal point was created in the Planning Commission (now NITI Aayog) to implement the scheme (Chand et al. 2020). This chapter provides useful discussions about the MI and its potential as a demand management strategy in the near future.

9.2 Current State of Micro-irrigation in India

The potential of MI systems in terms of drip and sprinkler was assessed by several researchers using the state-wise data (Raman 2010). Key variables considered in the assessment were: state-wise and source-wise irrigated area, cropping pattern, cropped area, and crop-wise suitability for different micro irrigation systems. Based on these assessments, the potentiality of bringing more area under drip and sprinkler in the country has varied from 42 M ha as estimated by Raman to about 72 M ha as estimated by Raman (2010), Chand et al. (2020), Narayanamoorthy (2006) and GOI (2004) (Table 9.1).

MI coverage in the country can be examined in terms of (a) actual area covered compared to potential in each state, and (b) state-wise actual area covered compared to All India MI area (Table 9.2). It was observed that compared to the potential area, the actual area covered was higher in Andhra Pradesh followed by Karnataka and Maharashtra states. Among the drip and sprinkler coverage, the coverage under drip was about 28.2 and 12.9% under sprinkler (Fig. 9.1). Among the states, the per cent of drip coverage was higher in Andhra Pradesh state and the sprinkler coverage was higher in Karnataka State.

In terms of actual area coverage to the total MI area in the country, five states viz., Rajasthan, Andhra Pradesh, Maharashtra, Karnataka, and Gujarat have shared more than 10% each, three states viz., Tamil Nadu, Haryana, and Madhya Pradesh have 4.5–7.5% share each and the remaining states have less than 4.5% share each indicating the importance of up-scaling the MI in these low MI coverage states (Fig. 9.2). The states have also been ranked based on their MI coverage to MI potential as well as their share to total MI area in the country. It is seen that Andhra Pradesh, Karnataka, Maharashtra, Chhattisgarh, Tamil Nadu, Gujarat, and Rajasthan have similar positions under both the rankings, and Jharkhand, Himachal Pradesh, Madhya Pradesh, and Telangana differ slightly in their rankings (Table 9.2). Among other things, the higher coverage in these states might be due to better implementation of the MI programmes, water scarcity situation, crop pattern followed, subsidy schemes adopted, and the farmers' interest to adopt MI technologies.

While analysing the spread and adoption of MI, it is also important to examine the suitability of sprinkler or drip adoption which differs mainly according to land terrain, soils, and type of crops (Table 9.3). In all the estimates, more shares are

Table 9.1 Micro irrigation potential and coverage in India (M ha)

A: MI Potential				
1	MI potential estimated by:	Sprinkler	Drip	Total
	(a) GOI (2004)	42.5	27	69.5
	(b) Narayanamoorthy (2006)	51	20.9	71.9
	(c) Raman (2010)	30.5	11.7	42.2
	(d) Chand-ICAR-NIAP (2020)	51.01	21.16	72.17
2	Actual coverage in 2019–20#	6.56	5.95	12.51
3	% potential covered based on:			
	(a) GOI (2004)	15.29	22.04	18.00
	(b) Narayanamoorthy (2006)	12.75	28.47	17.40
	(c) Raman (2010)	21.31	50.85	29.64
	(d) Chand ICAR-NIAP (2020)	12.74	28.12	17.33
B: Irrigated area and MI area				
1	Total irrigation potential created			113.53
	Medium & major irrigation			47.97
	Minor irrigation including groundwater			65.56
2	Net irrigated area			75.5
	Gross irrigated area			112.2
3	% micro irrigation covered in relation to:			
	Total irrigation potential created			11.02
	Net irrigated area			16.6
	Gross irrigated area			11.2

\# *Source* GOI (2021)
Note It is possible to assume that about 80% MI area come under minor irrigation & 20% under major and medium irrigation (MMI) projects

accounted for sprinkler, covering crops like cereals, pulses, oilseeds, fodder crops than drip, which can cover crops like cotton, sugarcane, fruits and vegetables, spices and condiments; and some pulse crops like red gram, etc. Scenario 1 gives the MI spread against the MI potential estimated. Scenario 2 gives the MI spread against the total irrigation potential created, net and gross irrigated area in the country.

9.3 Benefits of Micro Irrigation

The benefits and economics of MI were well documented over the past several years, which justified the investment in MI systems by the farmers. The benefits in terms of yield increase, water saving, and energy saving are discussed briefly.

Table 9.2 Potential and actual area under MI in different states, 2019–20 ('000 ha)

State	Drip			Sprinkler			Total			Ranking based on	
	P	A	%	P	A	%	P	A	%	Potential to actual	Actual to total
Andhra Pradesh	1390.0	1387.8	99.8	1230.0	519.2	42.2	2620.0	1907.0	72.8	1	2
Arunachal Pradesh*	–	1.8		–	1.2			3.0		22	25
Assam	50.0	2.4	4.9	80.0	9.7	12.1	130.0	12.2	9.4	14	19
Bihar	1010.0	7.5	0.7	2920.0	107.0	3.7	3930.0	114.4	2.9	19	13
Chhattisgarh	400.0	27.5	6.9	530.0	316.5	59.7	930.0	344.0	37.0	4	9
Goa	10.0	1.3	13.4	20.0	1.3	6.3	30.0	2.6	8.7	15	26
Gujarat	2280.0	800.7	35.1	2840.0	728.8	25.7	5120.0	1529.6	29.9	6	5
Haryana	970.0	33.6	3.5	3170.0	592.2	18.7	4140.0	625.8	15.1	11	7
Himachal Pradesh	10.0	5.1	51.1	40.0	4.3	10.8	50.0	9.4	18.8	10	22
Jammu and Kashmir*	–	0.1		–	0.0		–	0.1		23	28
Jharkhand	40.0	25.1	62.7	130.0	13.0	10.0	170.0	38.1	22.4	8	16
Karnataka	1270.0	723.2	56.9	1190.0	1048.9	88.1	2460.0	1772.1	72.0	2	4
Kerala	190.0	24.6	12.9	70.0	8.5	12.1	260.0	33.1	12.7	12	17
Madhya Pradesh	930.0	318.7	34.3	6560.0	241.7	3.7	7490.0	560.3	7.5	16	8
Maharashtra	1460.0	1316.8	90.2	1810.0	560.0	30.9	3270.0	1876.9	57.4	3	3
Manipur*	–	0.0		–	4.2		–	4.2		24	24
Mizoram*	–	4.1		–	0.7		–	4.9		25	23

(continued)

Table 9.2 (continued)

State	Drip			Sprinkler			Total			Ranking based on	
	P	A	%	P	A	%	P	A	%	Potential to actual	Actual to total
Nagaland*	–	3.3		–	6.4		–	9.8		26	21
Odisha	290.0	24.6	8.5	330.0	105.1	31.8	620.0	129.7	20.9	9	12
Punjab	1600.0	35.1	2.2 mplePara>	4670.0	13.7	0.3	6270.0	48.8	0.8	21	15
Rajasthan	1220.0	264.3	21.7	7240.0	1685.0	23.3	8460.0	1949.3	23.0	7	1
Sikkim*	–	6.5		–	5.1		–	11.6		27	20
Tamil Nadu	1440.0	686.6	47.7	1270.0	252.6	19.9	2710.0	939.1	34.7	5	6
Telangana	1120.0	196.0	17.5	1160.0	71.0	6.1	2280.0	267.0	11.7	13	10
Tripura*	–	0.1		–	0.4		–	0.5		28	27
Uttar Pradesh	4300.0	32.4	0.8	13350.0	178.6	1.3	17650.0	211.1	1.2	20	11
Uttarakhand	150.0	11.1	7.4	260.0	7.9	3.1	410.0	19.1	4.7	17	18
West Bengal	950.0	10.3	1.1	2050.0	78.2	3.8	3000.0	88.5	3.0	18	14
Total	21,080.0	5950.8	28.2	50,920.0	6561.2	12.9	72,000.0	12,512.0	17.4		

Note P = potential area taken from ICAR-NIAP estimate (Chand et al. 2020). A = Actual area worked out from the data available with GOI Directorate of Economics and Statistics, New Delhi; GOI, National Committee on Plasticulture Applications in Horticulture. GOI, Ministry of Agriculture and Farmers Welfare, Pradhan Mantri Krishi Sinchayee Yojna (PMKSY), 2021

Area under mini sprinklers, etc., are not included because of data inconsistency. Rough estimate of their coverage will be about 0.7 M ha (or about 5% of the MI area)

* Potential area is not available

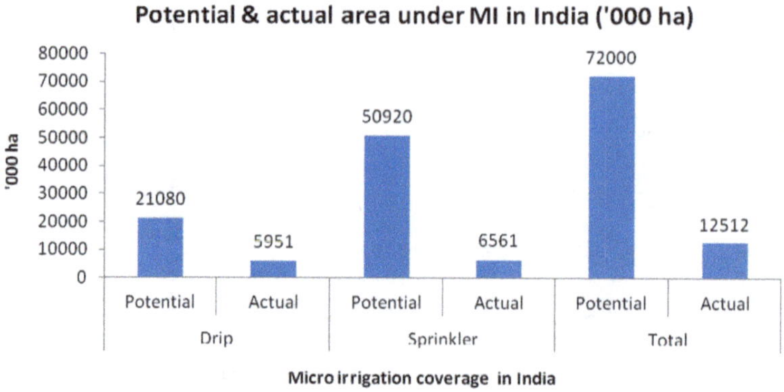

Fig. 9.1 Potential and actual area under MI in India ('000 ha)

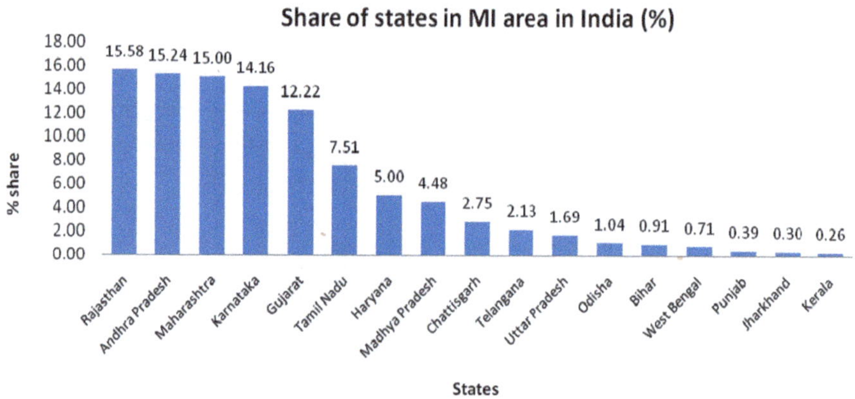

Fig. 9.2 Share of states under MI area in India

Enhanced Crop Yield and Water Saving

One of the significant benefits of adopting drip irrigation is the crop yield increase for almost all the crops. The increase varied from 33% in coffee (Robusta) to 362% in cardamom crop. Mango, guava, and capsicum output has increased by 150%, and mango and pomegranate reached higher yields in the fifth year of cultivation as opposed to the 12th year in the conventional methods. Likewise, groundnut and cotton yield increased by 200% and 186% respectively. Pigeon pea, onion, and soybean increased yield by 133%. Most of the vegetables under drip irrigation had increased yield from 50 to 86%. Oil palm and cashew nut increased yield by 71% and 85% respectively due to drip irrigation. The above information demonstrated the benefits of drip irrigation as a yield enhancing technology besides water saving. However, depending upon the soil conditions, crop management, and fertilizer applications the yield may vary across farms and between years (Soman 2023).

Table 9.3 Suitability of micro irrigation (MI) in India

Factors	MI Suitability	Remarks
1. Canal projects		Long-term trends in water supply
Severe water scarcity (>50%)	Fully	Possible to convert into canal-based MI irrigation; Structural and non-structural changes possible
Moderate water scarcity (20–50%)	Partially	Selected location (such as tail end command) with change in crop pattern
Low water scarcity (<20%)	No	Not attractive for water saving given the existing crop pattern
2. Pumping schemes		
Existing pumping schemes	Possible	Possible to shift from existing surface irrigation methods to micro irrigation with suitable crops
New pumping schemes	Possible	Possible to plan for micro irrigation from the beginning with high-value crops
3. Soil types		
Sandy and porous	Possible	Water losses can be minimized significantly with micro irrigation
Loam, alluvial	Partially	Water losses can be minimized moderately with micro irrigation
Clay	No	Micro irrigation not suitable due to low percolation
4. Water quality		
High sedimentation	Not advisable	High sediment level may clog the drip or sprinkler irrigation systems
5. Crop		
Wide spaced crops, perennial and high-value crops	Possible	Both water use and cost of systems can be reduced
Closed space crops like rice, wheat	Possible	Cost is very prohibitive
6. Terrain		
Plain	Possible	Depending upon the crops, it can be planned
Undulating	Highly possible	Land levelling not required and hence land debt cost minimized
7. Pumping and energy cost		MI investment cost will be high but when annualized, it will be comparatively less. Hence the pumping cost is considered here as a factor. Energy saving with MI is significant, relatively when compared to conventional irrigation

(continued)

Table 9.3 (continued)

Factors	MI Suitability	Remarks
High pumping cost/ha	Not attractive	With high pumping cost not financially feasible with resulting externalities of over pumping
Moderate to low cost/ha	Less attractive	With efficient crop pattern, marketing, etc., it is manageable
Low pumping cost/ha	Attractive	Possible to expand MI irrigation with solar power, etc., yet to establish popularity though solar pumping is sustainable only with subsidized investment
8. Farmers' socio-economics profile		
Moderate to high: education, skills, assets	Possible	With education, skills, and financial resources, farmers can manage MI systems
Low to moderate: education, skills, assets	Highly risky	Poor operation & maintenance results in inefficiency in water use, crop failure, etc.

Reduced Energy Consumption

As the volume of water applied to irrigate a crop under drip irrigation is lower compared to that applied in conventional method, the power consumption for pumping also decreases proportionately. Taking the example of sugarcane, flood irrigation method uses 2200 mm water compared to drip irrigation which uses only 1200 mm water. Thus the percentages of water and power saved are 45.5% and 58.3%, respectively. In addition, drip irrigation saves 29.9% in costs. Hence, drip irrigation is the most water-efficient irrigation system and saves energy and water in Indian sugarcane production (Narayanamoorthy 2005).

Economics of Micro Irrigation

Using the crop yield increase and the prevailing market prices, it was observed that investment in MI system was repayable in six months to 2 year period. In the last several years, a number of analyses by economists have been in public domain demonstrating the financial viability of drip and sprinkler irrigation (Palanisami et al. 2011). Further, if the benefits of water saving and power saving are considered, the returns were found to be still high and repayment period was also much less. The benefit cost (BC) ratio worked out for different crops covering several states has shown that MI for most of the crops was financially viable.

9.4 Adoption of MI by Different Farmers

Evidences show that by adopting the drip irrigation, water can be saved up to 40–80% compared to 30–40% under conventional practice (Palanisami et al. 2011, 2014). The successful adoption of MI systems requires, in addition to technical and economic efficiency, both technical knowledge about the use of the MI systems and its accessibility through institutional support systems such as subsidies (Namara et al. 2005). However, available empirical evidence is comparatively limited with respect to its level of adoption and economics under different farm categories. Hence, the key questions are: who has access to MI and what is the economics of MI under different farm groups (i.e., marginal, small, and large farmers).

Using the farm-level data, the following regression equation was fitted to study the factors influencing the area under MI.

$$Y_i = \beta_0 + \beta_1 X_i + \beta_2 D_{1i} + \beta_3 D_{2i} + \varepsilon_i$$

where,

Y_i = Area under MI by ith farmer (ha), X = Farm size of ith farmer (ha), D_{1i} and D_{2i} = Dummy variables for ith farmer representing marginal and small farmer category respectively, ε_i = error term.

Table 9.4 revealed that majority MI farmers in the case of Kerala state (52%) were marginal and in the case of Andhra Pradesh (71%), Karnataka (66%), Orissa (63%), and Punjab (55%) states, majority of them were small farmers. In the case of Maharashtra (63%) and Tamil Nadu (65%) states, majority of the farmers were large farmers. The study by Namara et al. (2005) has observed that majority of the MI farmers were rich farmers in Gujarat and Maharashtra states. In spite of the efforts by the GOI for the promotion of MI, the spread and adoption of MI is rather slow than expected. The MI adoption study conducted in nine states of the country had indicated the following as the main reasons for the slow adoption: farmers were reluctant to expand the area due to other constraints like high initial capital cost, lack of technical knowledge in the operation, and maintenance of the systems and crop preference by the farmers due to their past experience. The story is similar to the adoption of System of Rice Intensification (SRI) practices, which was low, despite the fact that SRI provided higher yields and income. This was mostly attributed to operating constraints like lack of skilled labour, basic knowledge on the practices, high management intensity, etc.

It was also observed that the internal rate of return (IRR) has ranged from 3 to 35% in case of marginal farmers, 14–88% for small farmers, and 15–128% for large farmers indicating financial viability of the MI adoption. Among the states, the IRR was higher in case of large farmers of Kerala and Maharashtra due to their diversified inter-cropping pattern in the orchard/plantation crops, ensuring higher rate of returns. In addition, the plantation crops are widely spaced and cost of investment is comparatively low. Overall, the study results indicated that the adoption of MI has

Table 9.4 Farm size and average area irrigated by MI systems

State	Farmer category	% of farmers	Average farm size (ha)	Average area under MI (ha)	% of area under MI
Andhra Pradesh	Marginal	6.00	0.82	0.76	92.68
	Small	70.67	1.7	0.90	52.94
	Large	23.33	14.08	2.96	21.02
Tamil Nadu	Marginal	13.33	0.62	0.48	77.42
	Small	22.00	1.72	1.31	76.16
	Large	64.67	4.67	2.41	51.61
Kerala	Marginal	52.00	0.54	0.15	94.44
	Small	28.00	1.44	1.25	86.80
	Large	20.00	2.38	2.22	93.27
Karnataka	Marginal	6.00	1.89	1.33	70.37
	Small	66.00	5.71	1.82	31.87
	Large	58.00	18.12	6.59	36.37
Maharashtra	Marginal	20.00	1.80	0.90	50.00
	Small	16.67	3.75	2.25	60.00
	Large	63.33	6.60	3.40	51.52
Orissa	Marginal	23.33	0.51	0.07	13.72
	Small	62.67	1.74	1.23	70.44
	Large	14.00	15.52	9.56	61.60
Punjab	Marginal	5.33	0.8	0.40	50.00
	Small	55.34	2.7	1.30	48.15
	Large	39.33	8.2	4.30	52.44
Rajasthan	Marginal	14.00	0.43	0.4	93.02
	Small	35.33	1.16	0.95	81.90
	Large	50.67	3.41	2.54	74.49
Gujarat	Marginal	02.00	0.8	0.58	72.50
	Small	20.67	1.75	1.13	64.57
	Large	77.33	3.65	3.0	82.19

The experiences of the GGRC in Gujarat indicated that in the recent years more number of small and marginal farmers have adopted the MI
Source Survey data. Palanisami et al. (2011)

boosted the crop yield by 20%, water saving by 25–30%, and energy saving by 15% (Palanisami et al. 2011).

The regression results (Table 9.5) showed that the coefficients of farm size were significant at 1% level, whereas coefficient of dummy variable for small farmers was significant at 10 per cent level and dummy variable for medium farmers was not significant The average farm size of small farmers in the 9 states was 0.91 ha, 2.41 ha, and 8.51 ha for marginal, small, and large farmers respectively. On average,

Table 9.5 Relationship between area under MI, farm size, and category of farms

Variables	Coefficients	Std. error	t-stat	*P-value*
Intercept	1.43	0.57	2.5	0.02
Farm size in ha	0.32	0.056	5.7	0.00
Dummy variable—marginal farmers (D1)	−1.15	0.62	−1.9	0.07
Dummy variable—small farmers (D2)	−0.8450	0.56	−1.5	0.15

Dependent Variable: Area under MI in ha, $R^2 = 0.81$, Adj. $R^2 = 0.79$

each farmer could allot about 0.32 ha of every additional ha of land to MI irrespective of the farm size categories.

9.5 Farmers' Suggestions for Better Adoption of MI Systems

The major suggestions to improve the adoption level of MI by the farmers include: (a) provision of technical support for MI operation and maintenance after installation, (b) relaxation the farm size limitation in fixing MI subsidies, (c) supply of liquid fertilizers at affordable prices, (d) improved marketing facilities, and (e) access to credit to expand the area under MI. In general, small farmers from Andhra Pradesh and Punjab and large farmers from Tamil Nadu state are in need of more technical support for the operation and management of MI systems. Karnataka state farmers demand more liquid fertilizers. Tamil Nadu and Punjab farmers need market facilities for the purchase and repair of MI components (Table 9.6).

9.6 MI and Subsidy

Given the growing water scarcity, state agricultural universities and the agricultural research stations in the country have started conducting research on MI in terms of water saving and yield increase. This has encouraged the state agricultural departments to popularize MI adoption for selected crops. However, the MI system cost is high particularly for the marginal and poor farmers to invest. Realizing this, the central and state government apart from announcing subsidy schemes, coordinate with the MI manufacturers and suppliers from time to time and insisted them on keeping the unit cost as low as possible. Central government also has launched a massive country wide scheme to promote MI, i.e., Centrally Sponsored Scheme (CSS) on MI which came into effect from 2005 to 2006. Now, MI subsidy is an integral part of MI expansion in the country which varies from state to state. In fact, the MI subsidy provided by both the Centre and the States aims to encourage farmers to adopt improved irrigation practices like MI systems on a large scale on a sustained

Table 9.6 Suggestions by the farmers for better adoption and management of MI systems

State	Farmer category	Percentage of farmers reported					
		More Technical support	Supply Liquid fertilizers	Providing Marketing facilities	Credit to cover more area under MI	No farm ceiling	Scientific knowledge of crop production
Andhra Pradesh	M (10)	100.00	11.15	0.00	11.18	33.38	0.00
	S (92)	96.75	5.55	0.00	0.00	10.96	6.62
	L (51)	10.20	0.00	2.05	0.00	56.20	0.00
Tamil Nadu	M (22)	90.40	50.30	99.5	98.00	90.30	50.40
	S (32)	90.90	42.45	60.65	96.95	96.90	48.52
	L (98)	92.75	30.95	97.97	97.89	97.96	49.55
Kerala	M (79)	50.02	7.70	29.54	24.41	20.55	7.72
	S (43)	66.69	9.55	30.97	33.30	4.70	9.50
	L (31)	70.30	0.00	10.40	63.30	73.30	3.35
Karnataka	M (10)	11.15	88.80	11.10	66.65	0.00	0.00
	S (99)	5.08	19.20	22.25	44.44	5.15	4.05
	L (43)	4.77	21.45	23.85	50.05	59.50	0.00
Maharashtra	M (25)	20.20	24.20	16.20	32.20	8.05	88.30
	S (21)	25.05	30.30	90.40	98.5	40.40	70.50
	L (103)	5.751	21.95	54.30	53.30	55.25	50.48
Orissa	M (16)	80.30	40.30	40.05	40.50	6.65	0.00
	S (115)	47.35	12.30	32.50	14.96	25.45	7.91
	L (22)	99.4	85.70	66.65	90.50	80.98	9.50
Punjab	M (9)	0.00	0.00	0.00	99.80	0.00	0.00
	S (85)	93.93	0.00	97.60	95.30	30.15	91.55
	L (60)	89.83	38.98	96.61	94.92	54.25	89.85
Rajasthan	M (26)	40.00	0.00	56.00	20.00	0.00	48.20
	S (51)	46.00	60.00	86.00	64.00	24.50	58.50
	L (76)	1.33	20.00	74.67	80.00	64.40	24.30
Gujarat	M (5)	66.67	33.33	0.00	33.33	0.00	66.65
	S (32)	19.35	19.35	25.81	19.35	12.92	38.72
	L (115)	23.28	11.21	18.10	10.34	39.60	37.10

M = marginal farmers; S = small farmers; L = large farmers
Figures in the parenthesis indicate number of farmers reported under each farm category
Source Survey data. Palanisami et al. (2011)

basis. In the course of time, many states have supplemented the given CSS subsidy budget with state funds from the other programmes. Several states also targeted the MI allocation towards specific regions and beneficiary groups to expand the MI areas. For example, in the case of Tamil Nadu, state government scheme provides 100% subsidy to small- and medium-scale farmers and 75% subsidy to large-scale farmers for micro irrigation (GoI 2021). The subsidy norms followed in different states are given in Table 9.7.

Table 9.7 Subsidy norms followed in different states, India

State	Beneficiary category		Subsidy (%)	
			Drip	Sprinkler
Andhra Pradesh	SC/ST farmers in small and marginal category		100	50
	Small and marginal farmers in other category		90	50
	Medium farmers of Rayalseema and Prakasam districts		90	50
	Medium farmers of coastal districts		70	50
	Other farmers		50	50
Gujarat	SC/ST farmers	* Dark zone area	90	90
		Non dark zone area	85	85
	General farmers** (Small and marginal)	Dark zone area	80	80
		Non dark zone area	70	70
	General farmers*** (above 2 ha)	Dark zone area	70	70
		Non dark zone area	70	70
Maharashtra	Small and marginal farmers of Vidarbha region		75	75
	Small and marginal farmers		60	60
	Other farmers		50	50
Punjab	SC/ST farmers in small and marginal category		90	90
	Other farmers		70	70
Rajasthan	DPAP/DDP	Small and marginal farmers	60	60
		Other farmers	45	45
	Non DPAP/DDP	Small and marginal farmers	45	45
		Other farmers	35	35
Tamil Nadu	Small and marginal farmers		100	100
	Large farmers		75	75
Utter Pradesh	Small and marginal farmers		90	90
	Other farmers		80	80

Note * indicates subsidy rates of MIS or Rs. 1,00,000 per ha, whichever is less; ** indicates subsidy rates of MIS or Rs. 80,000 per ha, whichever is less; *** indicates subsidy rates of MIS or Rs.70,000 per ha, whichever is less
Source Chand et al. (2020)

9.7 The Subsidy Syndrome

Farmers in general still expect a higher level of subsidy for MI systems because of the following reasons: (a) MI is capital intensive as the initial costs could vary from Rs 80,000 to 1.5 lakh per ha depending upon the crops, spacing adopted and type of MI systems (drip or sprinkler) and farmers are reluctant to make their own investment, (b) farmers skills and knowledge in the effective operation and maintenance of the MI systems are poor as often the systems are facing problems like clogging of the filters and drippers; also the required pressure from the pumps is not always maintained due to low pump discharge; (c) MI is not adopted for all crops except for wide spaced and high-value commercial crops, (d) farmers don't see that MI is an immediate need in their farms due to adequate water supplies except in locations where groundwater overexploitation is a problem.

Interactions with different stakeholders showed that actual subsidy realized by the farmers is always less than the prescribed subsidy, as the terms and prices are revised with a 2–3 year lag period and so do not account for inflation. Costs are also higher for closely spaced crops. Around 10% of components are not covered by the subsidies, which have to be paid separately by farmers. In fact, 25–30% of the actual costs of the MI systems are not accounted for in the subsidy calculations. Inefficient design of MIS also inflates the MI installation cost by 5–10%.

It was also observed during discussions with different stakeholders that subsidy syndrome is a major concern among the farmers. Without the subsidy, overall adoption of the MIS would not be able to meet the targets. At present, farmers are unwilling to invest in the systems on their own even if they have the capital, as subsidy is unavailable. Given the delays with receiving the subsidies, adoption of MI has thus slowed down. Further, the delays in the processing of MI subsidy applications, field inspection of the installed systems by the government department officials, and the release of the subsidies could pave the way for corruption at different levels. Hence it is important to see how best these issues could be addressed in the future. In this context the **Gujarat Green Revolution Company's (GGRC)** MI implementation is worth examining here.

9.8 Gujarat Green Revolution Company Ltd (GGRC)

The micro irrigation scheme in Gujarat is implemented by the **Gujarat Green Revolution Company (GGRC)** since May 2005. The GGRC (a Special Purpose Vehicle (SPV) was promoted by Gujarat State Fertilizers & Chemicals Limited, Gujarat Narmada Valley Fertilizers Company Limited and Gujarat Agro Industries Corporation Limited) for the implementation of the MI scheme. The subsidy norms adopted by GGRC are given in Table 9.8, which indicate that dark zones are given more subsidies as an incentive to adopt MI technologies.

Table 9.8 Subsidy norms adopted by GGRC

Sr. No	Category of farmer	Non dark zone area	Dark zone area
1	General Farmer: (2 ha and more)	Upto 70% of MIS cost or Rs. 70,000 per ha, whichever is less	Upto 70% of MIS cost or Rs. 70,000 per ha, whichever is less
2	General Farmer: less than 2 ha	Upto 70% of MIS cost or Rs. 80,000 per ha, whichever is less	Upto 80% of MIS cost or Rs. 80,000 per ha, whichever is less
3	Scheduled caste (SC)/ scheduled tribes (ST) farmers	Upto 85% of MIS cost or Rs. 1,00,000 per ha, whichever is less	Upto 90% of MIS Cost or Rs. 1,00,000 per ha, whichever is less

The scheme provides the hand holding of the farmers by facilitating bank loan as per need, insurance coverage for the system as well as for the beneficiary farmers for one year. Further, farmers can select the type of MIS as per their need, the extent of area to be covered under MIS, and the registered supplier who would install the MI system in their fields which will be subsequently inspected through third party inspection agencies. The MIS supplier, besides providing the MI equipment and agro-inputs, also renders agronomical support services to the farmer for one year to enhance agricultural productivity and also maintain the system for a period of five years. Since the implementation of the scheme by GGRC, (from May 2005 to March 2023) a total number of 13,87,613 farmers have adopted the MIS covering a total area of 21,83,507 ha. Out of this, 18,21,084 ha have been covered under agriculture crops and 3,62,423 ha under horticulture crops. The major crops covered were groundnut, cotton, sugarcane, castor, potato, banana, mango, papaya, and vegetables.

Use of Information Technology (IT) in MIS Implementation

As a part of Digital India, and the aim to provide agricultural services to farmers of the state through SMS, Voice Call, and Door Darshan live programmes, GGRC developed in-house software for implementing Micro Irrigation Scheme in the State.

GGRC has introduced the following methods under IT for more transparency in the scheme:

(1) Barcode: for easy online excess of physical files of beneficiary.
(2) Bio-metric method: for authentication of beneficiary farmer.
(3) Geo-fencing: for authentication of field of the beneficiary farmer before and after the trial-run of the MIS System.
(4) QR code-based GPS enabled mobile application: for verification of Trial run done at the farmer's field.
(5) Colour code: format for identification of MIS type as well as sharing cases.

Whole procedure from registration of MIS application to final payment under the Scheme is through software. Farmer can know the existing stage of his application at his door step through network.

Table 9.9 Impact of MI technologies in water and energy saving and increased yields

Sr. No	Details	Result
1	Saving in irrigation water	30–50%
2	Saving in labour cost in crop cultivation	35–40%
3	Saving in fertilizers **cost**	20–25%
4	Increase in crop yield	20–25%
5	Saving in electrical energy	940 kWh/ha
6	Increase in farm net return Rs./ha	Rs. 15500
7	Average pay back period	1.2–4 crop seasons

Source Author's compilation from the GGRC notes and reports (2023)

(6) GGRC has developed a mobile application for farmers to pre-register their intention for availing benefit under PMKSY—Per Drop More Crop to adopt Micro Irrigation. Through this portal farmer can show their interest in Micro Irrigation Scheme by entering few details, i.e., Name, Village, Taluka, and District by sending SMS or online portal "khedut.ggrc.co.in".

Impact of Micro Irrigation Adoption in Gujarat

The impact of the MI technologies is shown in Table 9.9. The payback period for the MI investment has been within 1.2–4 cropping seasons depending upon the crop income.

9.9 MI Expansion—Challenges and Expectations

In several states, it was observed that changing the subsidy levels has, in fact, encouraged more farmers to go for MI but at the same time, issues relating to fund disbursement and waiting time for availing the subsidy is creating new problems in MI expansion. Farmers have been waiting for the MI subsidy approval and it takes almost 2–3 years in some states by which time farmers lose interest in MI adoption. This long waiting process also encourages corruption at different levels. In states like Tamil Nadu, with 100% subsidy and 5 acres limit, many districts could not achieve the target area due to declining farmers' interest and prolonged delays from implementing departments. However, to address these issues, the government departments such as agriculture and horticulture department started to collect details of the interested farmers who are willing to take up MI, compile the farmers list and allot to different MI suppliers/firms for follow up. There is also the problem of supply where the suppliers/firms find it difficult to cover all the farmers in the list provided by government departments. Many farmers do not show interest in following up with the requirements for MI installation, and even if supplied, they were not prompt in installing them, thus making it difficult for the suppliers to collect their material costs. In some cases, even if the firms are able to supply all the target farmers with the MI equipment, it takes longer time to get the money sanctioned for them. In the long run, the MI installation process is getting distorted and delayed and thus making the MI expansion really challenging.

Challenges in implementation of MI are also due to lack of dedicated and trained personnel that can help the farmers in installation and maintenance. Periodical tracking of installation of the MI system from initiation of work order to installation and payment is a challenging task. Delays in release of changed guidelines/ government orders also affect the speedy implementation of the MI systems. Lack of easy financing mechanisms for farmers is another important constraint in MI adoption as collateral security is needed to avail the finance from private money lenders. Further, lack of knowledge on fertigation, scheduling of irrigation and fertigation, lack of availability of low-cost water-soluble fertilizers and other agrochemicals, and poor quality MI materials are also emerging as big challenges in MI adoption by the farmers.

Poor water control, over pumping, and free electricity further complicate the MI adoption as farmers have no incentives to save irrigation water. However, interactions and feedback from different stakeholders indicated that even though quantum of subsidy will be reduced over years, it is still possible to cover more area under MI as farmers will be interested to adopt MI due to increasing water scarcity. However, these challenges can be easily addressed through appropriate policy interventions which are discussed in the subsequent sections below.

At the macro level, the government agencies are of the opinion that micro irrigation technologies will result in overall water savings thus addressing the water scarcity problems in a region. At the micro level, farmers are interested in maximizing their crop yield, quality of product and income, and minimizing the labour use by adopting the MI.

Further, even though the adoption of MI can result in water savings at the field or farm level, it may not result in same level of savings at aggregate (basin) level. Hence, the mere adoption of MI technologies may not show total water saving at the basin level and it is important to initiate enabling institutional and economic policy instruments for effective management of the water resources (Molden et al. 2001). In this context, it is relevant here to refer to the concept of Jevons paradox where the efficiency not always leads to resource conservation rather it may lead to increase in resource use to produce more number of units (Jevons 1865). The rebound effect of MI technology can be interpreted as if the efficiency increased by "x" per cent then the resource use may also increase or decrease by "x" percent. For example, using the water use case here, if the water use efficiency increased by 6% it led to increase in water use by 4% (Yorka and McGeeb 2015). This is termed as rebound effect. Furthermore, rebound effect of technological efficiency will lead to counterproductive results of the technology's real purpose, which is called Jevons Paradox. The Paradox occurs, when rebound effect is 100%, for instance, if the energy efficiency raises by 10% it leads to increase in energy consumption by 2%. Furthermore, the energy-efficient technologies have not reduced the energy consumption rather they increased the use by 2% than the earlier level. Thus, the economic loss of benefit is 120% (Yorka and McGeeb, 2015; Young et al. 1998).

Micro irrigation (mostly drip and sprinkler irrigation) will reduce water consumption only if it is in phase with restriction on extension of irrigated area (Berbel and Mateos 2014). For example, in the case Northern Basin of Australia, a study concluded that greater the technology change in irrigation higher will be the consumptive demand for land and water because farmers switch to perennial crops with government incentives (Loch and Adamson 2015). However, in the case of India, area under drip irrigation has not increased proportionately to the water saving, thus there is no existence of rebound or Jevons paradox in drip irrigation technology in India (Patil et al. 2015).

9.10 Conclusions and the Way Forward

The poor MI adoption can be attributed to number of factors such as high cost of the MI systems, complexity of the technology, and other socio-economic issues such as lack of access to credit facilities, fragmented land holdings, localized crop pattern, etc. Among other things, reducing the capital cost and increasing the technical know-how will help the spread of the MI in a bigger way. The following cost reduction and capacity building options are important:

i. **Field Level** (*Affordability by the farmers–possibility for cost reduction)*
   The system cost can be reduced by slight modifications in the agricultural practices like paired row planting. Soil texture is important in fixing the emitter spacing which can reduce the system cost significantly.

ii **State Level**

   a. Many a time there is lot of time lag between the decision taken about the subsidy per cent and actual implementation. Any increase in the raw material prices during the lag period will reflect on the actual cost of the system, thus decreasing the actual subsidy per cent at the end users' level. Hence, periodical review of the unit cost is important as is done in few states.
   b. One of the major suggestions rendered by the farmers is improving the technical support. The capacity building of the implementing team is important, which in turn can train the farmers in the use of MI systems including routine operation and maintenance. Fertigation is not done in most of the farms due to lack of knowledge and hence to increase the crop productivity and income, fertigation scheduling should form a component in the MI installation and operation. Capacity building units should be encouraged in each region of the state. The recent experiences indicated that in Tamil Nadu state, the introduction of the TNDRIP capacity building programme covering 100 villages and 1000 farmers has resulted in 17% yield increase and 23% water saving under different crops compared to drip farmers without capacity building activities (Palanisami et al. 2014). The post-installation maintenance of MI

system and training of the farmers should be ensured so that the system works regularly without interruption. The firms supplying the system must be made responsible for the maintenance and supply of spares at least for five years. Training unemployed village youths to reduce time lag in installation and for entrepreneurship development is also warranted.

c. Administrative and operational procedures being followed in few states like Andhra Pradesh, Gujarat, and Maharashtra states are effective and satisfactory in terms of clarity, transparency, and disbursement of subsidy. The functioning of dedicated department promoting MI in these states has been found to be satisfactory and is the main reason for speedy progress of MIS in these states. A special purpose vehicle (an autonomous body) in each state should be created to handle the MI implementation following the success of the GGRC model. Follow up with the existing subsidy norms but adopting fully the CGRC model as a special purpose vehicle with core budget and full autonomy is important. The key elements of such special purpose vehicle will include funding authorities, implementing agency, MI firms, and farmer beneficiaries. In addition, other players include banks and other credit agencies, third party monitoring and evaluating agencies, agriculture extension personnel, MI dealers, and marketing agents (who in some cases represent multiple MI firms).

d. Inferences elucidate that the policy instrument like pecuniary benefits acts as a major determinant for enhancing adoption of MI technologies. Therefore, it is important to relook at the subsidy schemes. Also, other subsidy schemes already implemented will have a higher influence on MI adoption. Discussions with the MI companies and officials indicated that the differential subsidy pattern for different crops being followed in different regions is disturbing the farmers and the implementing agencies. Currently different government departments or agencies are involved in the implementation of the subsidy-oriented MI schemes. Due to the variation in the norms of different schemes which are implemented by different agencies, it is difficult to get the full details as and when required. Incorporation of the MI subsidy in the production itself (like fertilizers) and availability of the materials in the open market comparatively at a lower price than the present rates should be ensured. Many farmers suggested that the subsidy scheme can be removed and instead waiving taxes/providing concessions to manufacturing companies to reduce the purchase price paid by farmers in the open market. The quantum of the concessions needed for each manufacture must be decided based on the volume of MI materials produced, type of materials produced (accessories or main system materials), and quality of materials produced. Uniform standards can be fixed so that all the consumers can get the same quality materials at comparatively cheaper prices in the open market. Provide MI subsidies directly to the farmers, following the model of Direct Benefit Transfer (DBT). The DBT scheme was already launched by the Government of India in 2013. Responses of farmers, field implementing officials, extension officials, MI firms had shown that most of them like the ideas of direct money (subsidy) transfer to the farmers so that more farmers will be able to make investment in MI without waiting so long for the release of the subsidy. The unit price of the MI system components can be fixed and farmers can select their MI suppliers from the list of approved suppliers and install the MI systems in their fields by paying the full cost. Once the system installation is certified by the competent (Government department or third party inspection in terms of technical specifications, area, and costs), the eligible subsidy will be credited directly to the farmer's bank account by the concerned government department in the state. The new approaches show promising responses from the beneficiaries as well as implementing agencies. Appropriate norms including monitoring mechanisms can be set to implement the direct cash transfer model so that only eligible farmers can get benefits. Claiming such direct cash transfer benefits for additional area coverage or replacing the major components of the existing systems can be allowed once in 5 years.

e. Since fertigation scheduling will be an integral part of the MI systems to improve the productivity per unit of water, liquid chemicals and fertilizer should be made available easily at local level at affordable costs. Local production of liquid fertilizers should be encouraged.
f. The government should continue to provide additional incentives to the farmers for a large-scale adoption of MI in the dark or groundwater over-exploited zones along with an appropriate strategy for regulating the groundwater usage for sustaining the resources. While the subsidy rates vary across the states, the hard rock (dark) and over-exploited zone farmers can be entitled to get additional 10 per cent subsidy on adoption of MI. Further, investments in groundwater recharge structures like check dams and percolation and farm ponds will be highly complementary to the spread and adoption of MI systems in India.
g. MI can expand the scope of canal commands particularly where water scarcity is expected to increase in the coming years. Already, several programmes such as Ramthal in Karnataka, Indira Gandhi Nahar Pariyojna (IGNP) Project in Rajasthan have successfully been implemented in canal command areas. This will help to manage the irrigation systems with the planned coverage. In such cases, the solar energy can be appropriately harvested and used for pumping. Hence subsidy for installation of solar systems can be coupled with MI subsidy wherever possible. A sample worksheet is given in Annexure 1 to explain the MI introduction and calculation of water requirements in one of the major canal irrigation projects in Tamil Nadu state.
h. Given the above suggestions/interventions, farmers in India are more likely to go with the adoption of the MI technologies. The above interventions wherever possible could be extended for areas such as soil and crop management technologies as minimizing water and energy footprint is imperative given the goal of producing more with less. It is also important and timely for the researchers to conduct several empirical studies across few states and estimate the positive benefits of such interventions. Such studies could identify pros and cons of different interventions so that spillover effects could be measured,

and possible policy interface could be framed for wide scale diffusion of MI technologies in other states.

Annexure 1: Introduction of Micro Irrigation in the Canal Commands—A Sample Worksheet

Parambikulam Aliyar Project (PAP) is one of the major irrigation systems of Tamil Nadu; which serves the area of more than 4.30 lakh acres of the dry part of western Tamil Nadu. The PAP, interstate (Between Tamil Nadu and Kerala), a multi-valley and multi-purpose project was commissioned in the early 1960s and started functioning from the mid-sixties. Since, PAP is a supply-based irrigation system, the farmers' demand for adequate supply for better cropping is not feasible. This system follows a system of dry irrigated cropping which entails irrigation of a larger area with a given quantity of water (operating at a 120 duty basis). The rainfall in the command area is scanty with an annual average of about 700 mm, which is well below the state average of 960 mm. It is also concluded through different studies conducted so far that difference in productivity per unit of cultivated land with and without irrigation is much larger in this command. In recent years, there has been change in the water allocation management in the system, where the irrigation department controls the supply of water up to distributary/sluices after which farmers themselves manage water through Village Water User's Association (VWUA) functioning since 2000.

Water Realization for Irrigation System

The average quantity of water realized from the group of reservoirs over the past 35 years is around 15 TMC (Thousand Million Cubic feet). Out of 4.30 lakh acres in the entire PAP system, again this has been divided into two river basins namely Aliyar river basin with a command area of about 0.50 lakhs and the Palar river basin (Thirumurthy dam) with a command area of about 3.80 lakhs acres. The Palar river basin has been again divided into four zones system and each zone is being irrigated once in two years.

Existing Irrigation Method

Currently water is conveyed through concrete lined canals (main, branch, and distributaries) and from the sluice point to the field, water is flowing through earthen channel and in the field traditional irrigation methods like ridges and furrow, beds, and channels for annual crops and basins for perennial crops are used for irrigation.

All these traditional methods follow flooding irrigation. To provide irrigation to the 3.80 lakh acres of command area, water is released once in 2 years for 135 days dividing the entire area into four zones. The first two zones are in the first years and remaining 2 zones are irrigated in the next irrigation year (each zone's average command area is almost uniform ranging from 86,152 acres to 98,558 acres).

This is a unique project in the sense that the farmers get water once in 2 years for growing one dry crop with uncertainty in supplies as the supply is decided based on the many other factors including rainfall at catchments and command area, etc. However, according to the farmer's demand, from the year 2000 onwards, irrigation water is supplied by ***Alternative Sluice Irrigation System*** (ASIS) in order to keep the water flows in the main canal throughout the year so that wells get recharged. ASIS is nothing but regrouping of command area under branch canals, distributaries, and direct irrigation sluices, water is allowed in each zone through alternative sluices in main canal, branch canals, and distributaries having more than 1000 acres where main canal will be functional for the entire irrigation year to distribute the water for irrigation to different branches and distributaries.

Cropping Pattern

It is seen from the table below that 29% area is under coconut, 31% under millets viz., maize, sorghum, and cumbu, 11% under pulses, about 14% under oil seeds, 12% under vegetables, and remaining (around) 3% is under cotton.

S. No	Crop	Area (acres)	Percentage
1	Coconut	97,540	29.05
2	Maize	65,270	19.44
3	Sorghum	38,470	11.46
4	Pulses	36,920	11.00
5	Groundnut	32,550	9.69
6	Gingelly	14,495	4.32
7	Onion	10,470	3.12
8	Cotton	9,755	2.91
9	Vegetables (tomato, Brinjal, Bhendi and others	30,315	9.03
Total		3,35,785	100.00

Proposed Cropping Pattern Under Changed Irrigation Method

The quantity of water realized at Thirumurthy dam was reduced from 18 to 14 TMC in the last 20 years and sometimes water realization may even go less than 10 TMC in drought years. The non-agricultural demand as well as 15–20% conveyance losses should also be accounted for within the available supplies. Rainfall received in the command area over the years is also around 650–700 mm. In order to improve the water use efficiency in the water-scarce irrigation system and also to enhance the income of farmers, micro irrigation with the following crops in the canal command is suggested.

- **Low Water Consuming Crops**
  Mango, Sapota, Guava, Amla, Tamarind, Jetropa, and medicinal plants.
- **Medium Water Consuming Crops**
  Coconut, papaya, onion, all vegetables, maize, oil seed crops like groundnut, gingelly, and pulses.

Since, irrigation system is facing shortage of water for distribution to the entire command, water intensive crops like banana, turmeric, and sugarcane cultivation in the command may be discouraged.

Worksheet

Water Use with Proposed Irrigation Method

Water availability for utilization:

Surface water availability in the Thirumurthy reservoir	= 14 TMC
Groundwater availability in the entire command area	= 12 TMC

Assuming that command area as on date is around 3.0 lakhs acres with reduction of 25% area due to urbanization and converted into industrial use, etc.,

Quantity of water available per acre per year = 14,000 M.cft/0.30 M acres

= 50000 cft

Assuming this water is supplied for about 300 days (assuming the rainy days of three months—effectively around 75 days) the quantity available per day

= 50000/300

= 166 cft (or) around 5000 L/acre/day

Irrigation Methods

The water required for matured coconut tree through micro irrigation system is about 60–100 L per day. Variation may be due to season to season and soil moisture content, etc. Among the proposed crops, coconut may require more water. If coconut may be restricted to the 40% of the command area and remaining with other crops which consumes less water compared to the coconut. This calculation has been made without considering the groundwater potential of the location or region or zone. However, it is assumed that 12 TMC of groundwater potential is expected to be available from the command area which can be utilized in the drought years and whenever there is failure of monsoons.

Farmers' Preference for Micro Irrigation in the Command

It is interesting that majority of the farmers in the PAP system are well aware about the success and efficiency of micro irrigation techniques like drip and sprinkler and also drip irrigation method was already extensively adopted for cultivation of coconut crops (currently almost more than 40% of the coconut area is under drip irrigation in the PAP system) and also many other tree crops (mango, sapota, amla, and wood value tree crops) are cultivated under drip irrigation. Hence there may not be much problem in convincing the farmers and shifting the entire land under micro irrigation methods.

It is also high time to use the support extended by central and state governments for drip and sprinkler irrigation systems. In addition to this, financial sectors (agricultural banks) are also interested in funding for these types of activities.

Agricultural and Non-agricultural Benefits

By changing the crop/cropping pattern under micro irrigation method, farmers can cultivate the crops throughout the year instead of one season in two years irrigation cycle under existing water release pattern. Those areas brought under complete micro irrigation technique will be given water throughout the year and this arrangement will definitely help the farmers in managing their livelihood without much uncertainty in the agriculture since water is a crucial input. Groundwater level will improve due to continuous canal water supply. Equity can be achieved in time and quantity of irrigation water distribution among farmers.

Cost of Conversion to Micro Irrigation Technique

The entire command area will come under micro irrigation system and the ratio of drip and sprinkler may be 70 and 30% respectively.

The cost of drip irrigation per hectare for horticulture crops including water storage tank and pumpsets, etc., (well owning farmers may be educated to use their wells as such for storage of canal water and pumped for irrigation) will be around Rs. 40,000 per acre and Rs. 40,000 for development of water storage tank at farm level.

Initially, this modern irrigation water distribution will be implemented in one of the distributaries with a command area of 100–150 acres. The cost of this programme may be about Rs. 1–1.5 crores and based on the experience of this case, further expansion may be taken up. Already some success stories are done at micro level using drip irrigation in the command and hence this project will be successful if implemented properly. Successful implementation schedule is given below.

Case Study

• Sluice		1® of veruvedampalayam Distributary of PMC
• Ayacut area		21.30 Acres
• Proposed Area		21.30 Acres
• No. of Beneficiary		4 Nos
• No. of plants		1600 Nos
• Length of Hytrohole		1600 × 6 M = 9600 Nos
• No of Dripper/emitter		9600/0.50 M = 19200 Nos
• Total Discharge		19200 × 2 Lph = 38400 LPH
• Peak water requirement		1600 × 100 LPD = 160000 LPD
• Operating hours		4.16 h/Day
• Storage tank capacity		((10 + 8)/2) × (10 + 8)/2 × 2.25) = 182250 L
• Total head		(A) Operating head 10 m
		(B) System Head Loss17 m
		(C) Well Depth (Sump depth)2.25 m
	Total Head	29.25 or 30 m
Discharge		Q = 38400 L/3600 = 11 lps
HP 11 * 30/75 * 06(Q * H/ 75m)		7.3 HP 7.5 hp

Sluice (1R) of Veruvedampalayam Distributary of Parambikulam Main Canal

Total Ayacut Area	20.33 Acres
Total Number of Farmers	4 Nos
Total Number of Coconut Trees	1600 Nos
Total Number of Wells	4 Nos
Total Water Requirement/Tree/Day	100 L
Total Water Requirement/Day	1,60,000 L

References

Berbel J, Mateos L (2014) Does investment in irrigation technology necessarily generate rebound effects? A simulation analysis based on an agro-economic model. Agri Syst 128(C): 25–34

Chand S, Kishore P, Kumar S, Srivastava SK (2020) Potential, adoption and impact of micro irrigation in Indian agriculture. Policy Paper 36, ICAR-National Institute of Agricultural Economics and Policy Research (NIAP), New Delhi

GGRC (2023) Note on the current performance of the GGRC Ltd. Gujarat

GOI (2004) Report of task force on micro irrigation, Ministry of Agriculture and Farmers Welfare, Government of India

GOI (2021) Ministry of Agriculture and Farmers Welfare, Pradhan Mantri Krishi Sinchayee Yojna (PMKSY)

Jevons WS (1865) An inquiry concerning the progress of the nation, and the probable exhaustion of our Coal-mines [1865]. Macmillan and Co, London

Loch A, Adamson D (2015) Drought and the rebound effect: a Murray Darling basin example. Nat Haz 27(9):1429 1449

Molden DR, Sakthivadivel R, Habib Z (2001) Basin-level use and productivity of water: examples from south Asia. IWMI Research Report 49, International Water Management Institute (IWMI), Colombo, Sri Lanka

Namara RE, Upadhyaya B, Nagar RK (2005) Adoption and impacts of micro-irrigation technologies: empirical results from selected localities of Maharashtra and Gujarat of India, Research Report 93. International Water Management Institute, Colombo, Sri Lanka

Narayanamoorthy A (2005) Economics of drip irrigation in sugarcane cultivation: case study of a farmer from Tamil Nadu. Indian J Agric Econ 60(2):235–248

Narayanamoorthy A (2006) Potential for drip and sprinkler irrigation in India, Gokhale Institute for Politics and Economics, Pune

Palanisami K, Mohan K, Kakumanu KR, Raman S (2011) Spread and economics of micro-irrigation in India: evidence from nine states. Econ Polit Weekly XLVI(26 & 27)

Palanisami K, Ranganathan CR, Sureshkumar D, Malik RPS (2014) Enhancing the crop yield through capacity building programs: application of double difference method for evaluation of drip capacity building program in Tamil Nadu State, India. Agri Sci 5(1):33–42. https://doi.org/10.4236/as.2014.51003

Patil KK, Chandrakanth MG, Mahadev GB, Manjunatha A (2015) Sustainable path of extraction of groundwater for irrigation and Whither Jevons paradox in hard rock areas of India. In: AAEA & WAEA joint annual meeting, held at: San Francisco, California

Raman S (2010). State-wise micro irrigation potential in India: an assessment, Natural Resources Management Institute, Mumbai

Soman (2023) Field observations on impact of drip and fertigation in different crops. Jain Irrigation Systems, Ltd. Jalgoan, Maharashtra state

Yorka R, McGeeb JA (2015) Understanding the Jevons paradox. Envir Soc 2(1):1–11

Young KJ, Charles AS, Hall GL, Lopez L (1998) Resource use rates and efficiency as indicators of regional sustainability: an examination of five countries. In: Sandhu S, Jackson L, Austin K, Hyland J, Melzian B, Summers K (eds) Monitoring ecological condition at regional scales. Springer, Dordrecht

Chapter 10
Expanding Hill Water Management

Abstract The hilly region presents a paradoxical situation of scarcity amidst plenty on the water front. The hills and mountains are spread over five major regions viz., Himalayas (Western and North-Eastern ranges), Vindhya region, Western and Eastern Ghats, and Satpura ranges. Though sufficient rainfall is received in most parts of the region during monsoon season (June to September), most of it flows down the steep slopes as runoff and is not available for practical use. Improvements in hill agriculture will minimize the irrigation water demand to some extent. Of late, due to various developmental activities and degradation of catchments, water resources in the Himalayan region have been adversely affected. Deforestation, depleting fertility status of soil, erosion of some river banks, degraded common and grazing lands, inappropriate land use, fire resulting from shifting cultivation, fragmentation of land holdings, frequent droughts and floods, uncontrolled mining activities, and depleting groundwater resources are some of the major problems encountered in the hill agriculture. In view of the limited water resource availability in the hilly regions, soil and water conservation, rainwater harvesting and use of micro irrigation are important interventions. Suggestions are made to address the researchers and design engineers, local communities, implementing agencies, and policy makers in making needed interventions to improve the hill agriculture and water management.

Keywords Ecological regions · Hill agriculture · Jalkund · Water harvesting structures · Integrated Watershed Management Programme (IWMP) · Shifting cultivation · National bamboo mission

10.1 Introduction

India has several diverse agro-ecological regions, ranging from sub-temperate to tropical high mountains, plains, plateaus, deserts (hot and cold), hills, coasts, and islands. Hill regions are known for their: (a) monsoon season-based heavy and intense rainfall patterns with surface runoff leading to soil erosion and siltation of the downstream water bodies, and (b) summer season-based severe drought leading to acute

K. Palanisami and U. S. Nagothu, *India's Water Future in a Changing Climate*,
Advances in Geographical and Environmental Sciences,
https://doi.org/10.1007/978-981-97-1785-9_10

water scarcity. The spatial and temporal variations in the occurrence of rainfall and water availability further make the hill regions the most fragile ecosystems that need developmental interventions.

Areas with steep topographic slope, (Himalayan states, north-east regions, hilly tracts across central India) and specific geological structures, cause high runoffs, soil erosion, with little scope for rainwater infiltration due to thin soil layer. Groundwater potential is also limited to the intermountain valleys. Water requirement as well as its availability is wide-ranging and this needs to be considered while planning water management for a specified region, as one water management solution may not fit well to all other situations. Though the local water management practices to a greater extent consider the geological factors into account, the socio-economic profile of the regions have been rapidly changing putting more pressure on the hill ecosystems in the country.

Hilly and mountainous ecosystems in general are very fragile and have variable effects on the agro-climatic zones of the surrounding areas depending on the management of land. India has about 16.2% of total geographical area as the Indian Himalayan Region covering the 10 states of Jammu & Kashmir; Himachal Pradesh; Uttarakhand; Sikkim; Arunachal Pradesh; Nagaland, Manipur, Mizoram, Tripura, Meghalaya entirely, and covering partially the hilly regions of Assam and West Bengal. The variation in the average rainfall in the region is very high as shown in Table 10.1.

Table 10.1 Distribution of annual rainfall across ecological regions of India

Region	Annual rainfall (mm)
Eastern Himalayas (north-eastern states)	2000–3500
Western Himalayas (Jammu and Kashmir, parts of Himachal Pradesh, and Uttaranchal)	1600–2000
Lower Gangetic Plains (West Bengal)	1400–1600
Middle Gangetic Plains (parts of Uttar Pradesh and Bihar)	1100–1400
Upper Gangetic Plains (parts of Uttar Pradesh)	700–1100
Eastern Plateau and Hills (parts of Bihar, Jharkhand, Chattisgarh, Orissa, and Andhra Pradesh)	1200–1400
Central Plateau and Hills (parts of Uttar Pradesh, Madhya Pradesh, and Rajasthan)	500–1500
Western Plateau and Hills (parts of Madhya Pradesh and most of Maharashtra)	600–1100
Southern Plateau and Hills (parts of Karnataka, Andhra Pradesh, and Tamil Nadu)	650–1200
East coast Plains and Hills (eastern coastal regions of Andhra Pradesh and Tamil Nadu)	780–1300
West coast Plains and Ghats (western coastal strip of Maharashtra and Kerala)	2500–3500

Source Compiled by authors; Pangare et al. 2006

Table 10.2 Principal glacial-fed river systems of the Himalayas

River	Mountain area (km^2)	Glacier area (km^2)
Indus	268,842	7890
Jhelum	33,670	170
Chenab	27,195	2944
Ravi	8092	206
Sutlej	47,915	1295
Beas	12,504	638
Jamuna	11,655	125
Ganga	23,051	2312
Ramganga	6734	3
Kali	16,317	997
Karnali	53,354	1543
Gandak	37,814	1845
Kosi	61,901	1281
Tista	12,432	495
Raikad	26,418	195
Manas	31,080	528
Subansiri	81,130	725
Brahmaputra	256,928	108
Dibang	12,950	90
Lohit	20,720	425

Source Hasnain (1999)

The Himalayan glaciers are a source of water in the river systems in India (Table 10.2). The Himalayan ranges are well recognized for impacting the climate of several countries in Asia by creating idiosyncratic climate.

10.2 Hill Agriculture

The challenging environment creates certain disadvantages in terms of basic infrastructure (such as roads), communication channels, irrigation facilities, and other amenities. Yet, with these myriad challenges and limitations, hill regions differ from plains in terms of climate, natural flora and fauna, and cleaner air and water. Due to their geography, hilly regions could be richer in terms of natural resources. There are limitations as well, and the challenge lies in how the limitations can be overcome as we make efforts to promote economic growth in these regions.

Majority of the hill agriculture is rainfed farming based and irrigation practices are mostly supplementation oriented with limited scope for expansion due to socio-economic constraints. Land fragmentation, dispersed holdings, and terraced farming

Table 10.3 Comparison of agriculture in plains and hills

Agriculture in plains	Agriculture in hills
Characterized by commercial farming	Characterized by subsistence farming
Mono cropping is more common	Mixed cropping is more prevalent
Consolidated holdings common	Scattered and fragmented holdings
Irrigated area more (60–70%)	Irrigated area is less (<15%)
Length of growing period less priority	Length of growing period is more important
Seed replacement rate 15–20%	Seed replacement rate 3–4%
Average productivity of major crops high:	Average productivity of major crops low:
Rice 2.77 t/ha Wheat 3.06 t/ha	Rice 1.25 t/ha Wheat 1.35 t/ha

Source Compiled by authors

are the typical features of hill agriculture where failure of crops in one location due to micro-climatic conditions may not fully affect the crops in other locations. Traditionally, hill cropping pattern is based on indigenous knowledge and practices characterized mostly by crop rotation and crop diversity to retain soil fertility and minimize soil erosion risks where the length of growing period influences crop choices. Further, hill agriculture is predominantly organic in nature with minimum external inputs which is mainly due to: (i) small land holdings, (ii) family farming, (iii) indigenous cattle rearing, and (iv) limited accessibility and use of chemicals.

Agriculture in the hills and plains differ in several aspects (Table 10.3). Lands in the plains benefited greatly from the interventions of green revolution while the revolution did not reach the hills. Climate change also affects the hill agriculture to a larger extent. For example, in the hills, whenever *Kharif* season witnesses water stress, the area under rice cultivation falls while area under millets increases. In better monsoon periods, the situation is just the opposite.

10.3 Need for Water Management Technologies for Hills

While much of the water management research, experimentation, and government funding in the past in India focused on the plains, the hilly regions have in the past received comparatively less attention on water management-related issues. Currently, there are special initiatives targeting the hill regions and the most important is the sustainable development pathways of the water resources in this fragile ecosystem. This is particularly apparent in designing irrigation structures for hilly environments with emphasis on sustainability. In addition, mobility and communication is still a challenge and hence maintenance and operations of the small-scale irrigation systems can be difficult.

Type of water resources development and management practices in the mountains and hilly terrain will have an impact on the water availability in the downhill valleys and planes. Thus, there is a need to carefully design the future water management in the hills and valleys treating them as a continuum as any interventions taken upstream will have downstream impacts. While communities living in the hilly areas are susceptible to extreme situations from flash floods to droughts, they have over centuries, developed approaches that are local specific, to handle these extremities and utilize the available resources productively. Some community developed systems in addition to collecting water for irrigation purpose, also act as a controller of flash floods—thus reducing likely damages to life and property in the region. Hence, the community developed irrigation systems in fragile areas (hilly and coastal regions) must be fine-tuned to suit the present and future needs.

There is also a dearth of scientific data on the local water management practices in the hill regions. Hence, there is a pressing need to examine and document the successful practices or systems that have potential for expansion in the hill regions.

10.4 Factors Influencing Water Resources Investment and Management in Hills

The following factors have significant impact on developing and sustaining the water management in the hill regions:

- Climate change with increased spatial and temporal rainfall variability.
- Rainfed agriculture; traditional low input cultivation practices.
- Rugged terrain, steep slopes, alternate ridge and valleys, ever decreasing vegetative cover on hill slopes, prone soil slips.
- Typical geo-hydrology of hill regions with alternate steep slopes and valleys—thin soil layers, loose rocks vulnerable to surface runoff.
- Source of water (other than rain)—streams (perennial or semi-perennial) and springs. Stream water use for drinking and household uses must get priority over its use for supplementary irrigation.
- Springs are fed from sub-surface flow through sub-strata and rock fractures and flow pattern of which depends upon the amount of rainfall and recharge zone/basin characteristics in higher elevations.
- Streams not only vary with the flow rates but also change courses invariably in the event of land slides and cloud bursts.

10.5 Climate Change and Hill Agriculture

Closely related with water management are the climate change impacts in the hill regions. Almost all the regions of the world will experience a net negative impact of climate change with varied intensity and characteristics on water resources and freshwater ecosystems (IPCC 2007). Recent climate change has resulted in significant increases in terrestrial water storage (5.6 gigatonnes/year) in some regions of the Tibetan Plateau and significant decreases in terrestrial water storage (15.8 gigatonnes/year) in others. This is most likely because of the conflicting effects of glacier retreat, deterioration of seasonally frozen ground, and lake expansion (Paurush 2022). The impacts could be in the form of rising frequencies of the two extremes of natural hazards, such as droughts and floods, characterized by increasing number of dry spells and uncertainties in the monsoon arrival. Further, extremely high or low precipitation may increase the risk of water source contamination from sewage overflows, and runoff from agricultural lands and urban areas as well as increased sediment and non-point source pollutant loadings to watercourses. In the year 2009, peninsular India had experienced such a situation with higher uncertainties in the arrival of monsoon rainfall and its distribution pattern due to the "*El Nino*" effect in the Indian Ocean, resulting in crop failure, flood-related causalities, and increasing the risk of poverty and hunger. In reality, 40 million hectares of land in India are thought to be flood-prone and about 10 million hectares are genuinely impacted by floods each year on average. Floods frequently occur because of strong rainfall or persistent rain over an already sopped area (Murari 2001). The topographic undulations and varied altitudes of the hilly terrains further enhance the climatic uncertainties manyfold with multiple pockets of rain-shadow areas.

The hill agriculture is extremely susceptible to climate parameter changes. In the Western Ghats region, the yield of kharif crops such as irrigated rice is anticipated to alter from + 5% to −11% by 2030. Most of the region is expected to have a 4% decline in yields. In the case of rice grown using rainwater, the estimated change in production is 35% with a significant percentage of the region expected to experience yield losses of up to 10%. In the same way, by 2030, climate change is predicted to cut maize and sorghum yields by up to 50%, depending on the region. The North-Eastern (NE) region experiences a seasonal temperature range of 23/31 °C on average, with a little difference in temperature in the southern portion. The range of rainfall in this region, from 500 to 4000 mm, is quite variable. During the kharif season, much of the NE area experiences rainfall from 1000 and 2000 mm. The seasonal lowest and maximum temperatures throughout the month of April are approximately 9 and 24 °C. Winter rainfall in most of the NE region is about 200 mm. An increase of around 2 °C is anticipated in both the lowest and maximum temperatures during the kharif season. The maximum temperature is expected to rise by roughly 1.5 °C in the southernmost reaches of the NE region. It is predicted that the kharif seasonal rainfall would decrease by roughly 10%. With much of the NE area benefiting up to 5% in 2030, climate change may have a negative impact on irrigated rice production of from 5 to 10%. The expected decline in irrigation-related corn yields is roughly 40%. Wheat yields for rabi season harvests are expected to drop by up to 20%. Climate change is anticipated to cause a slight increase in potato yields of up to 5% in the upper NE region, but a projected yield loss of up to 4% in the middle and significantly greater negative effects in the southern parts of the region. It is also anticipated that the irrigated mustard crop will have yield losses of up to 10% (Naresh et al. 2011). Hence, the hilly agriculture in India is very much sensitive to the severity

Table 10.4 Trend in area and yield of major crops in hill regions of India Area in '000 ha, Yield in t/ha

Year	Potato		Orange		Carrot		Small millets*		Apple	
	Area	Yield	Area	Yield	Area	Yield	Area	Yield	Area	Yield
2011–12	614.80	15.58	51.50	7.57	26.35	13.39	162.60	0.84	321.91	2.92
2012–13	641.70	15.64	67.10	5.50	26.68	12.91	155.10	0.85	311.50	4.86
2013–14	673.90	15.59	74.20	5.63	26.96	14.47	145.70	0.85	313.10	6.28
2014–15	679.50	15.50	19.09	4.87	30.10	14.13	136.60	0.91	319.30	5.70
2015–16	673.71	15.83	125.24	6.03	35.94	13.87	141.50	0.87	277.15	6.15
2016–17	664.04	15.43	116.42	5.86	35.50	13.63	120.94	1.28	304.89	5.67
2017–18	662.46	16.87	115.86	5.68	37.90	13.14	132.21	0.89	301.03	5.33
2018–19	714.61	14.71	114.60	6.33	52.97	13.43	128.35	0.91	308.19	5.21
2019–20	706.37	14.99	114.33	5.82	47.23	13.49	133.79	0.98	309.68	6.01
2020–21	717.58	15.74	112.97	5.97	44.79	13.65	120.21	0.88	312.63	5.21
% change between 2021 and 2011	16.72	1.03	119.36	-21.14	69.98	1.94	-26.07	4.76	-2.88	78.42

* Small millets include mainly samai, foxtail, kodo, and barnyard millets
Source Ministry of Agriculture and Farmers Welfare, Govt. of India

of climate change. Thus, it is important to devise and implant strategies that help revive the hill agriculture in general and hill water management in the context of the changing climate. Trends in area and yield of major crops in the hill regions of India have clearly shown that area and yield increase over the past decade is not significant indicating the subsistence nature of hill agriculture (Table 10.4 and Fig. 10.1). It is observed that the area under small millets, apple, and tea has reduced over years, which might be due to changes in land-use pattern wherein other crops should have been cultivated. In the case of yield, except small millets and apple, all the crops had a reduction in yield indicating the vulnerability of the hill crops to climate change.

10.6 Exploring Water Management Options for the Hills

Efficient utilization and management of available rainwater in the hill regions are the core issue if the cropping intensity and production are to be enhanced for enhancing the livelihood options of the households in the region as much of the enormous water resources in hilly areas remain unutilized. As such all the rainwater is lost and leaves farmers with a situation, which occurs in water-stressed areas. However, farmers around the world have exhibited the skill of diverting water from surface streams into canals that deliver water to their fields.

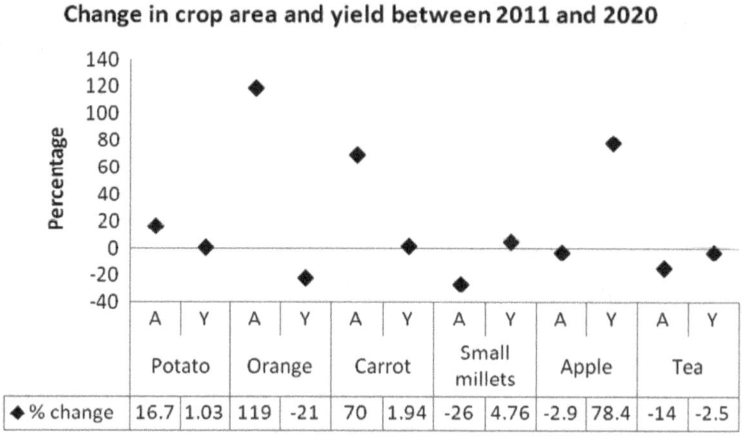

Fig. 10.1 Percentage change in crop area and yield between 2011 and 2020. *Note* A = Area; Y = Yield

The integrated watershed management concept, where community-based sustainable natural resource management is central in the context of hill areas, not only takes a holistic approach to the fast-depleting water resources but also helps in augmenting the resources through runoff management and harvesting techniques thereby ensuring groundwater/spring recharge in the hills. Shallow depressions for collecting rainwater are common in hills (for example, in Uttarakhand it is known locally as *chal/khal*) which are providing water during the dry season for irrigation and for use by animals, and recharging dry springs and streams. It is important to see how different water management practices have been developed in different regions in the past and to what extent they are relevant today. Good research results in water management technologies suitable to hilly regions must be combined with traditional knowledge for proper adoption in the hills.

(a) **Design Process**

In hilly or mountainous areas, the most common building materials are stones and boulders found on the river-bed, earth from the hill side, and forest products from the surrounding area, i.e., brush, branches, leaves, and timber. These materials are placed used to divert all or part of the discharges to one side of the stream, where a channel leads the water out of the riverbed.

The design process in irrigation development has traditionally been the domain of the civil engineers. However, in the hilly terrains, the domain has to be shared with local people with traditional wisdom. This makes the difference in developing proper design for irrigation in the hills. Hydraulic and physical design principles can be used in channeling water from supply source to the root zone of plants. Obvious information essential for carrying out the design process includes the crop water requirement, contribution from rainfall, losses by evaporation and seepage, soil hydrology, historical variations in the available water supply, etc. Less obvious and frequently neglected information is about water rights among systems and development of rules and plans for operation and maintenance of the systems. In the context of irrigation in hill regions, the basic factor that has to be addressed is sustainability. The factors affecting a proper and sustainable irrigation design and its process in the hill regions are listed below.

Cost Effectiveness

A cost effective design takes into account four types of cost (Yoder 1994*)*:

- *Initial construction costs including designing irrigation structures for hilly areas*
- *Recurring maintenance cost*
- *Rehabilitation or replacement cost*
- *Cost to the life or property if the structure fails*

Management Strategy

The failure of many structures is not due to their incorrect hydraulic or structural design but because they cannot be operated or maintained at certain times. In other words, the design should be as simple as possible so that people can understand how to replicate and operate it. A new structure can solve an existing problem but may create an even more challenging situation, if operation and maintenance rules relating to the structure are not properly followed.

Physical and Hydrological Data

Information intensive design procedures especially in hilly regions should be avoided, as it is very difficult to find out data in rough terrain with small command area. Other alternatives such as knowledge of local people should be utilized by involving them as partners in the design. Another option is that under conditions of insufficient data, the design staff should concentrate on structural improvements that are essential. For example, as the small steep gradient mountain streams are prone to flash floods, it may not be possible to design a reliable and permanent diversion weir using existing flood data. A low-cost temporary diversion may be best until further information is available. However, it would be possible to prepare reliable design for the accompanying control structure to limit the entry of floodwater into the canal (Coward et al. 1988).

Resource Availability

Delays in remote areas in terms of mobilizing funds mean production losses. An efficient design is one that helps to reduce the down time of the system. In subsistence economies, low cash needs may be one important element of an efficient design. Similarly, materials should also be available locally. But the replicability depends upon the future availability. For example, wood might have been an important material in irrigation structures of the past, but depletion of forest reserves or restrictions on access to forest products may determine whether wood should be used in new structures. On the other hand, designers tend to forget the material, which has an important role in the construction.

When it comes to human resources, designers should analyse the social conditions and culture. At times, it becomes very difficult to mobilize people for a concrete or major earthen dam. People who live in remote areas are totally unaware of the government flagship schemes. Further, getting skilled workers is anyway a problem in the hills. Hence, before proposing a structure, the local people must be trained. In case of migration, generally men shift to other places for additional earning and the question will be who will maintain the structures (i.e., burden would fall on women).

(b) **Community Participation**

It is important that community members should be involved in design process. Using local knowledge is important. This will enhance their ownership and responsibility to maintain the structures. Overall, a good design in hill areas should have the following five characteristics (Yodar and Thurston,1990): simplicity, equity, affordability, flexibility, and redundancy.

(c) **Types of Structures Used in Hill Irrigation Systems**

Structures for Water Acquisition

1. Gravity diversion

    a. Temporary—i. Brushwood with boulders, ii. Gabion, iii. Boulder filled
    b. Permanent—i. Masonry weir, ii. Bottom intake with trash rack, iii. Side intake

2. Intake discharge regulation—a. Fixed orifice, b. Gate, c. Weir, d. Siphon
3. Sediment removal and exclusion
4. Control device
5. Lift pump house, pumps, pipes, distribution from lift system, settling device before pumping
6. Groundwater/Bed flow

Structures for Water Storage

1. Before diversion: Large/small dam within stream reservoir, water harvesting
2. After diversion: Tanks, ponds, reservoirs fed canal

Structures for Conveyance

1. Canal cross section
2. Canal lining
3. Canal slope
4. Canal supports
5. Energy dissipaters—a. Drop structures, b. Chutes
6. Escape structures (for lining water discharge, not sediment)—a. Designed breach, b. Siphon, c. Gate, d. Weir
7. Cross drains—a. Level crossing, b. Super passage, c. Underpassage d. Inverted siphon, e. Aqueduct
8. Crossings on unstable slopes—a. Pipe, b. Tunnel, c. Covered canal
9. Slope stabilization—a. Retaining walls (upslope), b. Bioengineering, c. Catch drain

10. Drainage—a. To natural (storm) drain, b. To irrigation system,
11. Distribution—a. Open channel system, b. Pressure pipe system: Pressure regulator
12. Field irrigation and drainage—a. Surface (gravity) structures, b. Drip, c. Sprinklers, d. Drainage

The design of any engineering structure must be undertaken with due consideration of the terrain, farmer needs and capacity to manage, and costs. The basic approach should be to provide situation specific solutions and at the same time following the suggestions below:

1. Farmers to be consulted while planning about the type of structure, operation, and maintenance.
2. Traditional practices must be compatible with the new designs.
3. Prioritize traditionally proven and time-tested local design concepts, modify them to suit local situations.
4. Designs must be made with future repair and maintenance needs in mind. Transfer the technology and skills in the beginning itself that would be needed during maintenance of the system.
5. Build small, strong structures in the hills with minimal disturbance to the ecosystem.
6. To understand the real problems, system in operation under its most difficult situations should be monitored.

10.7 Hill Water Management—Case Studies Across India

Review of the hill water management works done elsewhere in the country indicates wider scope for expanding such investment options for enhancing the farmer's income and livelihood opportunities. A brief description of the successful interventions is given in the following sections.

(a) Erosion Control for Enhancing Productivity in Tea in Southern High Hills

Land degradation due to soil erosion is a serious problem in the hills of southern India due to low adoption rate of recommended soil and water conservation measures due to high initial investment costs (Shiferaw and Holden 1998). Effective soil and water conservation (SWC) measures are needed to control soil erosion for not only sustained agricultural production but also for ecological balance of hills. (Fig. 10.2).

Financial efficiency of tea plantation with SWC measures: The result of cost benefit analysis (CBA) over 25 years of tea plantation with SWC measures shows that CST, cover crop and combination of both were found to be financially attractive (Table 10.5).

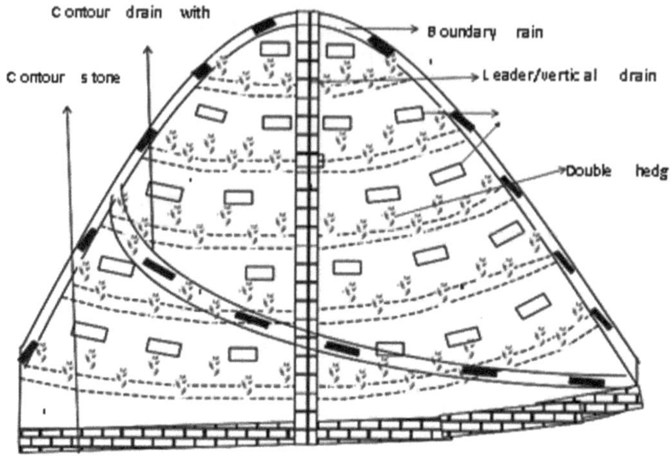

Fig. 10.2 Schematic diagram of conservation measures in tea plantation area

Table 10.5 Financial efficiency of tea plantation with SWC measures

Treatment	BCR *	IRR (%)	Pay Back Period (Year)
Control	1.46	41.25	6
CST	1.48	42.10	5
Cover crop	1.58	49.73	4
CST + cover crop	1.52	47.41	4

*discount rate 10%. CST = Contour Staggered Trenches. BCR = benefit cost ratio, IRR = internal rate of return

Soil and Water Conservation Efficiency

Soil conservation efficiency (SCE) and water conservation efficiency (WCE) measures are presented in Fig. 10.3. Higher WCE was recorded in CST where vegetative measures of *Eragrostiscurvula*(75%), *Cymbopogon Flexuosus*(70%), cover crop of beans, and *Tripsacumlaxum*(56%) were planted. Similarly, higher SCE was recorded in all the conservation measures except cover crop of beans alone and CST with *Tripsacumlaxum* where SCE was the lowest. Overall, the highest soil water conservation efficiency was observed in CST alone, CST with *Cymbopogon Flexuosus, Eragrostiscurvula, and with* cover crop of beans (Madhu et al. 2001).

(b) Water Resource Development and Management in the Indian Himalayan Region

The erosion rate in the Himalayan region covering about 39% area is more than 40 t/ha/year and about 68% of the area has an erosion rate of more than 10 t/ha/year. This calls for appropriate investment programmes to reduce the level of erosion. The cost of micro irrigation systems using small water harvesting structures has varied from Rs. 28,000 to 93,000 per ha. The BC ratios of various water harvesting structures

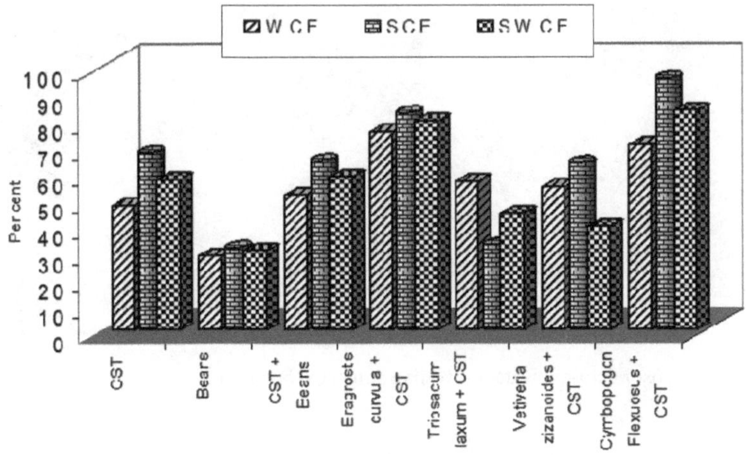

Fig. 10.3 Soil and water conservation efficiency of different conservation measures

varied from 1.48 to 3.89 (Babu 1987). The Sukhomajri Project where an embankment type pond (5.5 ha-m capacity) was constructed in a mini watershed has the BC ratio 1.63 for 30 years of project life at 10% discount rate (Palanisami et al. 2013). The water harvesting project constructed at Kalimati and Bhopalpani villages in Doon Valley of Uttarakhand where interflow was harvested in a pond and conveyed to fields through HDPE pipeline, has the BC ratio of 1.42 considering project life as 30 years and discount rate as 10%.

(c) **Multiple Water Use System for Sustainable Agriculture in North-Eastern Hilly Regions**

Among the NE hilly regions, water scarcity during the post-rainy season particularly from November to April is high in Nagaland. In the absence of major and medium irrigation facilities the minor irrigation provides more scope. A low-cost rainwater harvesting structure called *Jalkund* (Fig. 10.4) of varying capacity from 6000 to 30,000 L has been developed for the hilltops for effective use of water. The *Jalkund*is made up of clay and cow-dung plastering followed by 3–5 cm cushioning, laying down of 250 μm LDPE (low density polyethylene) black agrifilm and covering with 5–8 cm bamboo thatch. For example, a *Jalkund* with 30,000 L of stored water can support about 200 tomato plants, rear five piglets, or two ducks, or 50 poultry birds from November to April.

Economics of Construction of *Jalkund*

Following the costs estimated by Saha et al. 2007, the present cost of the harvested water in *Jalkund*is was worked out (Table 10.6). Considering the total expenditure of Rs. 5356 for preparing a 30,000 L capacity *Jalkund,* the cost/litre of stored water worked out to Rs. 0.18 in the first year and for the three years, the average cost was Rs. 0.06/L owing to less maintenance cost.

Fig. 10.4 Low-cost micro-rainwater harvesting structure (Jalkund)

Table 10.6 Construction cost of a *Jalkund* of 30,000 L capacity

Particulars	Unit	Unit cost* (Rs)	Quantity	Total cost (Rs)
Soil excavation	m^3	40	31	1240
Clay plastering	m^2	3.5	48	168
Cushioning with pine/banana leaf	m^2	3	48	144
Lining with LDPE black agrifilm	m^2	50	72	3600
Thatching	m^2	3	30	90
Fencing	M	3	38	114
Total				*5356*
Unit cost of stored water (1st year)				0.18/L
Unit cost of stored water (2nd year)				0
Unit cost of stored water (3rd year)				0
Unit cost of stored water (average for 3 years)				0.06/L

Source Survey data. *Cost relates to 2020 prices

Similarly, economics of different integrated farming systems utilizing 30,000 L of water from the Jalkund during dry spells is illustrated in Table 10.7.

(d) **Water Harvesting Technologies for Sustainable Farm Production in the Shivalik Region**

The technology of rainwater harvesting from hilly catchment by constructing earthen dam and providing supplemental irrigations to command area under integrated watershed management has proved very effective, sustainable, and eco-restorative. The technology implemented at *Sukhomajri* and *Bunga* in Haryana and *Relmajra* in Punjab was evaluated economically assuming a project period of 30 years. The BC ratios of the project varied between 1.58 and 1.81 proving their economic fitness. Many intangible benefits like eco-restoration, leisure, social security, etc., were not taken into account for calculating BC ratio.

Table 10.7 Economics of different integrated farming systems utilizing 30,000 L of water from the *Jalkund* during dry spells*

Farming system	Tomato-pig integration		Tomato-poultry integration		Tomato-duck-fish integration	
Total water requirement (litre)	Tomato	18,000	Tomato	18,000	Tomato	18,000
	Pig	10,000	Poultry	5,000	Duck-Fish	–
	Misc	2,000	Misc	7,000	Misc	NIL
Total expenditure (Rs)	Tomato	Hybrid seed:500 Other input: 250	Tomato	Hybrid seed:500 Other input: 250	Tomato	Hybrid seed: 500 Other input: 250
	Pig	Piglet:5,000 Feed:4,000	Poultry	Chicks: 750 Feed: 5,000	Duck-Fish	Duckling:30 Fingerlings: 100 Feed: 500
	Jalkund	Construction: 4205 Polypipe:250	Jalkund	Construction: 4205 Polypipe:250	*Jalkund*	Construction: 4205 Polypipe:250
	Total	14,205	Total	10,955	Total	5,835
Total income (Rs)	Tomato	3700	Tomato	3,700	Tomato	3,700
	Pig	20,000	Poultry	Eggs:15,000 Meat:12,000	Duck-Fish	Eggs:1600 Fingerlings: 2,000 Mature fish:1,500
	Total	23,700	Total	Eggs: 18,700 Meat: 15,700	Total	8,800
Profit (Rs)		9495		Eggs: 7,745 Meat: 4,745		2,965
BC ratio		1.67		Eggs:1.71 Meat:1.43		1.51

*The economics is according to the local rates prevailing during 2003–2006 in Shillong, Meghalaya

Economic feasibility of rainwater harvesting structures in *Shivaliks* was done using a village pond of 2 ha-m capacity renovated with proper irrigation network at village *Mandhala*, District *Solan*, *Himachal Pradesh.* A ten hectare of land was brought under supplemental irrigation and crop diversification index of the area has increased from 0 to 0.418 during *rabi* (winter) season. The cost of water storage in the renovated water harvesting structure was found to be Rs. 28.5 m^{-3} whereas, a net return of Rs. 38.2 m^{-3} was obtained (Singh et al. 2008).

Economic analysis of rainwater harvesting based on renovation of ponds for small agricultural watershed was worked out for *Johranpur* project in *Solan* district of *Himachal Pradesh.* The project life was considered as 20 years and the BC ratios were worked out. For the whole project, the BC ratio was 2.38 indicating the economic viability of rainwater harvesting based on ponds in small agricultural watershed (Yadav et al. 2005).

(e) **Potential Water Management Technologies for Sustainable Agriculture in Sikkim and Darjeeling Hills**

This region is completely land-locked and entirely exhibiting steep mountains and deep valleys. The trend of the mountain systems viewed generally is in east–west direction. The area being a part of the inner ranges of mountains of Himalayan systems has elevations varying from 300 to 8598 m. Agriculture in this area is practised on bench terraces designed and structured of varying widths and vertical intervals on such mountain slopes. Interestingly all the agri-horti crops and fodder trees are cultivated on the terraces and on the risers of the terraces depending on the crop. Agriculture in the north-east barring Assam is almost entirely rainfed. The natural slope gradient cannot be greatly altered and has to be used to divert water either directly to a storage tank, or water from the natural springs is tapped to be stored in a reservoir. ICAR Research Complex for NEH Region, Sikkim Centre has undertaken a project on Development of Model Watershed in Sajong, East Sikkim wherein an RCC reservoir of 1000 cu.m of water was constructed by the Watershed Development Society comprising the SHGs of the local community with an expected lifespan of at least 20 years. The reservoir was constructed at a cost of Rs. 2,21,156 (Table 10.8). The cost of construction is higher because all materials for this purpose were purchased and transported to the site including stones and sand thus increasing the overhead charges. The impounded water was used for fishery, irrigation of winter (rabi) crops such as field peas, cabbage, cauliflower, broccoli, mustard, radish, carrot, coriander, and leafy mustard in an area of 10 ha which earlier was winter fallow.

(f) **Enhancing Water Use Efficiency Through Micro Level Water Resources Management in the North Western Himalayan Regions (NWHR)**

Large-scale development of water resources and water distribution in hill regions is difficult. So, the best way is to develop micro-water resource management. Development of water harvesting structures and irrigation needed skills. The cost of pond construction varied with size. The construction cost of 20 and 80 m^3 LDPE lined tanks was around Rs 9,450 and Rs 18,400, respectively. The cost of construction reduced as the size increased. One cubic metre cost comes to around Rs 473 in case of 20 m^3 tank whereas it is only Rs 230 in case of higher size tanks (80 m^3).

The different irrigation systems were evaluated by Srivastava et al. (1994), Srivastava and Bisht (1984 1985) and reported economics of different irrigation systems in NWHR (Table 10.9). The lowest cost is Rs 0.09/m^3 of water in LDE lined tank as against Rs 0.33/m^3 in cement concrete when spring discharge is 20 L/minute. The water cost ranged from Rs 1.58 to 4.21/ m^3 in lift irrigation system. It showed that water collection in the LDPE lined tank is economical in comparison to other system of water collection and delivery.

The irrigation based on different types of catchments revealed that the mini watershed base catchment irrigation system is beneficial followed by the impermeable surface water harvesting (Table 10.10).

Table 10.8 Details of construction of RCC water harvesting tank of 1000 cu.m capacity, Sikkim

S. No	Works	Area (m^3)	Rate/10m^3 (Rs.)	Amount (Rs.)
1	Excavation in foundation trenches in mixed soil, waste disposal, etc	110	790	8,688
2	Plumbing 1:2:4 concrete with 50% clean, hard stone of size not exceeding 15 cm incl. shuttering, compacting, and curing	68.516	17,674	1,21,395
3	Hill cutting in mixed soil incl. waste disposal for 30 m lead for all lifts	41.60	323.9	1,347
4	Plumb line hand-packed stone soiling in building work with clean, hard stone	8.8	3032	2,668
5	Plumb line in 1:3:6 c.c. of specified grade including compacting, curing, etc	13.2	1908	25,185
6	P/L15mm thick cement plaster of specified mix in single cost incl. finishing even and smooth and curing	165.86	741	12,290
7	Carriage of stock materials	154 q	16.7/q	2,572
8	Carriage of non-stock materials-sand and stone	109.72	4312	47,311
			Total	2,21,156

(g) **Drip Irrigation in Citrus Cultivation in Meghalaya**

Majority of the orange cultivation is under rainfed situation in Meghalaya. Farmers are growing Khasi mandarin in the slopes of Khasi and Jayantia hills of the state. Orange plantation is considered a cultural aspect and a traditional plant that has been practised for long by the people. The growing water scarcity during winter and pre monsoon period causes dieback of citrus in the region. Bamboo drip which is a traditional method of irrigation operating with gravity force has provided supplemental irrigation to citrus plantations and also helped in conserving forests and natural resources in hilly areas. Farmers with bamboo drip irrigation systems do not resort to shifting cultivation which involves heavy deforestation. In general, farmers who possess the orchards in the proximity of a stream can very well avail the efficacy of this system. The water utilization efficiency for harvested citrus fruits is about 2–5 kg/m^3 with a moisture content of the fruits of about 85%, except for lime which contains about 70% moisture. A summary of the interventions in different hill regions of India is given in Table 10.11.

Table 10.9 Economics of different irrigation systems in North Western Himalayan Regions (NWHR)

Sl. No	Irrigation system	Investment (Rs/ha)	Annual cost (Rs/ha)	BCR
1	Cement concrete tank utilizing perennial spring with discharge			
(a)	20 lpm	7,095	448	0.33
(b)	10lpm	11,350	740	0.54
(c)	5lpm	22,280	1,450	1.06
(d)	1lpm	54,000	18,970	13.98
2	Lift irrigation with electric pump			
(a)	Lift 50 m	46,800	7,098	1.58
(b)	Lift 80 m	100,000	19,533	4.21
3	LDPE tank utilizing perennial spring with discharge			
(a)	20 lpm	2,023	127	0.09
(b)	10lpm	2,700	173	0.13
(c)	5lpm	3,240	227	0.17
(d)	1lpm	10,800	730	0.53
4	Runoff recycling through LDPE tank mini watershed basis (0.1 ha command unit)	20,000	1,350	1.80
5	Runoff recycling through LDPE Tank impermeable surface based (0.02 ha command unit)	25,000	1,350	1.76

Note lpm-litres per minute

Table 10.10 Evaluation of selected irrigation systems in North Western Himalayan Regions (NWHR)

Irrigation system	Investment (Rs/ha)	Increase in crop production (%)	BCR
Intra terrace based	28,571	25.85	1.33
Mini- watershed based	22,857	41.05	12.44
Impermeable surface based	25,000	27.71	3.54

Source Field survey

10.8 Need for Strengthening Water Management in Hills

It is needless to emphasize that increasing population growth and unplanned development for accelerated economic growth have caused enormous pressure on the ecology and water resources of hilly regions leading to irreversible changes in the land and

Table 10.11 Summary of the interventions in different hill regions of India

Location	Interventions	BCR	IRR %	Cost of stored water
Southern high hills	Contour staggered Trenches (CST)	1.49	42.09	
	Cover crop	1.57	49.72	
	Both	1.51	47.40	
	Control	1.45	41.24	
Himalayan region	Minor irrigation	1.48 to 3.89		
Doon valley	Unlined farm pond	1.85		
Sukhomajri	Embankment pond	1.63		
Kalimati and Bhopalpani, Uttarakhand	Pond, HDPE pipeline	1.42	94.5	
Nagaland	Jalkund	Tomato + poultry: 1.67		Rs. 0.14/L
		Tomato + duck fish: 1.51		
Haryana	Earthen dam	1.58		
Punjab	Earthen dam	1.81		
Shivalik	Village pond	2.38		Rs. 2.85/m^3
North Western Himalayan Region	LDPE lined tank			Rs 1.58 to 4.21/m^3
	Intra terrace-based irrigation	1.33		
	Mini watershed	12		
	Impermeable surface	3.54		
Meghalaya	Drip irrigation	Orange: 400–550 fruit/tree/ha; 25–40 t/ha/yr		
		Water utilization efficiency for harvested yield: 2 to 5 kg/m^3		

water resources in many parts of India and the world. These impacts of human activities may get further accentuated with plausible impacts of climate change which is becoming exceedingly visible.

Changes in land-use and land-cover conditions together with other developmental activities have brought several modifications/alterations in hydrologic cycle leading to uncertain water yields, nutrient loss, sedimentation loss, and loss of productivity and biodiversity and consequently poor socio-economic conditions almost everywhere in the hilly regions. Management of hilly ecosystems is not only crucial for people living in these areas but is also vital in determining the ecology and sustained livelihood of people living in plains. It is rightly said that the wealth of the plains very much depends on the health of the hills.

As indicated, hilly areas frequently experience water scarcity in the lean periods. Though average annual rainfall in the hills is generally higher than the national average of about 1200 mm, water availability after the cessation of monsoon is a matter of serious concern. Even in the North-Eastern hills of India which has the highest rainfall in the country, shortage of water during the non-monsoon period is a regular feature. The perennial water springs in the hills are drying up due to deforestation, excessive grazing, and many developmental activities.

The role of natural processes and their complexities and diversities, especially the impacts of human interferences/activities on the availability of water resource, its sustainable use, and upstream–downstream linkages are poorly understood. Although research institutes and developmental agencies have developed many technologies for water resources management, their wider adaptability and dissemination is still a matter of great concern, especially in the hilly areas. It is mainly attributed to the poor socio-economic conditions, small and fragmented land holdings, and the rugged terrain prevailing in the hills.

10.9 Challenges in the Design and Implementation of Water Management Options in Hilly Regions

Due to high gradient and difficult terrains irrigation in hilly areas has always been a challenge. The rainwater either has to be stored for irrigation at later stage or the stream water/natural spring water has to be properly diverted to meet the irrigation requirements. Therefore, design for structures on the hills is critical and it plays an important role in changing the livelihood pattern of the region. The basic challenges that a designer generally faces or the challenges that should overcome are as follows:

Small Systems

Compared to irrigation systems in the plains, systems in the mountains are relatively small and this is a serious constraint on their economic viability. High construction costs due to difficult access and technical problems associated with unstable mountainous terrain led to high costs per unit of irrigated area.

Water Source

Though water is frequently abundant in the large rivers draining glaciers and snow pack on high mountains, it is seldom accessible on high mountains without pumping. Large rivers tend to be deeply incised and require inordinately long feeder canals for gravity delivery of water to irrigable land. The multitude of cross drains and unstable hill slopes limit the possibilities of tapping the larger, more reliable water sources.

Frequent Floods and High Bed Loads

Hilly watersheds face severe pressure as growing populations remove forest and other vegetative cover. As vegetation is reduced, rainfall runoff is more rapid, increasing the frequency of severe floods. Erosion also increases and the eroded material is moved by floods, causing damage to irrigation structures and clogging canals. Though steep gradient mountain streams have heavy bed loads during floods, the gradient also provides the necessary head for hydraulic removal of the material.

Drainage

Terminal structures for canals are an important design challenge. Excess water leaving a river terrace, for example, frequently cuts a deep gully and damages the fields.

Cropping Pattern

Planning an irrigation system for market-oriented crops may promise the best economic opportunities for a community but in the transition period, if the system does not support crops traditionally grown, it may not be flexible. Hence the design of structures must be flexible enough to take this into account.

Road Access

Unless roads are used for other purposes, their maintenance becomes a costly burden, and they frequently fall into disrepair. Consequently, structures that require materials for maintenance that are not locally available may fail due to the lack of timely repairs.

Design Data

When access to a system is difficult, possibly requiring walking, repeated visits are time consuming and hence avoided. In these cases, collection of design data becomes difficult. Sometimes, it happens that very critical information is missed out. For example, if a stream being used for irrigation has low discharge, except during heavy rainfall, the violent nature of floods that move huge boulders along the bed may be missed and the designed structure may be destroyed in the first flood itself.

10.10 Government Programmes in Hill Water Management

(a) **Watershed Development Project in Shifting Cultivation Areas (WDPSCA)** The programme was put into effect beginning in 1995–1996. The Government of India's Ministry of Agriculture & Co-operation is providing this 100% centrally funded assistance with the goals of protecting the hill slopes of jhum areas through watershed-based soil and water conservation measures and helping jhummia families develop jhum land for productive uses through improved cultivation and appropriate set-up practices that lead to settled cultivation.

(b) **WDC-PMKSY Erstwhile Integrated Watershed Management Programme (IWMP):**

The scheme's central and state shares are 90:10. The main aspects of the IWMP are to preserve and conserve the ecology, restore and develop degraded natural resources by reducing soil loss, improving soil health, promoting water harvesting, recharging ground aquifers, enhancing crop production and strengthening livelihood opportunities.

(c) **NABARD Loan—Soil and Water Conservation Scheme Under RIDF:** The Scheme's primary focus is to increase the productivity of agriculture. In addition to these, the plan aims to encourage sustainable development by conserving and managing the soil and water resources.

(d) **Har Khet Ko Pani Erstwhile Accelerated Irrigation Benefits Programme (AIBP)**
The Central Assistance programme comes in the form of a central grant (90% of the project cost), with the state's share of 10%. The goal of the initiative is to expand the area irrigated, which will raise crop productivity and enhance the socio-economic situation of the people.

(e) **National Bamboo Mission (NBM)**
The Mission aims to enhance the area under bamboo cultivation by encouraging the establishment of new nurseries with high-quality planting material.

(f) **Springshed Development Works**
The Rural Management and Development Department (RMDD), Government of Sikkim, in collaboration with WWF-India and other NGOs, developed the Springshed Development or Dhara Vikas programme in 2008 with the aim of revitalizing and developing the state's mountain springs in a sustainable manner. A multidisciplinary approach is highly needed because springshed management is a trans-disciplinary notion with expertise from different but related fields such as groundwater, watersheds, hydrology, etc.

(g) **Rooftop Rainwater Harvesting Structure in School**
The school's rooftop rainwater collection system aims to offer a source of water for maintaining the kitchen garden, flushing the toilet, cooking, washing hands and feet before eating and after using the lavatory, keeping cleanliness, and eventually for drinking—if the water is treated well.

10.11 Future Options and the Way Forward

On the water front, the hilly region presents a paradoxical situation of scarcity amidst plenty. Of late, due to various developmental activities and degradation of catchments, water resources base in the Himalayan region has been badly affected. Even though sufficient rainfall is received in the hills during monsoon season (June to September), most of it flows down as runoff and is not available for productive use.

The terrace cultivation practices followed in the hills are mostly rainfed and irrigation facilities are very limited in valley areas. Diversion channels and tanks have limited potential since they can only irrigate lands below the water source. As most of the surface water sources in the hilly regions lie below the agriculture fields, technologies like hydraulic ram (hydram) which uses the energy of the falling water to lift it to a level much higher than the original source can be promoted. Properly designed check dams with loose boulder/gabion checks in their upstream for silt detention, ventilated weirs to tap the lean flow in streams, collection tanks and wells for subsurface water harvesting, diaphragm ponds in the mid reaches, and embankment cum dugout farm ponds in the lower reaches can be planned and adopted on watershed basis in the hills for effective water resource development.

In view of the limited water resource availability and low water use efficiency in traditionally adopted irrigation methods, micro irrigation systems such as drip and sprinkler need to be installed on a massive scale to improve the productivity and water use efficiency. Soil conservation measures such as terracing, trenching, bunding, and constructing gully plugs, check dams, etc., should be taken up in the catchment areas to control soil erosion which is a major land degradation problem in the hills.

Priority Setting for Different Stakeholders, Researchers, and Design Engineers

Climate change and extreme weather events is a reality. The current predictions are varying from low to higher levels particularly for temperature and precipitation. Compared to the plains, hill regions may be more susceptible to climate change, particularly, floods caused due to intense rainfall and subsequent impacts will be significant. Hence focused research on the impact of climate change on overall water supply, water use, and the type of storage structures needed should be done. The needed adoption strategies in terms of soil and water conservation, crop pattern, and irrigation methods should be identified and evaluated.

The impact of soil conservation measures, water harvesting methods, and land-use changes on water supplies and changes in flow regimes need to be quantified both at micro and macro-watershed scales. Strong field database on hill water resources and the management options should be created. Low-cost designs should be planned and implementation modes should be made easy involving the local communities.

Economics of different soil and water conservation and management technologies should be worked out for different slopes and soil types.

An irrigation investment plan for the hill regions should be prepared for possible implementation by the government departments.

Local Communities

Social capital will have significant impact in addressing climate change-related impacts both in the short and long runs. Hence how to build the social capital at the village or community level in the hill regions should be examined and needed efforts should be taken up.

Community participation in planning, execution, and maintenance should be ensured along with a community-based institutional mechanism (Water Use Society) for the success of any water resource development programme. As water resource developmental activities run under various governmental programmes, they should be harmonized through proper coordination among different departments to realise desired benefits.

Implementing Agencies

Mass awareness and capacity building programmes in a mission mode involving the local community are very important. The experiences from other locations have clearly indicated that water management technologies along with the appropriate capacity building programmes only will be successful for larger adoption (ITP, 2010). These programmes or interventions will not only increase the water saving at farm level but also enhance the farm productivity on a sustained basis. Hence, all the "technology packages" should have the inbuilt capacity building exercise.

Gender specific capacity building programmes for small-scale micro irrigation kits such as drip and sprinkler for kitchen gardens will have a higher pay-off in the hill regions as water storage and use will be a challenging task in the hills.

Policy Makers

Government programmes should aim at protecting the drying and dying springs in the hill regions. More specifically the *kuhls* should be rehabilitated through special programmes and water delivery should be streamlined from head to tail regions. Public–private sector participation can be ensured wherever possible.

Convergence of the various development programmes in the hill regions is very important as it will avoid the duplication of the investments. Assigning the responsibilities to a single and dedicated agency to implement all types of soil and water conservation technologies will have higher pay-offs through cross learnings.

Several research and development organizations are developing and implementing different types of technologies at micro level with moderate success. Hence the "cluster approach" where all relevant technologies can be piloted in a cluster of villages and based on their performance in 2–3 seasons, up-scaling programming can be taken up at a large scale.

Special incentive package can be provided to the micro irrigation firms who are committed to work and spread the technology in the hill regions along with the system repair and up-keep facilities.

References

Babu R (1987) Economic evaluation of water storage structures. Indian J Soil Cons 15(3):146–147

Coward W, Jhonson R, Walter M (1988) Improving public policies and programs for small scale irrigation development in Asia. Working paper, Irrigation studies group, Cornell University, NY, US

Hasnain SI (1999) Himalayan glaciers—hydrology and hydrochemistry. Allied Publishers, New Delhi, p 234

IPCC (2007) Summary for policymakers. In: Solomon S, Qin D, Manning M, Chen Z, Marquis M, Averyt KB, Tignor M, Miller HL (eds) Climate change 2007: the physical science basis.

contribution of working group i to the fourth assessment report of the intergovernmental panel on climate change. Cambridge University Press, Cambridge, United Kingdom and New York

Madhu M, Sikka AK, Tripathi KP, Ragupathy R, Singh DV (2001) Contour staggered trenching and cover crop for soil and water conservation in new tea plantation in Nilgiris. Indian J Soil Con 29(1):1–6

Ministry of Agriculture and Farmers Welfare (various issues), Government of India, New Delhi

Murari L (2001) Climatic change—implications for India's water resources. J Indian Wat Res Soc 21:101–119

Naresh K, Aggarwal PK, Swaroopa R, Surabhi J, Rani S, Nitin C (2011) Impact of climate change on crop productivity in Western Ghats, coastal and northeastern regions of India. Curr Sci 101(3):332–341

Palanisami K, Sharda VN Singh DV (2013) Water Management in the Hill Regions: evidences from field studies. Bloomsbury Publishing India Pvt. Ltd. New Delhi. ISBN: 978-93-82563-49-5

Pangare G, Pangare V, Das B (2006) Springs of life: India's water resources. Academic Foundation, New Delhi

Paurush O (2022) Freshwater shortage triggered by climate change threatens North India by 2060. https://www.livemint.com/news/india/freshwater-shortage-triggered-by-climate-change-threatens-north-india-by-2060-11660648274917.html

Saha R, Ghosh PK, Mishra VK, Bujarbaruah KM (2007) Low-cost micro-rainwater harvesting technology (Jalkund) for new livelihood of rural hill farmers. Curr Sci 92(9):1258–1265

Shiferaw B, Holden ST (1998) Resource degradation and adoption of land conservation technologies in Ethiopian highlands: a case study in AnditTid, North Shew. Agri Econ 18(3):233–248

Singh SS, Tiwari AK, Singh P Bhattacharya P (2008) Economic feasibility of renovating a water harvesting structure with package & practices of cultivation in Shivaliks a case study of Himachal Pradesh. Indian J Soil Con 36(1): 16–21

Srivastava RC, Bisht KKS (1984) Design specification of irrigation system to utilize low discharge water resources. In: Proceedings, national seminar of water resources management for rural development, Sultanpur, UP, India

Srivastava RC Bisht KKS (1985) Runoff management in hill agriculture. In: Proceedings, national seminar on formulation and appraisal of irrigation projects, Patna India, pp 62–70

Srivastava RC, Bhatnagar VK, Bhatnagar PR, Chandra, Subhash KD, Koranne (1994) Rainwater harvesting and utilization for sustainability of Hill agriculture in U.P. Hills. Paper presented in 8th ISCO Conference, New Delhi, pp 354–366

Yadav RP, Aggarwal RK, Singh P, Arya SL, Singh P, Prasad R, Bhattacharyya P, Tiwari AK, Yadav MK (2005) Rainwater harvesting and recycling technology for sustainable production in small agricultural watershed—Johranpur, Technical Bulletin No. T-50/C-11, central soil and water conservation research and training institute, research centre, Chandigarh Yoder, Thurston (1990) Design issues in farmer-managed irrigation systems. Int irr Manag Ins, Colombo, Sri Lanka

Yoder R (1994) Designing irrigation structures for mountainous environments. In: A hand book of experience. International Irrigation Management Institute, Colombo, Sri Lanka

Chapter 11
Reworking on Cost Sharing and Irrigation Subsidies

Abstract The irrigation subsidies are attributed almost exclusively to the irrigation sector even though many of these systems serve multiple sectors like domestic, industrial, fisheries, and recreation. It is therefore necessary to apportion the capital and operation and maintenance costs (O&M) so that actual subsidies to irrigation and other user sectors could be realistically assessed for suitable policy interventions. Case studies of three irrigation systems in Andhra Pradesh, namely, (i) Sri Ram Sagar Project (SRSP), (ii) Nagajunasagar Project (NRSP)-Right canal, and (iii) Tungabhadra Project (TBP)-Lower level canal were undertaken. The adjusted separable costs remaining benefits (SCRB) method was used to apportion the O&M cost to different uses(sectors). The results indicated that the current total O&M cost reported was Rs 177 per ha in the case of SRSP, Rs 98 per ha in the case of NRSP, and Rs 579 per ha in the case of TBP. However, using the SCRB method, the adjusted O&M cost per ha. for irrigation was only Rs 144, Rs 48, and Rs 274 respectively for these three projects. Accordingly, the actual subsidy per ha. with O&M irrigation component of the projects was Rs 55, Rs 48, and Rs 506 respectively. Hence, separating the irrigation part of the O&M cost and reworking the subsidy in different sectors are recommended in order to plan the irrigation water charges and design effectively the future canal investment programmes.

Keywords Multipurpose irrigation projects · Cost sharing · Separable Cost Remaining Benefit (SCRB) · Irrigation subsidy estimation

11.1 Introduction

The costs per hectare of creating irrigation potential in large and medium irrigation projects have increased from Rs 24,207 per hectare during the first plan period to Rs 156,378 per hectare during the ninth plan period (in constant 1999/2000 prices). Government outlays on major and medium irrigation during the ninth plan period (1997–2002) amounted to Rs 429.59 billion and cumulative expenditures since the first plan period (1951–61) to total Rs 2,557 billion (in constant 1999/2000 prices)

K. Palanisami and U. S. Nagothu, *India's Water Future in a Changing Climate*, Advances in Geographical and Environmental Sciences,
https://doi.org/10.1007/978-981-97-1785-9_11

(Planning Commission). While the costs of providing and expanding the irrigation services have increased, water rates for irrigation water have not increased proportionately. As a result, today the irrigation sector is facing major challenges: First, low irrigation water charges coupled with inefficiencies in the collection of these charges have resulted in irrigation revenues falling short of expenditures. Secondly, over time the accumulation of these financial losses has contributed to increasing fiscal crises in several states in India. Third, the growing deficits have also constrained the ability of state governments to expand and invest in new social and economic infrastructure.

Further, limited revenues from irrigation have resulted in inadequate budget allocation for operation and maintenance (O&M) of the irrigation projects contributing to the deterioration of existing irrigation infrastructures. Not only does dilapidated infrastructures result in lower yields and reduced incomes but it also leads to dissatisfaction among farmers who are more likely to delay or default on payments or resist revisions in water rates. This process perpetuates in a vicious circle compromising the sustainability of irrigation systems (World Bank 1999). Finally, subsidized irrigation water reduces incentives for conservation and encourages the cultivation of water intensive crops thereby contributing to substantial environmental degradation.

Available data on irrigation expenditures from 1985–2000 reveals a steady increase in total expenditures (the sum of plan and non-plan capital expenditures and revenue expenditures) in real terms. In 2000, irrigation expenditure totalled Rs 209 billion. On an average, total expenditure (constant 1999/2000 rupees) has increased by 20% per annum from a comparatively modest outlay of Rs 14 billion in 1985. The ratio of capital expenditures (plan and non-plan) to revenue expenditures has remained constant over time. Total capital expenditures in year 2000 on medium and major irrigation reached Rs 110 billion and accounted for approximately 40% of the total capital outlay on economic services, which includes capital expenditures on agriculture and allied activities, rural development, energy, transport, industry, science, and technology.

Recurrent revenue expenditures on irrigation accounted for about 18% of total revenue expenditures on economic services between 1985 and 2000, and in 2000 total revenue expenditure amounted to Rs 98.6 billion having increased from Rs 5.5 billion in 1985 (constant 1999/00 prices). The rapid growth in revenue expenditures could reflect an expansion in the number of staff employed in irrigation departments and a steep rise in staff salaries and allowances. Revenue expenditures per gross irrigated hectare have increased from Rs 202/ha to Rs 838/ha between 1988/89 and 1996/97 (constant 1999/00 prices). Revenue receipts accounted for Rs 10.3 billion in 2000. Although revenue receipts have grown at an average annual rate of 14% since 1985, they cover only a small fraction of total expenditures (5% in 2000) and revenue expenditures (10% in 2000). Inadequate revenues combined with large increases in revenue expenditures have resulted in large financing gaps.

Owing to increasing costs of expanding irrigation projects to more marginal areas, and higher costs of borrowing, capital expenditures at the national level have increased from about Rs 7.6 billion in 1985 to Rs 110 billion in 2000 (constant 1999/2000 prices). Various Finance Commissions have made strong recommendations on how to price irrigation water and concluded that water rates should at least cover the cost of operation and maintenance (O&M) of irrigation projects plus a part of the capital cost. Despite these recommendations, water charges have generally been set at very low rates. Hence the irrigation subsidies are increasing over years showing the poor performance of the irrigation systems in the country (Agricultural Finance Corporation 2002).

Given the increasing subsidies, it is important to define the planned and actual subsidies where the "planned" subsidy for canal irrigation refers to the difference between O&M expenditures and irrigation revenue demand (assessed irrigation charges) and the "actual" subsidy for canal irrigation water refers the difference between O&M expenditures and revenue receipts (actual water charge collections). The planned subsidy in fact captures the subsidy arising from government policies and reflects the government's water charging practices whereas the actual subsidy captures the subsidy arising from collection inefficiencies and the political economy of the irrigation sector.

Hence, more accurately estimating the level and assessing the targeting of public subsidies to surface irrigation in different states are critical in several important respects. First, a more accurate assessment of who the beneficiaries are will assist policy makers in formulating a more appropriate and equitable cost recovery/water pricing strategy for surface irrigation systems. While many states are moving forward towards increased recovery of operations and maintenance costs and the phasing out of subsidies, these programmes tend to exclusively recover costs from farmers. They generally do not take into account the multi-functionality of these systems (water conveyance, power generation flood control, etc.) nor the appropriate allocation of costs among various other water users (residential drinking water, government, industry, private power generating enterprises). Developing longer term, sustainable, and politically acceptable cost recovery and water pricing strategies, therefore, will require factoring these aspects more effectively.

Second, since the increasing level of subsidies to surface irrigation is contributing to the fiscal crisis in many States in India, the rationalization of public expenditures in irrigation (especially recurrent) will immediately contribute to improved fiscal health in these states. Appropriate cost recovery measures are key elements of such a programme and thus require more careful evaluation. Third, there is a growing debate on the inequities between farmers using canal irrigation and those using electricity to pump groundwater in India, in particular the perceived larger subsidies to canal users. This perception, in turn, is identified as an important obstacle to raising electricity tariffs and therefore power sector reforms. To make appropriate comparisons, however, there is a need to more accurately disentangle how much of the estimated "canal irrigation subsidy" really reaches farmers, what goes to other water users and other beneficiaries of the system (e.g. flood control), and how much serves as "social security" transfers to the irrigation bureaucracy (World Bank 2003).

The process of allocation of costs and subsidies, however, is theoretically complex due to the multiple functions of the canal infrastructure. Developing a practical methodology for the Indian context to sort this out is important not only in assisting state governments in formulating appropriate cost recovery/water pricing policies (for canal and groundwater), but also to pushing forward the dialogue on institutional restructuring/downsizing in the water sector.

11.2 Major Irrigation Projects Covered

Keeping this in mind, a detailed analysis of three irrigation systems in Andhra Pradesh state namely, (i) Nagajunasagar Project (NRSP) Right Bank Canal (mini-hydros, tanks/drinking water, irrigation) situated in the coastal region, (ii) Sri Ram Sagar Project (SRSP) (irrigation, industrial water supply, urban water supply, rural water supply, system tanks, conjunctive use) in the Telengana Region, and (iii) Tungabhadra Project (TBP) Lower Level Canal (drinking water, industrial, conjunctive use) in the Rayalseema Region was done mainly to (a) identify the multiple users of water delivered by the surface systems, and (b) apply the developed methodology to quantify and allocate irrigation water among the multiple users of the irrigation system to quantify and allocate the distribution of irrigation subsidies across the multiple users of irrigation water.

11.3 Methods of Calculating Project Benefits

Each of the methods described and evaluated below are ways of approximating willingness to pay for water project benefits (Young and Haveman 1986).

Method No. 1. Residual Valuation Approaches-Residual valuation follows the task of shadow-pricing by allocating the total value of output among each of the resources used in a production process. It applies to cases where water is an intermediate good and used by producers in the production of other goods and services and is not used for final consumption by households. The method assumes first, that the total value of production can be divided into shares, such that each resource is paid according to its marginal value productivity, and that the share allocation exactly exhausts the total value of output, and second, that market prices of each resource (excepting the one to be shadow-priced) are equal to returns to the resource at the margin.

Consider a simple example where four productive factors, capital (K), labour (L), land (5), and water (W) are used to produce a crop (Q). By the first of the above assumptions, we can write:

$$\text{TVPq} = (\text{MVPk} \times \text{K}) + (\text{MVPl} \times \text{L}) + (\text{MVPs} \times 5) + (\text{MVPw} \times \text{W})$$

where TVPq is the total value of output Q; MVPI represents the marginal value product of the i-th resource, and L, K, S, and W refer respectively to the quantities of labour, capital, land, and water employed in production. By the second assumption, market prices for K, L, and S are substituted for the respective marginal value products. Then the expression can be solved for the remaining unknown, MVPw; which is the estimated shadow price of water.

The limitations of the residual approach should be acknowledged. First, if an input variable and its costs are omitted, its value will be assigned to the residual input, water, thereby overestimating the latter variable's shadow price. In agricultural contexts, land rents and family labour and management costs may be omitted or are at least difficult to quantify. Second, distortions in either prices of non-water inputs or of outputs will bring about a distorted water shadow price.

Method No. 2. Net Incremental Return—This approach is a variation from method 1. It is the method recommended by the U.S. Water Resources Council for calculating irrigation project benefits. It is called the "change in net income" method. Benefits are calculated as the increment in net income or profit from the project based on a with/without comparison often called net primary returns. The calculations are identical to method 1 in the case of irrigation investments, because the "without" case net income is the pre-project net return or rent to the land resource. This approach is used in the present study for irrigation benefits.

Method No. 3: Cost of Next Best Alternative—In this approach, (sometimes called the "alternative cost" method), the willingness to pay is limited by the cost of the most-likely (i.e. least-cost) economically feasible alternative. The method is useful in cases where direct willingness to pay cannot be readily estimated, but where a substantially different economically feasible alternative exists. The alternative must be economically feasible in its own right, or any project purpose becomes justifiable by applying the alternative cost approach.

The benefits are taken to be the cost which would be incurred if the next best alternative were to be used. For example, the thermal power plant might be the next best alternative to hydroelectric power production. The incremental cost of energy produced by a thermal plant of equal capacity is the benefit ascribed to the hydropower facility. In this study, the alternative cost approach is heavily relied upon. It is found to be the method most appropriate for hydropower, navigation, and urban water supply.

Method No. 4. Avoided Damages—In certain cases, where a project permits avoidance of potential economic damages from either natural or manmade hazards, benefits are taken to be the expected value (i.e. the probability-weighted value) of monetary damages or costs prevented by the project. For example, a flood control project permits floodplain dwellers to avoid or reduce the probability of damages to homes or businesses. The benefit of avoiding these damages is infrequent, so the monetized damage-avoided measure of benefits must reflect the probability of damaging events.

Application of Methodologies for Estimating Benefits

The selection of the particular method of benefit calculations for allocating the costs depends on three aspects: (1) the project purpose and the type of benefits realized, (2) the needs and preferences of those performing the cost allocations, and (3) the availability of required data.

Method 3 is chosen for estimating system benefits from the use of dams for navigation, water supply, and hydroelectric power generation. This method is based on the cost of providing the service compared with the next best available alternative. The method requires data which is easy to obtain.

Method 2 is used for evaluating agricultural benefits because net returns can be calculated from the available data.

Method 4 suits situations where flood control is the focus because damage estimates are a measure of the value of the project to the users.

Method 1 is used for commercial fish harvests, which are valued on the basis of market prices. Due to a lack of cost data, the gross value of fishery revenues will be used.

Required cost data included the operation and maintenance expenditures (past 3–5 years) as well as the capital, interest, and depreciation costs since the inception of the three projects. The other large data is related to the benefits of each project that is identified as being significant. The major uses identified are: (1) domestic water supply (rural and municipal); (2) water supply for mini-hydropower stations; (3) industrial water supply; (4) irrigation; (5) flood control; (6) fisheries; (7) groundwater recharge.

Other Uses of Irrigation Water

In addition to irrigation, the irrigation projects also provide water supply for drinking (rural as well as urban), industry and for generating power through mini-hydel plants established on the canal systems. While drinking water supply is common in all the three systems, mini-hydel power is generated only on NRSP and SRSP. Unlike the water used for drinking and industry purposes, water used for hydel power generation is not consumptive. Water is diverted to generate power and released back into the canal systems for irrigation purposes. Fishing is a complementary activity in most of tanks and reservoirs (Table 11.1).

Benefit Quantification

Irrigation: increase in average net income from use of canal irrigation.

Domestic water supply: well and incremental cost of a deep aquifer supply; cost of water delivered by tankers from other locations.

Mini hydros: cost savings over thermal energy alternative.

Commercial fishing: average net income from fish harvests.

Industrial water supply: well and incremental cost of a deep aquifer supply.

Table 11.1 Allocation of water for different uses in the three projects (per year) (TMC)

Use	NRSP	SRSP	TBP
Drinking Water Supply (Rural + urban)	5.6	1.5	2.32
Industry	0.36	5.0	3.05
Irrigation	255	119	139*

* Total allocation for Andhra Pradesh and Karnataka states is about 230 TMC

Flood control: % of total irrigation benefits.

Groundwater recharge: additional area irrigated due to groundwater recharge and the resulting net returns from the area irrigated.

In addition to these benefits other potential benefits identified during the field visits included roads (canal banks), tourism (boating in reservoirs, amusement, and wildlife parks), water for livestock, water for bathing and washing clothes.

Regarding the alternative costs, the annualized capital cost of the project is used as this will reflect the probable expenditure that could be made in providing such facility on annual basis. Also 3% of the O&M cost is added with the annualized capital cost to arrive at the alternative costs for each multiple use type.

11.4 Allocation of Costs

Application of Separable Cost Remaining Benefit (SCRB)

While discussing the need to allocate costs, a distinction should be made between the cost allocation and cost sharing. Cost allocation is mostly concerned with the distribution of total project costs among the various users served by the project. Cost sharing on the other hand is the manner, in which costs actually are shared by the users of the project, with costs not necessarily being allocated in accordance with benefits. In general, the differences between cost allocation and cost sharing plans mainly arise from administrative policies.

The procedure designed to distribute the costs required to build, maintain, and operate the joint aspects of a modern resources system among those who benefit from its use should be based on sound economic principles of efficiency and equity. Once it is determined that a multipurpose project is economically justified, the costs of the project should be allocated equitably among the sectors benefiting from the project.

A review of the literature on cost allocation for multipurpose irrigation projects prepared for this study (Gittinger 1982; Sur et al. 2002; Easter 2003) revealed three most commonly discussed and used cost allocation methods: (1) use-of-facilities (UOF) method (where joint costs are allocated based on the relative use of the common facility for each purpose); (2) alternative justifiable expenditure (AJE) method (joint costs are allocated based on remaining benefits after subtracting out specific costs); and (3) separable costs remaining benefits (SCRB) method (joint costs allocated based on remaining benefits after subtracting out the separable costs). Separable costs are the difference between the cost of a multiple-purpose project with the purpose included and the cost without it. Joint costs for the AJE method are equal to the total project costs minus specific costs, whereas for the SCRB method they are equivalent to total project costs minus separable costs. In selecting the most appropriate methodology three criteria have to be set. First, the cost allocation should be reasonable, understandable, and transparent for all users. Second, required data should be available to apply the method at a reasonable time and cost. Finally, the allocation should be reasonable and also efficient in that: (a) total costs allocated for each purpose should not exceed total benefits; (b) total costs allocated for each purpose should not exceed the cost of a single purpose alternative providing equivalent benefits; and (c) the separable cost of adding each purpose as the last increment should not exceed its benefits (Loughlin 1977).

Although the AJE method could not necessarily meet the efficiency criteria, separable costs and specific costs are basically the same in the case of the irrigation projects in Andhra Pradesh state. Further, all the costs related to canals and reservoirs can be viewed as joint costs. The canals convey water for irrigation and also supply water to protected tanks for drinking water, water to industries, and hydropower plants. The reservoirs store water for irrigation and drinking purposes as well as provide flood control benefits and support fisheries.

Further, the adjusted separable costs remaining benefits method was developed mainly to adjust for the potential inequity in the SCRB formula by applying a credit to the separable costs so that separable costs are subtracted from the justifiable costs on a greater than a 1: 1 basis. The justification for applying a credit to the separable costs in allocating joint costs is that the separable costs share in the benefits provided by the joint costs. This improved method provides better results than the SCRB method for meeting the equity criterion.

After evaluating the existing methodologies, the Adjusted SCRB method was applied for the cost sharing study.

Categorizing Costs

The allocation of the project costs implies that it is important to attribute the costs precisely to the user sectors where benefits are expected. Thus, as far as possible, specific and separable costs are assigned to those sectors using a project service, and the remaining joint costs then are allocated. Specific costs are those which are clearly linked with a particular user group or beneficiary category of the project such as domestic, industry, fishery, etc. For example, costs associated with canals which supply only irrigation water are specific costs to irrigated agriculture. Similarly, a power generation is clearly a specific cost to energy sector. Separable costs normally refer to those incremental investments in a multipurpose project which result from the inclusion of a particular use. Separable costs for a particular use/service include needed changes in the project design to include that service as a specific purpose.

Relationship Between Cost Allocation and User Fees

The main purposes of the cost sharing in irrigation projects are: (1) to generate revenue that could pay for the investment, and (2) to increase the project efficiency. The water allocation could reach optimum level, when prices charged in each use equal the marginal cost of the services provided by the project. In a multipurpose irrigation project, an equitable allocation of the costs to all users is important for efficient system operation and maintenance. The usual norm is to formulate water rates or fees which will distribute costs among all users more or less in proportion to the benefits.

The results of the adjusted SCRB procedure adopted for the three projects are given in Tables 11.2, 11.3 and 11.4.

In the case of SRSP, out of the total capital and O&M expenditure of Rs 11,415 lakhs, irrigation component alone accounts for Rs 9512 lakhs (83.3%) followed by industry (4.35%), flood control (4.07%), hydropower (3.57%), drinking water supply (2.27%), and fishing (2.39%). When only O&M component is considered, irrigation component accounts for Rs 201 lakhs out of the total O&M expenditure of Rs 247.6 lakhs (81%) followed by flood control (6.8%), industry water supply (3.7%), and hydropower (3.67%).

In the case of NRSP, the capital and O&M for irrigation component accounts for Rs 4326 lakhs against the total O&M of Rs 6917 lakhs (62.58%), followed by hydropower (29%), drinking (4.13%), flood control (3.27%), and fishing (0.52%). However, when only the O&M component is considered, the irrigation accounts for Rs 462 lakhs against the total O&M expenditure of Rs 951.7 lakhs (48.6%). The share of hydro, flood, drinking water, fishing, and industry supply is 43%, 4.85%, 2.45%, 0.77%, and 0.30% respectively.

In the case of TBP, the capital and O&M for irrigation component accounts for Rs 4106 lakh against the total of Rs 8693 lakhs (47.2%). The share of industry water supply is 22.91%, drinking water supply is 17.7%, hydropower is 4.94%, flood 4.86%, and fishing 2.32%. In the case of total cost represented by only O&M expenditure, the share of irrigation component is Rs 564.7 lakhs against the total O&M expenditure of Rs 1195.6 lakhs (47.2%). The share of other multiple uses is drinking water 17.7%, industry supply 22.91%, hydropower 4.94%, flood 4.86%, and fishing 2.32%.

In summary, it could be mentioned that the actual share of total O&M cost related to irrigation is ranging from 47% in TBP to 83% in SRSP, which helps to identify the other multiple use sectors which enjoy the sizeable share of the total O&M expenditure of the irrigation projects. This will be useful in analysing the exact subsidy that could be attributable to irrigation sector. It is also important that other multiple use sectors such as hydropower, drinking water supply, industry water supply, and fishing could be proportionately addressed to examine the subsidy level and to recover the portion of the total O&M cost spent.

11.5 Calculation of Irrigation Subsidies

The estimates implicitly define irrigation subsidy as the difference between cost of supplying water for irrigation (O&M costs) and the revenue received as payment from the users of irrigation water; it is presumed that the losses of the irrigation department are on account of supplying water at concessional rates.

Quantifying the Irrigation Subsidy

The concept of irrigation subsidy adopted in this study is based on the difference in the cost of supplying irrigation water and what the farmers pay for irrigation water as its

Table 11.2 Cost allocation using adjusted SCRB method—Sri Ram Sagar Project (SRSP). SRSP-capital and O&M costs (Rs lakhs)

Items	Irrigation	Hydro	Drinking	Industry	Flood	Fishing	Total
Benefits	18,683.9	629.3	270.82	925.35	934.19	338.52	21,782.08
Alternative costs	11,502	504.19	271.41	519.12	574.74	1149.48	14,520.94
Justifiable costs	11,502	504.19	270.82	519.12	574.74	338.52	13,709.39
Separable costs	7258.55	0	214.88	417.12	0	0	7890.55
Cost for all other purposes	4156.45	11,415	11,200.12	10,997.88	11,415	11,415	3524.45
Justifiable cost for purpose	2207.39	13,205.2	13,438.57	13,190.27	13,134.65	13,370.87	
Justifiable cost for all other purposes	2207.39	11,415	11,200.12	10,997.88	11,415	11,415	
Adjustment factor	1.21	1.044	1.00	1.01	1.05	1.03	
Adjusted separable costs	8717.50	0	215.93	420.85	0	0	
Remaining benefits	2784.50	504.19	54.89	98.27	574.74	338.52	4355.11
Joint cost proportions	0.64	0.12	0.01	0.023	0.13	0.078	1
Allocated joint costs	2253.40	408.02	44.42	79.53	465.12	273.95	3524.45
Total allocated costs	9511.96	408.02	259.30	496.65	465.12	273.95	11,415
%	83.33	3.57	2.27	4.35	4.07	2.40	100

SRSP- O&M ONLY

Items	Irrigation	Hydro	Drinking	Industry	Flood	Fishing	Total
Benefits	18,683.9	629.3	270.82	925.35	934.19	338.52	21,782.08
Alternative costs	11,167	504.19	271.41	519.12	1149.48	1149.48	14,760.68
Justifiable costs	11,167	504.19	270.82	519.12	934.19	338.52	13,733.84
Separable costs	0	0	0	0	0	0	0

(continued)

Table 11.2 (continued)

Items	Irrigation	Hydro	Drinking	Industry	Flood	Fishing	Total
Cost for all other purposes	247.6	247.6	247.6	247.6	247.6	247.6	1485.6
Justifiable cost for purpose	2566.84	13,229.65	13,463.02	13,214.72	12,799.65	13,395.32	
Justifiable cost for all other purposes	247.6	247.6	247.6	247.6	247.6	247.6	
Adjustment factor	46.10	3.04	2.09	3.10	4.77	2.37	
Adjusted separable costs	0	0	0	0	0	0	
Remaining benefits	11,167	504.19	270.82	519.12	934.19	338.52	13,733.84
Joint cost proportions	0.81	0.04	0.02	0.038	0.07	0.02	1
Allocated joint costs	201.32	9.09	4.88	9.36	16.84	6.10	247.6
Total allocated costs	201.32	9.09	4.88	9.36	16.84	6.10	247.6
%	81.31	3.67	1.97	3.78	6.80	2.46	100

direct price. Like the method adopted in the estimation of power subsidies, irrigation subsidies represent the perspective of the supplying agency. Accordingly, the focus is on estimating the cost of public irrigation water through major and medium irrigation schemes and the payments made by farmers for irrigation water. The cost of irrigation through major and medium schemes comprises the following: (a) the capital cost and (b) the working cost of operation and maintenance (O&M). The O&M costs are added to capital cost, and then payments made by farmers for irrigation water are deducted to obtain relevant public irrigation subsidy, for major and medium irrigation. The adjusted figures were derived using the adjusted Separable Cost Remaining Benefit (SCRB) procedure adopted for each of the three projects studied.

Table 11.3 Cost allocation using Adjusted SCRB method for Nagarjunasagar Project. NRSP-capital and O&M costs (Rs lakhs)

Items	Irrigation	Hydro	Drinking	Industry	Flood	Fishing	Total
Benefits	136,212.66	5760.00	1011.05	66.62	6810.60	98.42	149,959.35
Alternative costs	6144.26	5438.40	311.47	37.90	614.40	614.40	13,160.83
Justifiable costs	6144.26	5438.40	311.47	37.90	614.40	98.42	12,644.85
Separable costs	3993.77	0.00	272.10	30.20	0.00	0.00	4296.07
Cost for all other purposes	2923.23	6917.00	6644.90	6886.80	6917.00	6917.00	37,205.93
Justifiable cost for purpose	6500.59	7206.45	12,333.38	12,606.95	12,030.45	12,546.43	
Justifiable cost for all other purposes	2923.23	6917.00	6644.90	6886.80	6917.00	6917.00	
Adjustment factor	1.31	1.79	1.01	1.00	1.09	1.01	
Adjusted separable costs	5235.43	0.00	273.65	30.23	0.00	0.00	
Remaining benefits	908.83	5438.40	37.82	7.67	614.40	98.42	7105.54
Joint cost proportions	0.13	0.77	0.01	0.00	0.09	0.01	1.00
Allocated joint costs	335.23	2005.99	13.95	2.83	226.63	36.30	2620.93
Total allocated costs	4329.00	2005.99	286.05	33.03	226.63	36.30	6917.00
%	62.58	29.00	4.14	0.48	3.28	0.52	100.00
NRSP- O&M ONLY							
Items	Irrigation	Hydro	Drinking	Industry	Flood	Fishing	Total
Benefits	136,212.66	5760.00	1011.05	66.62	6810.60	98.42	149,959.35
Alternative costs	6144.26	5438.40	311.47	37.90	614.40	614.40	13,160.83
Justifiable costs	6144.26	5438.40	311.47	37.90	614.40	98.42	12,644.85
Separable costs	0.00	0.00	0.00	0.00	0.00	0.00	0.00

(continued)

Table 11.3 (continued)

Items	Irrigation	Hydro	Drinking	Industry	Flood	Fishing	Total
Cost for all other purposes	951.70	951.70	951.70	951.70	951.70	951.70	5710.20
Justifiable cost for purpose	6500.59	7206.45	12,333.38	12,606.95	12,030.45	12,546.43	
Justifiable cost for all other purposes	951.70	951.70	951.70	951.70	951.70	951.70	
Adjustment factor	7.46	6.71	1.33	1.04	1.65	1.10	
Adjusted separable costs	0.00	0.00	0.00	0.00	0.00	0.00	
Remaining benefits	6144.26	5438.40	311.47	37.90	614.40	98.42	12,644.85
Joint cost proportions	0.49	0.43	0.02	0.00	0.05	0.01	1.00
Allocated joint costs	462.44	409.31	23.44	2.85	46.24	7.41	951.70
Total allocated costs	462.44	409.31	23.44	2.85	46.24	7.41	951.70
%	48.59	43.01	2.46	0.30	4.86	0.78	100.00

11.6 Calculation of the O&M Cost Components

Four scenarios are used to derive the O&M cost components, viz., (I) full O&M costs of the projects, (ii) O&M Cost of the projects adjusted to irrigation component alone. These adjusted figures are derived using the adjusted Separable Cost Remaining Benefit (SCRB) procedure; (iii) 10% of the capital cost and full O&M cost and (iv) 10% of the capital cost plus O&M Cost of the projects adjusted to irrigation component alone. The magnitude of the O&M costs derived is given in Table 11.5.

Quantification of Irrigation Subsidies

The estimates of irrigation subsidy so derived for the three projects have differed among the projects. Since data were available only on selected WUAs and DCs, the subsidy was calculated on per hectare basis. Then total subsidy was worked out using the weighted subsidy (per hectare basis) multiplied by the total command area. These estimates of irrigation subsidy are presented in Tables 11.6, 11.7, 11.8 and 11.9.

Table 11.4 Cost allocation using adjusted SCRB method for Tungabhadra Project. TBP-capital and O&M costs (Rs lakhs)

Items	Irrigation	Hydro	Drinking	Industry	Flood	Fishing	Total
Benefits	22,356.90	1025.00	418.86	564.46	1117.84	32.20	25,515.26
Alternative costs	654.00	68.50	245.14	317.24	67.36	67.36	1419.60
Justifiable costs	654.00	68.50	245.14	317.24	67.36	32.20	1384.44
Separable costs	425.10	0.00	123.20	184.80	0.00	0.00	733.10
Cost for all other purposes	8267.90	8693.00	8569.80	8508.20	8693.00	8693.00	51,424.90
Justifiable cost for purpose	730.44	1315.94	1139.30	1067.20	1317.08	1352.24	
Justifiable cost for all other purposes	730.44	1315.94	1139.30	1067.20	1317.08	1352.24	
Adjustment factor	0.16	0.16	0.16	0.16	0.16	0.16	
Adjusted separable costs	67.70	0.00	19.62	29.43	0.00	0.00	
Remaining benefits	586.30	68.50	225.52	287.81	67.36	32.20	1267.69
Joint cost proportions	0.46	0.05	0.18	0.23	0.05	0.03	1.00
Allocated joint costs	3681.41	430.12	1416.05	1807.17	422.96	202.19	7959.90
Total allocated costs	4106.51	430.12	1539.25	1991.97	422.96	202.19	8693.00
%	47.24	4.95	17.71	22.91	4.87	2.32	100.00

(continued)

Table 11.4 (continued)

Items	Irrigation	Hydro	Drinking	Industry	Flood	Fishing	Total
TBP- O&M ONLY							
Items	Irrigation	Hydro	Drinking	Industry	Flood	Fishing	Total
Benefits	22,356.90	1025.00	418.86	564.46	1117.84	32.20	25,515.26
Alternative costs	654.00	68.50	245.14	317.24	67.36	67.36	1419.60
Justifiable costs	654.00	68.50	245.14	317.24	67.36	32.20	1384.44
Separable costs	0.00	0.00	0.00	0.00	0.00	0.00	0.00
Cost for all other purposes	1195.60	1195.60	1195.60	1195.60	1195.60	1195.60	7173.60
Justifiable cost for purpose	730.44	1315.94	1139.30	1067.20	1317.08	1352.24	
Justifiable cost for all other purposes	730.44	1195.60	1139.30	1067.20	1195.60	1195.60	
Adjustment factor	1.16	1.06	1.16	1.16	1.06	1.03	
Adjusted separable costs	0.00	0.00	0.00	0.00	0.00	0.00	
Remaining benefits	654.00	68.50	245.14	317.24	67.36	32.20	1384.44
Joint cost proportions	0.47	0.05	0.18	0.23	0.05	0.02	1.00
Allocated joint costs	564.79	59.16	211.70	273.97	58.17	27.81	1195.60
Total allocated costs	564.79	59.16	211.70	273.97	58.17	27.81	1195.60
%	47.24	4.95	17.71	22.91	4.87	2.33	100.00

Table 11.5 Comparison of the O&M Costs in Different Projects (Rs / ha)

Cost type	SRSP	NRSP	TBP
O&M	177.7	98.42	579.46
O&M adjusted for irrigation component	144.49	47.82	273.72

Table 11.6 Details of demand, collection, and subsidies for SRSP

Details	Demand (Rs/ha)	Collection (Rs / ha)	O&M (Rs/ha)	Planned subsidy (Rs/ha)	Actual subsidy (Rs/ha)
O&M	354.79	89.49	177.71	−177.08	88.22
O&M IRRIGATION	354.79	89.49	144.49	−210.30	55.00
CAPITAL + O&M TOTAL	354.79	89.49	979.18	624.39	889.69
CAPITAL + O&M IRRIGATION	354.79	89.49	794.50	439.71	705.01

In the case of SRSP, the distributory committees (DCs) were compared and the actual subsidy with the O&M component was Rs 88 per ha among the DCs (Table 11.6). The proportion of the actual subsidy has been low when the O&M was adjusted for irrigation component. Similarly, the subsidy was on the higher side when the capital cost was added. In the case of NRSP, the subsidy was worked out using the WUAs located in the head, middle, and tail end of the system. Comparing the 192 WUAs in the NRSP-RC and the actual irrigated area, the planned subsidy was negative indicating that the existing water charges are already higher than the O&M expenditure per hectare. However, when comparing the actual subsidy which is the difference between the actual collection and the O&M expenditure, it is about Rs.98 per hectare indicating that the recovery is almost negligible among the water users association (Table 11.7).

However, since the total O&M cost is accountable not only to irrigation but to other benefits (multiple users), it is reasonable to workout the subsidy only taking into account the irrigation component. Using the separable cost and remaining benefit method, the actual O&M component attributable to the irrigation component was worked out. The planned subsidy is zero and the quantum of actual subsidy has reduced almost by half in the head, middle, and tail locations which is only about Rs. 47.6 per hectare.

Since it is argued that at least 10% of the capital cost be recovered along with the O&M expenditure (Gulati 2002), the planned and actual subsidy was calculated. Considering the localized command, the planned subsidy was ranging from Rs.430 per hectare in the head to Rs.546 per hectare in the tail indicating the tail end farmers are enjoying comparatively more subsidy and the actual subsidy, on the other head was uniformly higher in all the locations (i.e., Rs.840 per hectare) indicating the

Table 11.7 Details of demand, collection, and subsidies for NRSP, right canal

Particulars WUAs [N = 192]	Irrigated Area (ha)	Water charges (demand in Rs)	Collection (in Rs)	Demand (Rs./ha)	Collection (Rs/ha)	O&M (Rs/ha)	Planned subsidy (Rs/ha)	Actual subsidy (Rs/ha)
O&M TOTAL								
Head	135,926.50	58,294,000.00	15,610.89	428.86	0.11	98.42	−330.44	98.31
Middle	88,835.85	31,396,000.00	18,153.31	353.42	0.20	98.42	−255.00	98.22
Tail	72,925.91	21,665,000.00	8283.33	297.08	0.11	98.42	−198.66	98.31
O&M IRRIGATION								
Head	135,926.50	58,294,000.00	15,610.89	428.86	0.11	47.82	−381.04	47.71
Middle	88,835.85	31,396,000.00	18,153.31	353.42	0.20	47.82	−305.60	47.62
Tail	72,925.91	21,665,000.00	8283.33	297.08	0.11	47.82	−249.26	47.71
CAPITAL + O&M TOTAL								
Head	135,926.50	58,294,000.00	15,610.89	428.86	0.11	160.10	−268.76	159.99
Middle	88,835.85	31,396,000.00	18,153.31	353.42	0.20	160.10	−193.32	159.90
Tail	72,925.91	21,665,000.00	8283.33	297.08	0.11	160.10	−136.98	159.99
CAPITAL + O&M IRRIGATION								
Head	135,926.50	58,294,000.00	15,610.89	428.86	0.11	44.35	−384.51	44.24
Middle	88,835.85	31,396,000.00	18,153.31	353.42	0.20	44.35	−309.07	44.15
Tail	72,925.91	21,665,000.00	8283.33	297.08	0.11	44.35	−252.73	44.24

Table 11.8 Details of demand, collection, and subsidies for NRSP, left canal

Particulars WUAs [N = 131]	Total (In ha)	Demand (in Rs)	Collection (In Rs)	Demand (Rs./ha)	Collection (Rs / ha)	O&M (Rs/ ha)	Planned subsidy (Rs/ha)	Actual subsidy (Rs/ha)
O&M TOTAL								
Head	72,623.6	36,311,800	21,801,305	500	300.19	98.42	−401.58	−201.77
Middle	44,799.6	22,399,800	11,740,115	500	262.05	98.42	−401.58	−163.63
Tail	36,292	18,146,000	1,645,423	500	45.33	98.42	−401.58	53.08
O&M IRRIGATION								
Head	72,623.6	36,311,800	21,801,305	500	300.19	47.82	−452.18	−252.37
Middle	44,799.6	22,399,800	11,740,115	500	262.05	47.82	−452.18	−214.23
Tail	36,292	18,146,000	1,645,423	500	45.33	47.82	−452.18	2.48
CAPITAL + O&M TOTAL								
Head	72,623.6	36,311,800	21,801,305	500	300.19	160.1	−339.9	−140.09
Middle	44,799.6	22,399,800	11,740,115	500	262.05	160.1	−339.9	−101.95
Tail	36,292	18,146,000	1,645,423	500	45.33	160.1	−339.9	114.76
CAPITAL + O&M IRRIGATION								
Head	72,623.6	36,311,800	21,801,305	500	300.19	44.35	−455.65	−255.84
Middle	44,799.6	22,399,800	11,740,115	500	262.05	44.35	−455.65	−217.70
Tail	36,292	18,146,000	1,645,423	500	45.33	44.35	−455.65	−0.98

Table 11.9 Details of demand, collection, and subsidies for TBP

Particulars WUAs (N = 32)	Total area irrigated (In ha)	Demand (In Rs)	Collection (In Rs)	Demand (Rs / ha)	Collection (Rs/ha)	O&M (Rs/ ha)	Planned subsidy (Rs/ha)	Actual subsidy (Rs/ha)
O&M TOTAL								
Head	10,085.51	2,009,310	782,000	199.23	77.54	579.46	380.23	501.92
Middle	9665.49	1,926,306	639,900	199.30	66.20	579.46	380.16	513.25
Tail	NA	NA	NA	NA	NA	579.46	0.00	0.00
O&M IRRIGATION								
Head	10,085.51	2,009,310	782,000	199.23	77.54	273.73	74.50	196.19
Middle	9665.49	1,926,306	639,900	199.30	66.20	273.73	74.43	207.53
Tail	0	0	0	0.00	0.00	273.73	0.00	0.00
CAPITAL + O&M TOTAL								
Head	10,085.51	2,009,310	782,000	199.23	77.54	942.5	743.27	864.96
Middle	9665.49	1,926,306	639,900	199.30	66.20	942.5	743.20	876.30
Tail	0	0	0	0.00	0.00	942.5	0.00	0.00
CAPITAL + O&M IRRIGATION								
Head	10,085.51	2,009,310	782,000	199.23	77.54	197.9	−1.33	120.36
Middle	9665.49	1,926,306	639,900	199.30	66.20	197.9	−1.40	131.70
Tail	0	0	0	0.00	0.00	197.9	0.00	0.00

collection mechanism is very poor. When comparing the actual command area, the same trend was observed both under planned and actual subsidy, even though the capital cost indicates the fixed cost of the project. However, the capital and O&M cost was adjusted for irrigation component for using the SCRB method, both planned and actual subsidies were considerably reduced. The planned subsidy was zero and the actual subsidy was about Rs 44 per hectare. The planned subsidy, however, will be higher when the localized area under the NRSP is considered.

In the case of NRSP-Left Canal, with the actual area irrigated, the WUAs located in the head, middle, and tail regions had shown no difference in the planned subsidy, where there is no subsidy experienced. However, with actual subsidy, only the tail end WUAs had a subsidy of Rs.53 per hectare indicating inefficiency in the collection mechanisms or poor water deliveries resulting in poor cost recovery. However, when the O&M expenses related to only irrigation are considered, only the tail end WUAs enjoyed a marginal subsidy of Rs.2.5 per hectares when WUAs in other location had no subsidy. When the capital and O&M total expenses were related to actual area irrigated, only actual subsidies were seen particularly in the tail end (Rs 114.7 per hectare). When the irrigation portion of the O&M expenses was considered, there was no subsidy enjoyed by the farmers in all the categories.

In the case of Distributory Committee (DCs), actual subsidy enjoyed by the farmers under the DCs located in the head, middle, and tail regions was Rs.22, Rs 17, and Rs.106 per hectare respectively. When the O&M expenses were related to irrigation component, the actual subsidy with actual irrigated area was ranging from Rs.3 in the head to Rs.6 per hectare in the tail region of the project, and this was same when the localized area of the project was considered. As indicated earlier, with the capital and O&M expenses added, the actual subsidy was ranging from Rs.40 to Rs.43 per hectare. The trend was same when the localized area under the project was used. In the case of WUAs, in the NRSP Right Canal, the actual subsidy was observed only in the tail end portions of the canal in the head, middle, and tail regions (Table 11.8).

In the case of Tungabhadra project, the planned subsidy in the head was Rs 380 per hectare and the actual subsidy was Rs 502 per hectare. When the irrigation component of the O&M was included, the actual subsidy was ranging from Rs 74 to Rs 196 per hectare. When the O&M and capital cost was included, the planned and actual subsidy was ranging from Rs 743 to Rs 865 per hectare (Table 11.9).

11.7 Reworking O&M Cost—The Way Forward

Since total O&M expenditures on canal irrigation include expenditures arising from inefficiencies and possible overstaffing in the irrigation department, the planned and actual subsidy capture both the subsidy received by users of canal water as well as the subsidy accruing to the state irrigation bureaucracy.

It is important that other multiple use sectors such as hydropower, drinking water supply, industry water supply, and fishing could be proportionately addressed in the future O&M calculations of irrigation projects. Accordingly, actual subsidy relating to irrigation sector should be re-examined taking into account the irrigation component alone. Using the separable cost remaining benefit method, the actual O&M component attributable to the irrigation component can be worked out for making policy prescriptions. The actual subsidy is expected to be lower when the O&M expenditure is adjusted for the irrigation component alone. This will give justification for the allocation of adequate budgets for regular O&M of the irrigation projects and revising the water charges under different uses accordingly.

References

Agricultural Finance Corporation (2002) Baseline survey- SRSP, Hyderabad

Gittinger JP (1982) Economic analysis of agricultural projects, 2nd edn. EDI series in economic development. World Bank and Johns Hopkins University Press.

Gulati A (2002) A subsidy syndrome. Oxford Publishing Company, New Delhi

Loughlin JC (1977) The efficiency and equity of cost allocation methods for multipurpose water projects. Water Resour Res 12(1).

Sur M et al (2002) Draft report prepared by the incidence of canal irrigation subsidies in India - A Policy Note. South Asia Rural Development Unit, World Bank, Washington D.C

William Easter (2003) Cost allocation and the distribution of costs for irrigation in Andhra Pradesh. Draft report

World Bank (1999) The irrigation sector. The World Bank, Washington D.C

World Bank (2003) The incident of canal irrigation subsidies in India. 'A Policy Note'. South Asia Development Unit Washington D.C

Young RA, Haveman R (1986) Economics of water resources: a survey and the evaluation of state or regional water plans. Water Resour Res 21(12):1819–1823

Chapter 12
Learning from Success Stories and Best Practices in Water Management

Abstract Water resources development and management are taking many turning points in terms of their size and scale of investment, approach and methodologies used, varying levels of interventions and their impacts. However, past experiences indicate not all is going well with the investment and management aspects of the irrigation projects. There are many best practices and success stories realized in different regions over years. Adoption of such practices is expected to minimize the cost of wrong implementation decisions and shows the ways for higher project performance. This chapter gives a briefing on the new initiative viz., Support for Irrigation Modernization Program (SIMP) in MMI modernization and outlines several best practices at national and international levels. Also the cases of private sector participation in irrigation project modernization under five different irrigation typologies are presented.

Keywords SIMP approach · Telemetry Systems for Data Acquisition (TSDA) · Design Build Operate (DBO) · Tamil Nadu Drip Irrigation Project (TNDRIP) · Mission Kakatiya · Underground taming of floods · Turpan water management model · Barind model

12.1 Introduction

Over years, improvements in the design, construction, and operation of the canal system have been made and both the central and state governments have spent substantial amount of funds, time, and energy on command area development and improving the overall performance of the irrigation projects. However, they are not successful in utilizing 100% of the potential created. Further, with all these improvements, still irrigation projects remained as supply-based system with less focus on demand management aspects. Automation of the canal systems has helped to improve operational schedules to some extent but it failed to answer major issues like equitable distribution and low water productivity. Efforts are still made with new insights and approaches in modernizing the MMI projects in the country where mostly the

K. Palanisami and U. S. Nagothu, *India's Water Future in a Changing Climate*, Advances in Geographical and Environmental Sciences,
https://doi.org/10.1007/978-981-97-1785-9_12

blueprint approach is followed without giving flexibility for modification. This status quo or somewhat rigid approach is mainly due to lack of evidences to show that such modifications will result in better implementation and improved performance efficiency. Hence getting access to cross learning is important particularly in irrigation programmes because of more money, time, and energy involved. This chapter brings out some of the success cases of canal irrigation modernization with improved efficiencies covering both national and international cases.

12.2 The SIMP Approach for Canal Modernization

Support for Irrigation Modernization Program (SIMP): To address the issues of low project performance and low water productivity, the Support for Irrigation Modernization Program (SIMP) has been launched by CWC. The programme is supported by the DOWR, GOI and financial and technical assistance of the ADB. The programme will support the states with preparing new irrigation investments by applying national and international best practices when preparing feasibility studies, detailed project reports, detailed designs and tender documents and sustainable asset management plans. In a way, the SIMP will be focusing more on demand management aspects as nationally and internationally proven technologies and best practices will be included in improving the performance of the systems (Asian Development Bank 2021a, b). For example, digital technologies can play a key role both at system level and farmer level.

The key features of the SIMP approach include:

- Establishing a Central Irrigation Management Office (CIMO) in the Central Water Commission (CWC) to coordinate modernization projects at state left.
- Establishing State Irrigation Modernization Units (SIMUs) to plan, design, and oversee the implementation of the modernization projects in the state.
- Careful selection and prioritization of modernization projects based on stipulated criteria.
- Preparation of Irrigation Modernization Plans (IMPs) for each project tailored to the individual conditions and needs of that project, with full engagement in the process with water users, field and office irrigation and drainage staff.
- Adoption of a "cafeteria" rather than blueprint approach to modernization whereby interventions for each individual project are selected from a wide range of possible modernization interventions. Knowledge and awareness programmes to improve the understanding of the modernization process and requirements, together with strengthened knowledge for more flexible and site-specific designs for modernization.
- Formulation of procedures for monitoring and evaluation of project implementation, and scheme performance following modernization.
- Introduction of procedures for preparing asset management plans to ensure the sustainability of the irrigation and drainage schemes following modernization.

It is expected that modernizing MMI schemes by following the SIMP approach will increase land and water productivity and farmer income thus bringing in much needed financial and social security to the farmers.

12.3 Current Best Practices in Irrigation Management—National Success Cases

The national-level best practices in this chapter mainly include selected management interventions implemented in the past to fulfil the objective of MMI modernization in the country. Further, performance assessment tools like Rapid Appraisal Process (RAP) developed by FAO and Mapping System and Services for Canal Operation Techniques (MASSCOTE) performance assessment tools, telemetry for flow measurements in canals are also important as part of the best practices. The SIMP report covers many success cases (Asian Development Bank 2021b). Some of the best practices are discussed below

(a) **Management approach to increase irrigated agriculture area and production, Madhya Pradesh**

Madhya Pradesh is one of the leading states interested in improving the performance of the irrigation systems. However, the performance till 2010 was poor mainly due to lack of optimal operation and maintenance resulting in a large gap in the utilization of the created command area. The net sown area was only about 34% of the total land area (Julaniya et al. 2016).

In order to improve the performance of the systems, the state Water Resources Department (WRD) has initiated several management interventions such as: (i) speeding up the ongoing projects and (ii) rehabilitation of existing projects by adopting the following measures in 2010 (Julaniya et al. 2016):

- monitoring reservoir storages.
- setting targets for irrigating areas based on stored volume.
- regular and transparent progress monitoring using web-based MIS.
- regular inspections.
- Improved maintenance of canal systems.
- empowering and orienting the staff towards service delivery.
- rewarding good performance, and
- eliciting active support from WUAs.

Adoption of the improved water management practices supported by modern technology (Burton and Stoutjesdijk 2017), rigorous monitoring, and highly motivated staff led to a much greater utilization of the designed irrigation command area. As a result of the improvement measures undertaken, the utilization of the command area has increased from 34% to 85% and a three-fold increase in irrigated area from 0.94 m.ha in 2010–11 to 2.91 m.ha in 2015–16. In order to meet the increased O&M expenditure, the budget for maintenance and repair of the irrigation systems has substantially increased from Rs112/ha in 2009–10 to Rs 820/ha in 2015–16 (Julaniya et al. 2016). The substantial increase in the maintenance funding by the GoMP is to operate a canal system effectively and obtain sustainable high levels of production.

(b) **Drip-to-Market-Agro-Corridor (DMAC) in Ramthal Project, Karnataka**

This DMAC is a typical case of a Build Operate and Transfer (BOT) model commissioned by Krishna BhagyaJala Nigam Ltd. (KBJNL), a Government of Karnataka enterprise. It is a joint initiative of WRD, Karnataka through KBJNL with Agriculture and Horticulture departments. Installation of drip irrigation systems for 24,000 hectares during 2017 was undertaken from the backwaters of Narayanpur reservoir in Upper Krishna Project benefitting about 15,000 farmers under the Participatory Irrigation Management (PIM) programme and formed 51 Water User Committees (WUCS). Market linkages have been established through the DMAC (Drip-to-Market-Agri-Corridor) programme for effective marketing of agriculture produced by the participation of private sector in procuring high-value crops and building farmers' capacities in best agricultural practices. This initiative has transformed the WUCs into participants in agribusiness marketing. DMAC incorporated downstream linkage with agro-processing firms for better price realization.

Key features of the BOT drip irrigation model are:

- Design, supply, installation, operation, & maintenance (for 5 years) of fully automated drip irrigation system, formation of WUCS, training and capacity building of farmers, support to agro-produce, and marketing linkages are the responsibility of the implementing agencies.
- Entire system is operated from a single location with a centralized pumping system.
- The total project area is divided into 23 zones of average 500 ha area with one WUCS per zone.
- The 500 ha Zone is then further divided into 10 numbers of Water User Group (WUG) of 50 ha area.
- Irrigation is to cover both the seasons: 80% during Kharif season and 50% during Rabi season.
- Crops grown include: jowar, groundnut, maize, pulses in Kharif and jowar, cotton, and vegetables in Rabi.

A Project Implementation Unit (PIU) constituted jointly by the departments of WRD and DoA supervises the DMAC with a special focus on linking farmers to high-value markets. Another initiative under DMAC during the 2019 Rabi season was that payments to the farmers for the agricultural produce sold to the companies were credited directly to farmers' accounts. Within a year of establishment, the DMAC has demonstrated its impact by improving farm incomes, promoting demand-driven water management, and incentivizing adoption of water-efficient technologies through private sector market off-takers (World Bank Group 2020).

(c) **Telemetry Systems for Data Acquisition (TSDA), KISWRMIP, Karnataka**

The TSDA is a classical example of how advanced measurement tools can be applied for improving the overall irrigation project performance. The Karnataka Neeravari Nigam Limited (KNNL), a Government of Karnataka enterprise is implementing ADB-assisted Karnataka Integrated and Sustainable Water Resources Management Investment Program (KISWRMIP) since 2012. The programme aims to improve water availability to meet the competing water demands by adopting the Integrated Water Resources Management (IWRM) approach and to improve water use efficiency in irrigated agriculture.

Under the component of flow measurement and telemetry system of the KISWRMIP, 100 online flow measurement stations were installed in canals. The contract for the installation of telemetry system provided for five years of maintenance after the successful installation and commissioning. The scope of work included automated discharge measurement at remote canal sites in the six schemes, data logging, data transmission to Central Monitoring Station at Shivamogga for flow and volumetric accounting of the water utilization. The Contract covered supply, installation, testing, commissioning, and maintenance of 24 multi-path transit time flow meters for main canals, 76 non-contact surface velocity flow meters with non-contact ultrasonic water level transmitters for branches and distributaries, supply of 4 ADCPs for verification of the discharge in the canals, 6 fully automatic weather stations, data transmission and centralized monitoring and information sharing facility on turnkey basis.

The discharge and depth of flow at all the measurement locations is made available on web portal and to authorized smart phones of concerned officers using the mobile application developed for the purpose. This gives the officers a complete network view with real-time flow rate, the daily volume, the fortnightly and seasonal volumes of water supplied at the location. The real-time data availability from these telemetry installations in six schemes has been very helpful in supporting effective water management during the irrigation seasons.

(d) **Pressurized Irrigation Implementation Model, Karnataka, Madhya Pradesh**

The implementation model adopted here is the Design Build Operate (DBO) model mainly to improve water use efficiency in the irrigation projects where pressurized irrigation systems are adopted. DBO model facilitates Participatory Irrigation Management (PIM), provides agricultural and agronomic support to farmers, and develops market linkages to the farmers, O&M and support to the state WRDs in formation of WUCs. A synthesis of DBO-based micro irrigation contracts of four micro irrigation systems under implementation in Madhya Pradesh and Karnataka are discussed briefly below.

In the case of Madhya Pradesh Irrigation Efficiency improvement Project (MPIEIP), the designed command area was 125,000 ha. The cost per ha worked out to be Rs. 233,520. The system level efficiency was 80%. The project facilitated the formation of WUCs. In the case of micro irrigation, the project provides sprinkler/drip connection for 1 ha area & the farmers to connect their own systems (sprinkler/drip). Further, the smart water management system comprises of SCADA system for collecting information from remote sites in the command area and DSS with an irrigation database.

In the case of Karnataka state, two irrigation projects viz., Tarikere LIS and Singatlur LIS were designed. Command areas of these projects were 24,150 and 70,877 ha respectively. The cost per ha worked out to be Rs 4,33, 871 for Singatlur LIS, which was higher followed by Tarikere LIS (Rs 3,36,231). The system level of efficiency for all the projects was 90%. These irrigation projects facilitated the formation of WUCs and provided drip irrigation system right up to the farmers' fields including laterals (drip lines). Further these projects applied SCADA for automated operation.

(e) **Telangana Government's Mission Kakatiya on Tank Rehabilitation**

The State of Telangana after its formation has launched Mission Kakatiya programme in 2015 to restore the minor irrigation tanks and water bodies in the state and strengthen community-based irrigation management. Several interventions were initiated which include desilting tank beds, repairing feeder channels, repairing dilapidated sluices, weirs, sectioning of irrigation channels to standards, etc.

The government planned to restore all the 46,531 minor irrigation sources over five years by taking up 20% of the tanks each year. Since its launch, Mission Kakatiya has helped increase crop yields especially in paddy and cotton. Other aspects include: decrease in consumption of chemical fertilizers, increase in tank silt application, large-scale plantation of toddy trees on the bund slopes, and benefits to the fishermen community in the tanks.

(f) **Underground Taming of Floods for Drought Management**

International Water Management Institute (IWMI) in association with Climate Change, Agriculture and Food Security (CCAFS) has developed a new technique called "Underground Taming of Floods for Irrigation" (UTFI) in Rampur village of Uttar Pradesh which involves diverting high water flows from rivers or canals and recharging the groundwater via village ponds or small storage structures. Implementing partners include: Government of Uttar Pradesh, Indian Council of Agricultural Research, Central Soil Salinity Research Institute, Acacia Water, Livelihoods and Natural Resource Management Institute, The Energy and Resources Institute, KrishiVigyan Kendra and International Water Management Institute (CCAFS 2017). As a follow-up, this technique has been scaled up through detailed technical and non-technical investigations, stakeholder engagement, and communication activities and is now included in the district development plans.

(g) **Precision Farming as a Business Model, Tamil Nadu**

The pilot project on precision farming was a turnkey project implemented by Tamil Nadu Agricultural University (TNAU) during 2004–05 to 2006–07 in Dharmapuri and Krishnagiri districts as a state demonstration with the specific objectives of promoting hi-tech horticulture through application of precision farming technologies and to encourage the farmers to take up market-led horticulture by forming producers associations. The key PF technologies applied in the pilot project were: remote sensing technology (to understand the soil fertility), chisel ploughing to break the hard soil pan at the sub-surface level, crop geometry, drip and fertigation, growth management, plant protection, and postharvest management.

The project was spread over three years with coverage of 100 ha, 200 ha, and 100 ha respectively in each year beginning 2007–08 benefitting 400 farmers. The project size in terms of financial investment was Rs 720.60 lakhs of which Rs 160 lakhs allocated for cultivation expenses and Rs 300 lakhs for drip and fertigation. In the first year, each of 100 beneficiary farmers received a subsidy of Rs 40,000 for cultivation expenses and Rs 75,000 for investment on drip-fertigation system. In the second year, the subsidy was reduced by 10% to the 200 farmers and in the third year, it was reduced by 20% for the remaining 100 farmers. The entire budget was provided by the Department of Horticulture, Government of Tamil Nadu.

The cluster (of villages) approach was adopted for operational convenience. About nine clusters were formed in the study districts. A Nodal Officer from TNAU was in charge of the project and was assisted by the two Project Officers one for each district. The field-level workers were supervised by the Project Officers. The farmers adopted technologies as per TNAU's recommendations. The highly effective extension work done by the TNAU faculty can be attributed as one of the key factors for the success of the PF project. TNAU guided the farmers in all aspects including purchase of modern inputs and marketing of PF produce. The major crops covered were: tomato, brinjal, banana, musk melon, cassava, cabbage, rose, sugarcane, and cotton. The PF farmers organized themselves at cluster level and registered "Producers Associations" and used this collective action approach in purchase of farm inputs, organizing community nursery, and marketing of their produce.

The impact of pilot PF project was highly positive. The yields of crops in PF fields were 0.75 to 15 times higher than the average farm yields and farmers realized proportionately higher income from crops. The drip irrigation led to more than 50% in water saving with uniform distribution of water across the entire field. Since PF considered collective action as one of the features of the scheme, farmers organized community nurseries and resorted to group marketing. The Adhiyaman Precision Farmers Association launched a venture called Adhiyaman Agro Service Centre which has taken up dealership of fertilizers, seeds, pesticides, etc. and distributes quality inputs to their members and also non-member farmers.

On witnessing the success of pilot project demonstrated by TNAU, Government of Tamil Nadu planned to up-scale PF to the state level. Hence the project was launched covering an area of 12,800 ha in 28 districts during 2007–08 and an area of 13,400 ha was brought under PF during the subsequent two years. The scheme is continuing utilizing the funds provided by the central government under National Agricultural Development Project.

(h) **Tamil Nadu Integrated Agriculture Modernization and water bodies restoration and Management (TN-IAMWARM)**

The World Bank-funded TN-IAMWARM project is a multidisciplinary project involving water, agriculture, and livelihood sectors that had been successfully implemented between 2007 and 2015 covering 61 sub-basins in Tamil Nadu with a mantle of excellent note on many success stories and good learning. The follow-on project TNIAMP (Tamil Nadu Irrigated Agriculture Modernization Project) which commenced in January, 2018 with the overall budget out lay of Rs 2962 crores for covering 66 sub-basins is now nearing completion. The project has introduced many innovations and technologies as outlined below:

Augmenting water resources through partial reclamation of 6,621 tanks, 920 anicuts, 10,818 km supply channels and construction of recharging wells (101 Nos) in the tanks water spread areas. The increase in storage capacity is between 4 and 9% in tanks. The innovative Climate Smart, cutting edge technologies on improved agricultural practices aimed to increase the water productivity by 20% which include promotion of SRI (91,000 ha), SSI (3600 ha), micro irrigation (65,220 ha), maize improvement (87,000 ha), hybrid varieties of horticulture crops and mulching (77,958 ha).

About 700 Model Villages were created and redressal mechanisms through SWIKC (Single Window Information Knowledge Centers) in each Model Village were established. Introduced *e-velanmai* an ICT-based initiative for effective crop husbandry with timely action to control any pest and disease attacks. In the case of agri-marketing component, value addition activities through the marketing infrastructure like storage shed, drying yards, agri business centre, pack house, etc. 771 numbers, benefitting 1,73,422 farmers to transact additional quantity of 3.80 lakh tonnes of commodities. Besides 50 Farmers Producers Organisation registered and started transaction.

The silent but effective contributions in revolutionizing the Animal Care with increased Artificial inseminations by 51.7% against national average of 35% motivate the farmers on fodder cultivation leading to incremental milk yield of 7.06 lakh tonnes. The promotion of inland aquaculture by Fisheries Department in farm ponds constructed by Agricultural Engineering Department (AED) is another proven convergence by the project facilitating development of aquaculture in farm ponds (831 nos) and in irrigation tanks (804 nos) with good fish seeds increase in fish production 14,686 tonnes as additional income generation. Two success cases of changing canal operating rules are given in Box 12.1

(i) **Capacity Building Program for Improving WUE—Tamil Nadu Drip Irrigation Project (TNDRIP)**

The International Water Management Institute (IWMI) under its IWMI-TATA programme has initiated a farmer capacity building programme (called TNDRIP) in Tamil Nadu state in association with Tamil Nadu Agricultural University mainly to address those issues which lead to dis-adoption of drip technology by farmers. The TNDRIP was undertaken in selected districts in Tamil Nadu during 2010 in 100 villages covering about 800 farmers who had adopted drip-fertigation technology for various crops. The training was imparted through lectures, field demonstrations, and question–answer sessions between the resource persons and trainees. The programme focused mainly on the operation and maintenance practices of drip irrigation system besides fertigation and irrigation scheduling practices.

As a result of the programme, practices such as cleaning the filters, sub-mains and laterals, pressure regulation, acid treatment, thatching the drip system, placing the laterals in shade, etc. have been fully adopted by a majority of the farmers. The yield of banana was 45.9 t/ha for the control (non-participating) farmers before the capacity building training period. The yield of banana for the same set of farmers after the capacity building training period was 47.7 t/ha (Palanisami et al. 2014). Given this cost effectiveness of the capacity building programmes, this model has been adopted in several regions of the state. Sir Ratan Tata Trust (SRTT) has extended the TNDRIP programme for another 3 years to cover more districts in the state to cover around 10,000 farmers. The spillover impact (off shore impact) will be about 50,000 farmers in the long run.

At national level, the programme was given the award, "Rotary Up-town International, Coimbatore Chapter" in 2011. At international level, the TNDRIP was selected as one of the finalists in the category 1—Best Water Management Practices under "The 2013 Water for Life—UN Best Practice Award" (Palanisami 2014).

Box 12.1 Interventions in canal operating rules and system operation Two projects in Tamil Nadu state India are referred by the stakeholders as the best Performing projects. The major intervention made was the change in canal operation procedure/rules making more equitable and efficient water allocation and use.

1. Parambikulam Aliyar Project (PAP): PAP is a multipurpose, multi-reservoir irrigation system situated in Tamil Nadu with a command area of 172,256 ha. A four-zone water allocation was followed (each zone is approximately 25% of the total command area), each farmer will get water once in 24 months. Since 2000, alternate sluice irrigation system (ASIS) is followed which means in each turn, alternate sluices will get water along the canal or distributary where water is supplied for the entire length of the main canal and distributaries. Thus canal is run every year and sluices are opened in alternate years. The canal seepage will be used by the farmers in other zones for recharging the wells and groundwater irrigation during non-canal water supply seasons.

2. Lower Bhavani Project (LBP): The Lower Bhavani Project (LBP) is located in Tamil Nadu in Bhavani River, irrigates 83,980 ha in two seasons in a year. The canal water supply is provided for one wet crop from August to December (41,885 ha) and one dry crop from January to April (41,885 ha). Command area was divided into 2 zones, each year one zone gets water from August to December for wet crop and other zone gets water in January to April for dry crop. The zones irrigated during wet and dry season will be swapped during the next year, so that each farmer will get water once in a year to grow wet and dry crop alternatively. Water supply in the main canal is continuous during the crop season. Occasionally, water for special wettings (1–2 weeks) are also given before and after the August-December releases for land preparation and irrigation needs, depending upon the storage in the reservoir.

After the introduction of ASIS in PAP in the year 2000, the average percentage of target area irrigated has increased to 97% compared to 78% in the pre-ASIS periods. In the case LBP, after the introduction of alternate sluice (AS) irrigation in 1980s, the irrigated areas have increased to 89% compared to 74% in the pre-AS period. Increase in water productivity and water saving are also realized over years.

Source: field survey

12.4 Success Cases of Private Sector Participation

In the recent years, private sector is also playing a key role in irrigation project improvements. For example, the Jain Irrigation System Ltd (JIS) is promoting a major initiative to convert the existing canal command areas into "Jain Integrated Irrigation Solutions (JIIS)" wherein the irrigation efficiencies can be increased up to 75% by adopting improved technologies such as micro irrigation (Jain Irrigation Systems 2018). The Integrated Irrigation Solution concepts initiated by Jain irrigation in association with the state irrigation departments for some of the existing canal projects in India are discussed below (Tables 12.1, 12.2, 12.3, 12.4 and 12.5).

Table 12.1 Type—gravity-based irrigation

S. No	Scope & components of the project	Purna/Chandrabhaga project, Achalpur, Maharashtra
	Replacement of field irrigation canals with pipe network up to farmers field including silt trap chamber	√
1	Area of the Project: Acres	29,600
	Cost of the project: Rs. Million	210
	Cost of the project: per acre	Rs.7,095
2	Expected conveyance efficiency, (%)	66%
	Expected application efficiency (%)	60%
	Overall project efficiency (Anticipated) (%)	40%

Table 12.2 Type—canal-based/pressurised micro irrigation

S. No	Scope & components of the project	Sardar Sarovar Narmada Nigam Limited (SSNNL) Project, Gujarat	Indira Gandhi Nahar Pariyojna (IGNP) Project, Rajasthan	Narmada Canal Project (NCP), Rajasthan
	Replacement of field irrigation canals with pipe network up to farmers field	√	√	√
	Construction of Sumps and Storages	√		√
	Construction of secondary pumphouse	√		√
	Construction of secondary pumping machinery	√	√	√
	Area of the project, Acres	617	37,050	338,400
	Cost of the project: Rs. Million	25.7	220	2770
	Cost of the project per acre:	Rs. 41,654	Rs. 5,938	Rs. 8,186
	Expected conveyance efficiency, (%)	76%	76%	76%
	Expected application efficiency (%)	Up to 95%	75%	75%
	Overall project efficiency (Anticipated) (%)	71%	56%	56%

12.5 Current Best Practices in Irrigation—International Success Cases

Even though several success cases are demonstrated at international level, only few cases that are relevant to Indian situations interns of their future replicability are discussed below.

(a) **Turpan Water Management Model, China**

The Xinjiang Turpan Water Conservation Project is an example of the successful combination of advanced remote sensing (RS) technology, targeted improvements in irrigated agriculture and reforms to water rights and water pricing to achieve a "whole system" water resources management approach (World Bank 2018, 2019). There was a rapid expansion of the irrigated land in Turpan Prefecture from 60,000 ha to 80,000 ha during 1970–2000. To relieve the pressure on groundwater from this expansion and to conserve water, modern irrigation technologies including drip and sprinkler were introduced. However, the expected water savings were not realized

Table 12.3 Type—lift cum sprinkler irrigation

S. No	Scope & components of the project	LIMIP (Lift Irrigation Micro Irrigation Project), Kadapa, Andhra Pradesh	Shiggaon Sprinkler irrigation project, Shiggaon, Karnataka
	Replacement of distributary canal with pipe network		√
	Replacement of Minor Canal With Pipe Network		√
	Replacement of sub-minor canal with pipe network		√
	Replacement of field irrigation canals with pipe network up to farmers field	√	√
	Construction of sumps and storages	√	√
	Construction of secondary pumphouse	√	√
	Construction of secondary pumping machinery	√	√
	Shiftable sprinkler irrigation system	√	√
	Semi-permanent sprinkler irrigation system	√	
	Construction of substation works		√
	Construction of transmission line networks		√
	Area of the project, Acres	14,820	24,453
	Cost of the project: Rs. Million	460	1683
	Cost of the project per acre	Rs. 31,040	Rs. 68,704
	Expected conveyance efficiency, (%)	66%	99%
	Expected application efficiency (%)	Up to 95%	75%
	Overall project efficiency (Anticipated) (%)	62%	74%

as farmers used the "saved" water to increase their irrigated area. The groundwater levels had further declined by 1.5–2 m per year.

Under the World Bank-funded Xinjiang Turpan Water Conservation Project several measures were introduced to arrest and reverse the deteriorating situation including a policy decision to cap the water use by farmers by monitoring their actual crop water use (crop evapotranspiration) using remote sensing technologies. This initiative was attempted by adopting the following measures:

Table 12.4 Type—community tube well drip irrigation

S. No	Scope & components of the project	Gujarat Water Resource Development Corporation (GWRDC), Gujarat	Solar operated community Tubewell Kurani, (Lucknow) UP
	Piping network from community tubewell to farmer's field	√	√
	Semi-Permanent Sprinkler Irrigation System	√	
	Drip Irrigation System (NMMI/PMKSY)	√	
	Solar Pumping Systems		√
	Area of the Project, Acres	5000	2500
	Cost of the Project: Rs. Million	260	23
	Cost of the Project per acre	Rs. 52,000	Rs. 9200 (Approx)
	Expected conveyance efficiency, (%)	99%	99%
	Expected application efficiency (%)	Up to 95%	60%
	Overall project efficiency (Anticipated) (%)	90%	59%

(a) Conducting a water accounting assessment at the river basin level,
(b) Determining the target water allocations,
(c) Revising the suitable cropping pattern to fit within the allocated water right,
(d) Applying the remote sensing to monitor the actual ET,
(e) Charging for overuse of water, and
(f) Taking measures to bring down the actual ET consumed within the target ET value.

As a result of these measures, the groundwater overdraft was reduced and agricultural production was enhanced with the significant water savings.

(b) **Barind Multipurpose Development Authority (BMDA), Bangladesh**

The Barind Multipurpose Development Authority (BMDA) is a successful example of an autonomous authority under the jurisdiction of the Ministry of Agriculture in Bangladesh. The BMDA was established under the Ministry of Agriculture in 1992 and then formalized by an Act of Parliament in October 2018. The Authority is responsible for providing irrigation water to a total cultivable area of around 600,000 ha of which 84% is single cropped, 13% is double cropped, and the remainder triple cropped.

Table 12.5 Type—lift cum drip irrigation

S. No	Scope& components of the project	Sampatrao Deshmukh Sahakari Pani Purvatha Sanstha Maryadit (Cane Agro Energy), Sangli, Maharashtra	Ramthal Micro Irrigation Project, Hungund, Karnataka
	Replacement of main canal with piping network	√	
	Replacement of distributary canal with pipe network	√	√
	Replacement of minor canal with pipe network	√	√
	Replacement of sub-minor canal with pipe network	√	√
	Replacement of field irrigation canals with pipe network up to farmers' field	√	√
	Construction of secondary pumping machinery		√
	Drip Irrigation system	√	√
	Network level automation system	√	√
	Field-level automation system with complete SCADA	√	√
	Area of the project, Acres	2000	30,381
	Cost of the project: Rs. Million	274.7	3857.8
	Cost of the project per acre:	Rs. 137,350	Rs. 126,981
	Expected conveyance efficiency, (%)	99%	99%
	Expected application efficiency (%)	Up to 95%	Up to 95%
	Overall project efficiency (Anticipated) (%)	90%	90%

(a) The key features of the "Barind model" are:
(b) The BMDA is an autonomous, self-financing organization managed by a Board chaired by an appointee of the Prime Minister,
(c) BMDA is responsible for 16,072 individual irrigation systems comprising 15,553 tubewell systems and 519 water lift systems,
(d) Water distribution from the pump station is via buried pipes to the farm turnout,
(e) The Authority is totally financially self-sustaining, covering all its costs from charges to water users,

(f) Charges vary per system but range from US$7.75 per 1000 m^3 for larger capacity tubewells (29–56 l/s) to US$21.08 per 1000 m^3 for smaller capacity (14 l/s) tubewells. Charges for surface water lift schemes range from US$9.71 to US$15.50 per 1000 m^3,
(g) The irrigation charges are calculated for each scheme based on the electricity costs, pump operator costs, repair and maintenance costs for the physical assets (pump house, pumping equipment, buried pipe system, meters, etc.), the depreciation costs of the machinery and equipment, BMDA staff overhead costs, and taxes,
(h) The fees recovered from water users cover all costs, including staffing, O&M, electricity, and asset depreciation and repairs to the irrigation system, including the metering system is carried out by private contractors hired by BMDA,
(i) The Authority has installed a Pre-Paid Metering System whereby farmers pay for a volumetric supply of water using pre-paid cards,
(j) Fee recovery is 100% and no farmer can irrigate without using the pre-paid card.

This initiative has resulted in reductions in irrigation costs per hectare, average water use per hectare, and water requirements per kilogram of rice, and substantial (34–43%) increases in average yields and average earnings. Thus, the BMDA is providing a sustainable, on-demand irrigation service which is profitable to the farmers for which they are prepared to pay a realistic irrigation service fee. The fee level is set to cover the full-life costs of running the systems with income exceeding expenditure such that funding is secured to cover eventual asset replacement costs (Zaman 2019).

12.6 Application of Advanced Methods/Tools

These success stories can be better applied along with the modern tools of analysis as better use of such appropriate tools will help in the effective implementation and evaluation of the impacts of the modernization programmes. Most commonly used tools are: Mapping System and Services for Canal Operation Techniques (MASSCOTE), Rapid Appraisal Procedure (RAP), Remote sensing technology applying Surface and Energy Balance Algorithm for Land (SEBAL), and Bench marking which can be used to compare the performance of one irrigation scheme with another. Double difference (DD) method is used for studying the impact using panel data. It has the advantage to control for the time-invariant characteristics of farmers when comparing adopters and non-adopters as well as before and after adoption of a technology intervention.

12.7 Success Stores and the Way Forward

The lessons learnt from the success cases will be useful in: (a) planning the type of interventions needed, (b) selecting the type of methodologies that can be better used for effective implementation, (c) designing appropriate capacity building programmes, (d) identifying the type of public private partnership implementation models, and (e) proper monitoring and evaluation of the impacts of modernization. Using the success cases, a matrix of interventions, implementation strategies and evaluation methods can be prepared and used as guidance documents for each irrigation typology.

In addition to systems level interventions, farm-level interventions are also important. The selection of these farm-level interventions mainly depends upon: the project types (gravity or pumping), water availability, soil types, crop seasons, farmers interest and marketing aspects, the technology choices, etc. The farm-level successful practices otherwise called climate-smart technologies that are discussed in Chap. 7 can be examined and integrated with the systems level interventions to make the irrigation system modernization programme as holistic one. Capacity building programmes for different stake holders will be an integral part of irrigation modernization and stakeholders need to be familiarized with different modernization interventions. Ongoing government programmes and policies can be converged at project level to maximize the benefits from irrigation modernization. Implementation of the transformative policy framework encompassing all the relevant implementation protocols in irrigation sector improvement will have a higher pay-off.

References

Asian Development Bank (ADB) (2021a) Holistic approach to meeting the people's republic of China's agriculture challenges. https://www.adb.org/news/features/holistic-approach-meeting-prc-agriculture-challenges

Asian Development bank (ADB) (2021b). Strengthening capacity to design and implement water and rural infrastructure facility: support for irrigation modernization program, vol I. India. TA-9738 IND

Burton M, Stoutjesdijk J (2017) Increasing irrigated agricultural production using ICT and modern management: a case study from Madhya Pradesh, India. In: 23rd international congress on irrigation and drainage, Mexico

Climate Change, Agriculture and Food Security (CCAFS) (2017) Progressing towards climate resilient agriculture; top ten success stories from CCAFS in South Asia. CGIAR Research Program on Climate Change, Agriculture and Food Security (CCAFS) of South Asia, New Delhi, India. https://cgspace.cgiar.org/handle/10568/83477

Jain Irrigation Systems (2018) Resource to root. In: Integrated irrigation solutions, pp 55–62, Jalgaon, Maharashtra, India

Julaniya RS, Singh M, Choubey MG, Biswas S (2016) A management approach to increase irrigated agricultural area and production in Madhya Pradesh, India. In: Second World Irrigation Forum (WIF2), Chiang Mai, Thailand, 6–8 November

Palanisami K (2014) Making a differenced drop by drop, everything about water. https://www.iwmi.cgiar.org/Publications/Success_Stories/PDF/2013/Issue_18-Making_a_difference_drop_by_drop.pdf

Palanisami K, Ranganathan CR, Sureshkumar D, Malik RPS (2014) Enhancing the crop yield through capacity building programs: application of double difference method for evaluation of drip capacity building program in Tamil Nadu State, India. Agrisci 5(1):33–42. https://doi.org/10.4236/as.2014.51003

World Bank (2018) China Turpan water management model. Project innovations knowledge sharing series, Water global practice, World Bank, Washington DC. https://documents1.worldbank.org/curated/zh/128351554920790216/pdf/China-Turpan-Management-Model.pdf

World Bank (2019) ChinaTurpan water management model. Presentation by Liping Jiang (EAP, World Bank) to the regional workshop on irrigation modernization, Almaty, November 19–20

World Bank Group (2020) Finding new purpose: propelling agribusiness marketing through irrigators'cooperatives. https://www.2030wrg.org/finding-new-purpose-propelling-agribusiness-marketing-through-irrigators-cooperatives

Zaman A (2019) Participatory irrigation management: barind model a new sustainable initiative. Univ J Mana 7(6):220–228

Chapter 13
Strengthening Institutions, Policies, and Governance

Abstract Irrigation institutions in India are facing several constraints and challenges as the existing policies, laws/legislations, acts are complex to understand and implement. River basin planning and management through institutions has been emphasized in various policy papers and it is yet to become a reality. Water-related laws, acts are introduced from time to time but implementation aspects are weak. The water governance phase of the India's water future with embedded investment and management phases is expected to address the emerging challenges to address both the water security and sustainability. The risks of climate change, interstate conflicts and increasing level of water pollution are making the policy interface further a complex phenomenon. However, these are the key areas where the Centre and the states need to work in a partnership mode involving institutions and regulatory mechanisms. Good governance needs to strengthen the institutions, implementable programmes, relevant policies related to water resources development and management in the country. This chapter discusses the issues related to water institutions, river basin organizations, existing acts, agricultural and water policies, government programmes and schemes. Finally, the chapter indicates how good governance can be a reality to address India's water future with a positive note.

Keywords Water institutions · River basin organizations · Irrigation management transfer · Participatory irrigation management · Water legislation · National agricultural policy · National water policy · Water governance

13.1 Introduction

Water Institutions

Water institutions influence the irrigation development and management in the country, as governance with strong and effective regulatory and administrative mechanisms decides the quality of irrigation services available to the stakeholders. Irrigation institutions in India are facing several constraints and challenges as the existing

K. Palanisami and U. S. Nagothu, *India's Water Future in a Changing Climate*, Advances in Geographical and Environmental Sciences,
https://doi.org/10.1007/978-981-97-1785-9_13

policies, laws/legislations, acts are complex to implement. Further, the organizational arrangements with several water departments/ agencies/with varying roles and responsibilities are governed by the norms of both the central and state governments. In several states, the organizational structures are also changing to meet the requirements of the funding agencies or donors. In addition, the roles of non-government organizations (NGOs), universities and colleges and corporate organizations associated with water-related research, development, and outreach activities are also changing. This chapter brings in a comprehensive view of the organizations, institutions, programmes, policies, and types of governance related to water resources development and management in the country.

River Basin Organizations (RBOs)

Comprehensive management of river basins is very important to address the future challenges in water availability and this can be possible only through the promotion of River Basin Organisation (RBO) (Chitale, 1992). Following the success of the Tennessee Valley Authority (TVA) in the United States of America, the GOI has enacted the Damodar Valley Corporation (DVC) Act in 1948 mainly to improve the performance of the river basins in a holistic manner. Subsequently, the River Boards Act was enacted in 1956 to enable the development of other rivers in the country on the similar lines of DVC. The DVC Act has clearly indicated the importance of regulating and developing the interstate rivers and river valley projects, by setting up a River Board under the consent of the respective state governments (CWC 1997).

Though the initiatives and the follow-ups to set up RBOs are justifiable, however, in practice they faced several constraints due to the inability to anticipate, assess, and address the various challenges faced by the large-scale dam projects and to evolve acceptable management practice to the changing needs (CSE 1991). Even though the river basin planning and management aspects have been emphasized in various policy papers including the National Water Policy, 2012, and initiated, it is yet to become a success case in India. Some of the observed problems are: the reluctance of the state government to devolve powers to the people and local institutions; existence of complex institutional structures to plan and manage the river system, growing interstate river conflicts, etc. One of the obvious features of the RBOs in India is that they do not have autonomy similar to that of the TVA.

Hence, amending legislation is needed to enable: a) creation of the new water allocation and sharing mechanisms with appropriate institutions; b) strengthened regulatory powers, c) adoption of pollution control measures, d) establishment of water resource-based multi-sectoral stakeholders associations in the basin, e) participation of private sector and civil society for decision-making, and f) new approaches for revenue generation (World Bank, 1998). In order to manage the water resources for future needs, a "threshold approach" can be examined involving action-oriented analysis of the river management problems and efforts to plan and manage them efficiently (Newson, 1992).

Institutional Interventions in Water Resources Management

In the recent decades, it is observed that funding agencies like the World Bank and the Asian Development Bank are actively suggesting the state governments to promote institutions like water user association (WUA) and also supporting the states in renaming the existing irrigation departments as Water Resource Organisation (WRO) or Council. These quick initiatives are mostly seen as a step for scaling up the "participatory" approach to suit funding agencies, without giving due attention to the stakeholder needs. During the 1970s and the 1980s, the World Bank and USAID had invested more resources in establishing more than a dozen state level institutions like Water and Land Management Institutes (WALMIs). The infrastructure and training facilities in these WALMIs were so good and the planned programmes went on smoothly until their funding ended. Subsequently, the state governments had taken over them but due to fund shortages their activities were diluted and reduced resulting in poor performance of the WALMIs in several states. They also failed to deliver the needed handholding and capacity building support to the local WUAs. Currently, the WALMIs are managed by different departments with limited experience on land and water management resulting in poor knowledge sharing to field workers and farmers. However, given the emerging challenges in water sector, the roles and responsibilities of the local water users associations (WUAs) and panchayat raj institutions will be important in managing the water supply and farmer conflicts and they need to be encouraged for improving their effectiveness (GOI 2006).

WUAs and Panchayati Raj Institutions—the Linkages

It is often seen that in the major and medium irrigation projects, the management is on a hydraulic (canal) basis and not village based. In many cases, only a part of the village area is getting irrigation from canals due to the defined boundaries of the canal command. On the other hand, organizations comprising of water users are likely to be much effective particularly in minor surface irrigation projects where the area of operation is comparatively less with erratic water supplies. However, the panchayats who work on administrative basis can support the Water Users Associations (WUAs) who work on hydraulic basis by giving supplementary grants for maintenance and repair of the structures. They can also help WUAs in any outstanding disputes between them, between farmers and governmental agencies because of their commanding position in the village and administrative system. The interaction between the two local institutions can be strengthened if a representative of Panchayat can participate in the regular meetings of the WUAs. Similarly, a representative of WUAs can also join the meetings of the Panchayats as and when required. A similar arrangement can be planned between Zila Parishad (district level Panchayat) and distributory level WUA or a federation of WUAs at district level in case they exist so that cross-learning and needed follow-ups can be made easy.

Irrigation Management Transfer (IMT) and Participatory Irrigation Management (PIM)

In the process of Irrigation Management Transfer (IMT), it should be made pro-poor, which can ensure the viability of WUAs through assured labour contribution by small and marginal farmers. In many cases, small and marginal farmers are not part of the local committees and also not participating in important meetings indicating that they are not actually gaining the benefits that could be realized from IMT. Hence, raising awareness, especially among small and marginal farmers, is important.

For e.g., the Andhra Pradesh Farmers Management of Irrigation Systems (APFMIS) serves as a good model for IMT where farmers have equal decision-making powers within the WUA, regardless of their farm size (i.e., one farm and one vote). Hence, it is important that the government agencies/departments ensure that farmers easily understand the IMT process and their roles and responsibilities. In this context, reforming the election process within WUAs will be one of the ways to substantially improve participation of small and marginal farmers in the WUAs.

The National Commission on Farmers (NCF) had indicated that irrigation management could be transferred to WUAs for the effective distribution of water, fixing appropriate water charges, collection of water revenue and repair and rehabilitation of the structures thereby enhancing the efficiency of the irrigation systems. Recognizing the need for adequate legal support for Participatory Irrigation Management (PIM), the Ministry of Water Resources has introduced a Model Act which could be adopted by the States for better functioning of the PIM. Consequently, several states such as Andhra Pradesh, Bihar, Goa, Madhya Pradesh, Maharashtra, Karnataka, Kerala, Orissa, Rajasthan, & Tamil Nadu have either enacted new Acts or amended the existing Irrigation Act to facilitate the functioning of the PIM. In all, more than 55, 000 WUAs have so far been formed in 23 states covering an area of about 10 million hectares, where majority of these WUAs are active mostly in MI schemes. Evaluation studies of PIM done in Madhya Pradesh and Andhra Pradesh states have shown mixed outcomes about the PIM performance. Practices like introduction of warabandi system of water distribution, increase in crop acreage, reduction in water theft, timely availability of water even at tail ends, etc. have been reported as some of the benefits realized in these two states. However, many of the WUAs are not very active due to political interference, lack of funds, substandard work, poor participation of tenant and women farmers in these two states (GOI 2006).

Among the several factors that could influence the functioning of the PIM, actual quantity of water available mainly at farm level determines the level of functioning of PIM where PIM works better in command areas of low to moderate water deficit. Hand holding after transfer of irrigation management for at least five years also helps better functioning of the PIM indicating the importance of capacity building and training of the stakeholders. In this context, enabling legislation will be a prerequisite to implement the PIM on a large scale which will be possible through a strong support from village, political, and administrative circles. Unfortunately, the current focus of PIM is only on water distribution, and not on other factors that contribute for productivity enhancement and value addition. Innovative ideas and practices in a bottom-up manner could help the farmers to raise productivity of existing crops, diversify crop pattern towards more remunerative crops, and get more income.

Organizational restructuring of government agencies is also gaining importance in the recent years. The present structure of PIM under the state administration is looking unsuitable and weak for making needed policy decisions and their satisfactory implementation. Hence, there is a need to re-examine the possibility of a more efficient, productive, and accountable framework at the basin/project level particularly in large canal irrigation projects. After gaining adequate experience at the project level, restructuring can be considered at the State Irrigation Department level where it is possible to convert into an empowered authority that will have the responsibility, accountability, authority, and resources to ensure efficiency, equity, and sustainability in the functioning of the PIM (GOI 2006).

NGOs in Water Augmentation

In recent years there have been significant efforts to support locally based institutions to manage water. Several NGOs across the country have been successfully promoting water management with active participation of local communities. Three examples of NGOs' participation in tank irrigation rehabilitation are given below.

(a) The NGO PRADAN operates a rainwater conservation project in Alwar district, Rajasthan. The project mainly aims at reviving the traditional paal (bund) system of rainwater harvesting involving the local communities. It has helped villagers to build over 110 paals in several micro watersheds.

(b) Another NGO, Tarun Bharat Sangh (TBS) works with johads in 550 villages in the Ruparel river basin, Alwar district where their water harvesting works cover an area of about 6500 square kilometres. TBS had several hundred volunteers spread over in several villages, which have later acted as small grassroots organizations. People who benefitted have also contributed labour required in addition to some materials and cash. Each johad is designed with needed modifications to meet the unique needs of each site and group. TBS's works are low-cost compared to government structures. The check dams promote groundwater recharge and also support drinking water needs of human and cattle during post-monsoon periods. Construction of check dams is made in such a way they don't interfere with inflows to the tanks.

(c) DHAN Foundation, Tamil Nadu initiated an action research project in 1992 for rehabilitation of the tank irrigation systems. Later it was converted in to "Vayalagam Tank-fed Agriculture Development Program", with the focus on tank-based watersheds and reviving chains of tanks. DHAN foundation works in partnership with the state, private philanthropies, corporate institutions under corporate social responsibility, banking and financial institutions, and research institutions across the state. So for it has rehabilitated about 4000 tanks and village Ponds and took up soil and water conservation measures nearly in 100 watersheds, resulting in enhanced water availability. In addition to restoring the physical structures of the irrigation tanks, DHAN has facilitated proper maintenance of the tanks, efficient water management and improved cropping practices to ensure sustainable crop production and income.

13.2 Water Legislation in India

It is understood that water and irrigation development and management activities are the exclusive domain of the state government both in terms of water governance and institutions and also water infrastructure and water services development, except in the context of interstate river dispute and environmental aspects and pollution.

Article 246 of the Constitution of India deals with the subject matter of laws to be made by Parliament and by the Legislatures of the States. The allocation of responsibilities between the Centre and the States in respects of laws fall into three categories: (i) The Union List (List–I); (ii) The State List (List–II); and (iii) The Concurrent List (List–III). The subject of water is at Entry 17 of List–II, State List. The specific provision is: "State List 17. water supplies, irrigation and canals, drainage and embankments, water storage and water power subject to the provisions of Entry 56 of List–I". Hence, Entry 17 is subject to the provisions of Entry 56 of List–I, the Union List, which is: "Union List 56. Regulation and development of Interstate rivers and river valleys to the extent to which such regulation and development under the control of the Union is declared by Parliament by law to be expedient in the public interest".

The laws governing water resources and irrigation in India vary both at the central and state government levels. These state level acts vary and the implementation modalities are also different in each state. As the irrigation development is mostly governed by the state laws/acts, the water resources conservation and allocation, and development and management, operation, and maintenance of irrigation projects are important. The set of water legislations related to water allocation and conservation include the Rivers Board Act, 1956, the Interstate River Water Disputes (Amendment) Bill, 2019, the Water (Prevention and Control of Pollution) Act, 1974, and the Environment Protection Act, 1986. Further, at the state level, State Groundwater (Regulation) Acts, the State Irrigation Acts, State Water Tax/Rates/Charges regularization Acts, State WUA Acts, etc. play a key role in mobilizing the financial resources for O&M of the irrigation projects. To manage the interstate water sharing issues, the Tribunals have been established from time to time under the Interstate Water Disputes Act, 1956. The Water (Prevention and Control of Pollution) Act, 1974, the Environment Protection Act, 1986 give clear guidance in fixing standards and monitor water quality and permissible pollution levels in the water bodies and their uses. However, in practice, the implementation of the Act is not smooth, and has its own challenges.

Interstate conflicts are increasing over years. Tribunals have been constituted in the past for Narmada, Godavari, and Krishna water sharing claims between states. Although time limits have been prescribed for Tribunals, still the adjudication process is a long-drawn affair. Tribunal decisions are interpreted differently by Co-basin States resulting in further disputes in defining and implementing the Awards without any enforcement mechanism. This has led to a situation where in most cases there is no clarity about who can use what amount of water. In most cases, the tribunals follow the old norms of crop water requirement (duty) in allocation decisions thus creating a mismatch between actual requirement and allocation. Only recently, the Krishna Water Disputes Tribunal-II formed with reference to Section 89 of the A.P Reorganization Act 2014, has asked for reworking of project-specific crop water requirements using scientific approach.

13.3 Missions and Programmes

Several agriculture development programmes/schemes have been initiated from time to time by different Ministries/Departments which range from a single component/ commodity-based sectoral scheme to area-based integrated watershed development. Important schemes are: the Rashtriya Krishi Vikas Yojana (RKVY), the National Horticulture Mission (NHM), the National Food Security Mission (NFSM), the Rainfed Area Development Programme (RADP), the National Mission on Micro Irrigation (NMMI), Accelerated Pulses Development Programme (APDP), Integrated Scheme of Oilseeds, Pulses, Oil palm and Maize (ISOPOM) and National Initiative on Climate-Resilient Agriculture (NICRA). Similarly, various watershed development programmes are also being implemented by the Ministry of Agriculture and Ministry of Rural Development focusing on the development of rainfed areas and degraded lands including wastelands across the country.

Watershed Management: It is considered to be a vehicle for natural resources management (NRM) and rainfed area development where the participation of local community is high. Initially started as soil and water conservation programme, the watershed management programmes have undergone many changes over the years. Multiple ministries/departments and agencies have been actively involved in the growth and development of watersheds with an array of watershed schemes. Many such schemes have been observed for their overlapping activities over time and space. Three schemes, viz., Drought Prone Area Programme (DPAP), Desert Development Programme (DDP) and Integrated Wasteland Development Programme (IWDP) of Ministry of Rural Development have now been merged into Integrated Watershed Management Programme (IWMP) in 2008. Some of the schemes that are visible are: National Watershed Development Programme for Rainfed Agriculture (NWDPRA), River Valley Projects (RVPs), Catchment Area Programme in Flood Prone Rivers, Shifting Cultivation scheme of Ministry of Agriculture; Watershed Development Fund (WDF) administered by NABARD; Hill Area Development Programme (HADP) and Western Ghats Development Programme (WGDP) of Planning Commission; various afforestation programmes and other national and externally aided projects are among the various watershed-related schemes. Besides, the Mahatma Gandhi National Rural Employment Guarantee Act (MGNREGA) is mainly focusing on rainwater harvesting and NRM-related activities for creating durable assets to build production and livelihood system in the rural sector. Overall, the programmes are facing lot of challenges in terms of implementation protocols and management issues over space and time resulting in partial success.

Further, Government of India has formulated the National Agriculture Policy (GOI 2000), National Water Policy (GOI 2012), and National Policy for Farmers (GOI 2007) based on National Commission on Farmers (GOI 2006). The agricultural programmes, policies, and initiatives mentioned above can be broadly grouped under three categories: (a) technological reforms, (b) institutional reforms, and (c) marketing and prices. All these programmes and policies aim to increase crop productivity and overall farmer income. In order to enhance irrigation potential in the country, Ministry of Water Resources, Government of India has started implementing various development programmes over the years, viz. Accelerated Irrigation Benefits Programme (AIBP), Command Area Development and Water Management Programme (CADWM), National Project for Repair, Renovation and Restoration (RRR) of Water Bodies, and Scheme on Artificial Recharge (AR) of groundwater. Besides, there are many schemes being funded by state governments from time to time depending upon the need and they also complement the ongoing schemes in augmenting the water supplies.

13.4 Policies

13.4.1 Agricultural Policies

Food shortages in the early 1960s made the advent of the green revolution, which has raised the crop yield through improved crop technologies and seed varieties. To strengthen agricultural research and education, State Agricultural Universities were also been set up during the 1970s. With aim of enhancing the institutional lending to the agricultural sector, major commercial banks were nationalized, and new financial institutions such as National Bank for Agriculture and Rural Development (NABARD) and regional rural banks have also been established.

In the 1980s and 1990s the green revolution technologies were expanded to additional crops and regions, and emphasis was also given for high-value commodities such as milk, fish, poultry, vegetables, and fruits. Since 1991, after the economic reforms, the focus of agricultural policy shifted to improving the functioning of markets, reducing excessive legislation, and liberalizing agricultural trade.

National Agricultural Policy 2000 (NAP 2000)

Given the past development in agriculture sector and the need for further strengthening it, the National Agricultural Policy 2000 (NAP 2000) of India was formulated essentially to make efficient use of the available resources and the opportunities emerging as a result of advancement in science and technology applications in agriculture. The National Agricultural Policy (2000) aimed to attain a growth rate of more than four per cent per annum over the next two decades through efficient use of resources. Besides, the policy emphasized that the higher growth should be demand-driven that caters to domestic markets, and where possible maximize benefits from exports of agricultural products.

The key features of the National Agricultural Policy (NAP 2000) are: (1) four per cent annual growth rate over the next two decades, (2) optimum use of land water resources, (3) development of animal husbandry, poultry, dairy, and aquaculture, (4) updated national agricultural insurance scheme, (5) price stability and protection for farmers, (6) private sector participation through contract farming, (7) setting up of agro-processing units, (8) removal of restrictions on movement of agricultural commodities throughout the country, (9) capital inflow and assured prices and markets for crops, (10) protection of plant varieties through legislation, (11) adequate and timely supply of quality inputs to farmers (GOI 2000).

As a follow-up of the agricultural development programmes of the GOI, the National Commission on Farmers (NCF) was set up during 2004 by the GOI and some of the major observations of the NCF include farmers' immediate need for an assured access to and control over basic resources such as land, water, credit and crop insurance. Knowledge of crop choices, farming practices, and markets is also the key. It is also debated that agriculture and water can be dealt in the concurrent list.

The National Policy for Farmers, 2007, on recommendations of National Commission on Farmers, has provided for a holistic approach to development of the farm sector. The broad areas of its coverage include: (i) focus will be on the economic well being of the farmers in addition to production and productivity, (ii) asset reforms, (iii) water use efficiency, (iv) new technologies like biotechnology, information and communication technology (ICT), renewable energy technology, space applications, and nano-technology, (v) national Agricultural Bio-security System, (vi) seeds and soil health, (vii) support services for women, (viii) credit and insurance, (ix) Gyan Chaupals at village level with the help of ICT, (x) social security scheme for farmers, (xi) minimum support price (MSP) mechanism, (xii) food security basket. The NCF is indeed a derivative of the NAP with more focus on farmer-centric strategies and programmes.

Weakness of NAP 2000

The major weaknesses can be attributed mainly for not focusing on the following issues: (a) imbalances in input supplies with different quality inputs, (b) price variability and lack of timely credit supply in the use of some inputs, (c) no integration of agriculture and allied sectors resulting in inefficiency, (d) private sector domination in procurement, storage, and marketing, (f) poor coverage on small and medium farmers and local traders in marketing, (g) focus on profit-making enterprises only.

The impact of the policy couldn't be seen because of the following reasons: (i) Small and marginal farmers could not have easy access to processing and value addition; (ii) Processing and value added is done only for few commodities due to quality constraints and poor demand from consumers; (iii) The current extension system can't handle the latest technology aspects to farmers; (iv) Gap exists between the research stations and govt. extension departments; (v) Farmers lack skills to adopt high-tech technologies such as laser levelling, GIS and RS-based applications; (vi) Less awareness and confidence among the farmers on crop insurance schemes due to delays in payment of compensation; (vii) Climate change induces more risks in crop production which are not handled adequately in the policy; (viii) The magnitude of risks in crop production is changing every year; and (ix) Agriculture being the state level subject, lot of variations in marketing channels and market margin.

However, the measures indicated in the National Agricultural Policy 2000 are in conformity with the interventions needed for agricultural and irrigation modernization programmes where NAP will provide opportunities to the irrigation systems to deal with several of the modernization interventions through the central and state level implementable programmes. The provisions provided under different agricultural development programmes of the state will be fully exploited while making the recommendations for implementing the modernization strategies. The type and extend of the interventions to be implemented depend upon the irrigation project requirements of the region and the state with the stakeholders involved at all levels.

Agriculture reforms without much improvement in the market sector will be incomplete as most of the interventions are pro-agricultural and irrigation modernization. Farmer's access to market information through digital devices will be an added strength to these implementable interventions. The role of FPOs will be much significant as carriers of these market measures to reach out even the small and marginal farmers. The type and extend of these interventions will be decided based on the irrigation project priorities in the region.

National Water Policy 2012

The National Water Mission (GOI 2011) aims at conservation of water and ensuring its equitable distribution both across and within States through integrated water resources management thereby improving the present water use efficiency by 20%. As a follow-up, National Water Policy was formulated by the Ministry of Water Resources to govern the planning and development of water resources in a sustained manner and their optimum utilization. The first National Water Policy was adopted in September, 1987. Subsequently, it was reviewed and updated in 2002 and later released in 2012. The main emphasis of National Water Policy (2012) is to consider water as an economic good and to promote its conservation and efficient use. Over the years, there is a paradigm shift in approach from service provider of water to facilitator of service and delivery. The key points of the National Water Policy 2012 are summarized here:

1. Planning, development, and management of water resources should be within a common integrated perspective.
2. Management of the water as a common pool resource should be under a national-level legal framework. The existing water Acts can be revisited.
3. Minimum ecological requirements should be considered.
4. River basin concept is important for all hydrological planning, inter-basin transfers of water, etc.
5. Climate change adaptation strategies, adoption of climate-resilient technologies, etc. should be promoted.
6. Mapping of aquifers should be done to know the quantum and quality of groundwater, reducing the overexploitation of groundwater, etc.

7. Watershed development activities be integrated with the MGNREGA.
8. Water footprints and water auditing for efficient use of water are necessary.
9. Recycle and reuse of wastewater are important.
10. Water saving in irrigation and minimizing wastage of water and electricity be promoted.
11. Water should increasingly be subjected to allocation and pricing on economic principles.
12. Establishment of a Water Regulatory Authority (WRA) in each state to fix and regulate the water tariff. Water users associations (WUAs) should be given statutory powers.
13. Conservation of river corridors, water bodies, and infrastructure. Efficiency benchmarks to be prescribed in planning and implementation of projects.
14. Flood forecasting, coping mechanisms, and relevant measures to prevent floods and drought are important.
15. A permanent Water Disputes Tribunal at the Centre to solve all the issues in an efficient manner is important.
16. International agreements with neighbouring countries on a bilateral basis for the exchange of hydrological data on international rivers on a near-real-time basis.
17. All hydrological data, other than those classified on national security considerations, should be put in the public domain.

There is an urgent need to develop suitable institutional and policy frameworks at local, national, and international level for the expected changes to occur. However, the current policy does not adequately address the issues of agriculture. It is imperative to initiate efforts for designing such a policy which could provide a framework for sustainable water use in India by different sectors without a compromise on development and future needs of enormously growing human and livestock population. Some of the areas such as the rainwater harvesting, groundwater recharge, improving overall water use efficiency, climate change and multiple uses of water need to be revisited. Further, increased focus needs to be given to bridge the gap between the potential created and potential utilized as this will be comparatively cost effective than creating new irrigation potential at a higher cost (Table 13.1).

The state water policy and state water legislations (Acts) deal with issues such as irrigation development, command area development, irrigation tax/rates, PIM, groundwater regulation, improving efficiency, etc. In terms of institutional/organizational aspects of water resources and irrigation, reforms are lacking in the following areas:

- Greater participation by water users in irrigation system MOM promoting WUA participation in irrigation scheme MOM through trainings, O&M finance devolution, and bulk water allocation.
- Improved water resources information/data systems, which require reliable and timely data on various aspects of the irrigation systems to facilitate decision-making and practice in such areas as water accounting, benchmarking, performance assessment, modernization planning, scheme MOM, etc.

Table 13.1 Strengths, weaknesses, opportunities, and threats (SWOT) Analysis of the National Water Policy (2012)

Strength	Weakness	Opportunities	Threat
Includes policy principles and strategies Performance improvement linking investment options, Infrastructure development, O&M improvement, Cost recovery, Climate change and stakeholder Participation Overall emphasis for improved water use efficiency	The policy principles and strategies outlined are mostly theme areas and no specific focus on key or thrust areas No measurable key performance indicators, targets, or timelines for intervention in the short, medium and long term planning Entirely qualitative in nature Less focus on multiple uses, groundwater recharges, and climate mitigation	Some of the key areas such as climate change, stakeholder participation, water quality improvement can be taken up in the future investment programmes Possible to prepare detailed irrigation investment plans and implementable programmes using the framework outlined in the Policy Gives scopes for institutional reforms and refinement for preparing future water policy	Increasing interstate conflicts may affect the implementation of the some of the policy recommendations Socio-political issues may emerge as constraints in major investment programmes and technology adoption Intersectoral Impact of climate change may be serious in certain (vulnerable) regions

- Interstate water sharing and focus on multiple uses of water such as domestic water supply, hydropower, industry, and environmental uses.
- Private sector participation in irrigation management.
- Climate change mitigation and river pollution.

13.5 Water Governance

Water governance is widely acknowledged as an important factor for sustainable development of water resources in the country. Governance may be understood to be the traditions and institutions by which authority is exercised (Kaufmann et al. 1999). The governance in general can also refer to a government's capacity to formulate and implement sound and acceptable policies involving stakeholders and institutions. Good governance therefore is considered to be broader than institutions and includes the people, the state, and their interactions (United Nations Development Programme 2014).

A concerted effort is needed for resolving conflicts and developing a shared vision for the efficient use of water resources. Hence, a good water governance necessarily demands a systematic Centre's response where the Centre and the States work in a partnership mode with defined responsibilities. The three pillars of governance to achieve better performance are efficiency, effectiveness, and commitments which are influenced by several factors as explained in Fig. 13.1.

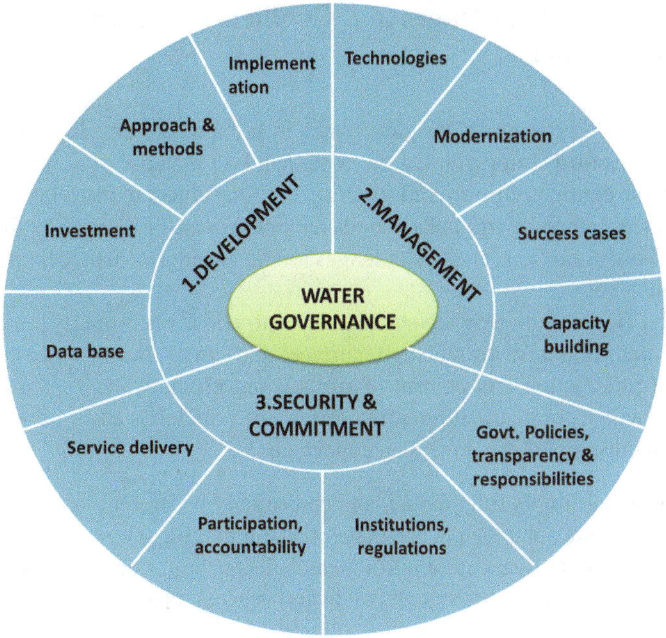

Fig. 13.1 Three pillars of water governance (*Source* authors' compilation)

It is further clear from the governance diagram, the three broad dimensions of governance are: (a) institutions and regulatory mechanism, (b) participation and accountability, and (c) service delivery to the stakeholders. An empirical study on irrigation governance in India had shown that two dimensions of irrigation governance viz., institutions and regulatory mechanism, and service delivery to the stakeholders had positively related in influencing the irrigation performance (Kannan et al. 2019).

Recently, two important bills were passed by the Lok Sabha viz. the Interstate River Water Disputes Amendment Bill 2019 and the Dam Safety Bill 2019. These bills mainly focus on how to deal with the emerging challenges of interstate water disputes, safety of existing/aging dams and their governance mechanisms. The Interstate River Water Disputes Amendment Bill 2019 therefore mainly aims to improve the interstate water disputes aspects through a permanent tribunal supported by an institutional mechanism, the dispute resolution committee. The Dam Safety Bill 2019 on the other hand aims to deal with the risks of old/ageing dams using an effective institutional framework involving committees and authorities both at national and state levels.

13.6 Emerging Challenges and Conflicts in Water Governance

- Centre vs state powers: Water governance in India is perceived and practised as the states' exclusive domain. On the other hand, in the case of major river water sharing, the centre is empowered to act as they are subject to interstate river water governance. Interstate tribunals awards are not properly followed by the riparian states. These conditions create challenges for the water governance both at Central and State level.
- The River Boards Act 1956 legislated under Entry 56 is not properly implemented.
- Fragmented policies: fragmented surface and groundwater policies due to different jurisdiction and control of surface and groundwater systems. Lack of political will in implementing some important policies is also a constraint for better functioning of the governance mechanism.
  - Data availability is fragmented, and controlled by different agencies. In many cases, data are not updated, and inadequate for making investment and management decisions. Data on environmental flows, pollution, salinity and water logging, interstate flows, cost recovery are not properly documented and utilized.
  - Increasing over-abstraction. No concerted efforts to deal with the over-abstraction and overuse of water in different regions which need the cooperation of both centre and state government departments.
- Pursuing National Projects: Lack of push on inter-basin transfers is creating an uncertainty in the effectiveness of governance mechanism.

13.7 Conclusions and the Way Forward

In sum, the water governance phase of the India's water future with embedded investment and management phases (as discussed in Chap. 1) is expected to address the emerging challenges such as long-term national water security and sustainability, the risks of climate change, interstate conflicts and the increasing level of water pollution. These are the key areas where the Centre and the states can work in an active partnership mode involving institutions and regulatory mechanisms. For better water governance, the following are important:

(a) Centre-States Dialogue: The Centre can work with the concerned states in building a credible institutional framework for collecting and sharing important water-related data and states can improve the delivery of this essential service to the stakeholders,

(b) Multi-stakeholder Approach: In view of multiple stakeholders (farmers, urban communities, industry, NGOs, researchers, and government) influencing the level of water use, an effective governance mechanism can aim for a joint decision-making for the efficient use of the water resources in the country. For example, the establishment of active stakeholder forums and councils can help to a large extent, where respective ministries, stakeholder groups representatives, and NGOs are included and an effective platform for consultations is established.

References

Chitale MA (1992) Development of India's river basins. In: Water resources development, vol 8, no 1

Centre for Science and Environment (CSE) (1991) Floods, floodplains and environmental myths. State of India's Environment: A Citizen's report. CSE, New Delhi

Central Water Commission (CWC) (1997) Legal instruments on Rivers in India. Constitutional provisions, central laws and notifications, vol 1. Government of India

Government of India (2000) National Agricultural Policy, New Delhi

Government of India (2006) National Commission on Farmers, New Delhi

GOI (2007) National Policy for Farmers. Ministry of Agriculture. New Delhi

Government of India (2011) Comprehensive Mission Documents for National Water Mission. Ministry of Jal Shakthi. New Delhi

Government of India (2012) National Water Policy, New Delhi

Kannan E, Bathla S, Das GK (2019) Irrigation governance and the performance of the public irrigation system across states in India. Agric Econ Res Rev 2019, 32 (Conference Number), 27–41

Kaufmann D, Kraay A, Zoido-Lobaton P (1999) Governance matters. Policy Research Working Paper 2196, World Bank, Washington, D.C

Newson M (1992) Land, water and Development-River Basin systems and their sustainable management. Routledge

United Nation Development Programme (2014) Governance for sustainable development: integrating governance in the post-2015 development framework. Discussion Paper, United Nations Development Programme (UNDP)

World Bank (1998) Report on inter-sectoral water allocation, planning and management under India water resources management sector review

Chapter 14
India's Water Future—A Way Forward

Abstract Problems and challenges faced by the Indian water sector warrant both potential supply and demand-side interventions. Several interventions have been suggested and these interventions have also been ranked by experts to know their importance in bridging the future supply–demand gap. A matrix of main as well as a sub-set of interventions highlighting their specific focus/purpose, their coverage in the past, and their potential in bridging the supply–demand gap covering in the short and long-term implementation phase is also presented. The supply-side interventions could bridge the supply–demand gap by about 40% and the demand-side (mostly technology and institutions related) interventions could bridge the gap by about 60%. Selection of these interventions to suit different segments of the river basins is emphasized. Need for a portfolio of implementable policies and programmes that could integrate both supply and demand-side interventions supported by well-functioning water institutions (governance) is also emphasized.

Keywords Canal modernization · Crop diversification · Desalination · Inter-basin transfers · Water governance · Digital agriculture

14.1 Introduction

India's water sector faces a turbulent future with emerging climate challenges that need to be addressed on a priority basis. Unless the policies and technical interventions are adopted and implemented as planned, providing water to the different sectors and managing water sustainably will not be easy. Many drivers will continue to influence India's water availability and use in the future. As discussed in Chap. 2, the traditional drivers are the increasing population, urbanization, increasing food demand, growing industrial and commercial activities, new technologies, and fast changing lifestyles. The non-traditional drivers are the climate extremities, steady economic growth, and change in consumption patterns which also influence the future demand for water. Under this backdrop, this chapter focuses on two key issues: a) the need and priority to address the key challenges facing the irrigation sector both

K. Palanisami and U. S. Nagothu, *India's Water Future in a Changing Climate*,
Advances in Geographical and Environmental Sciences,
https://doi.org/10.1007/978-981-97-1785-9_14

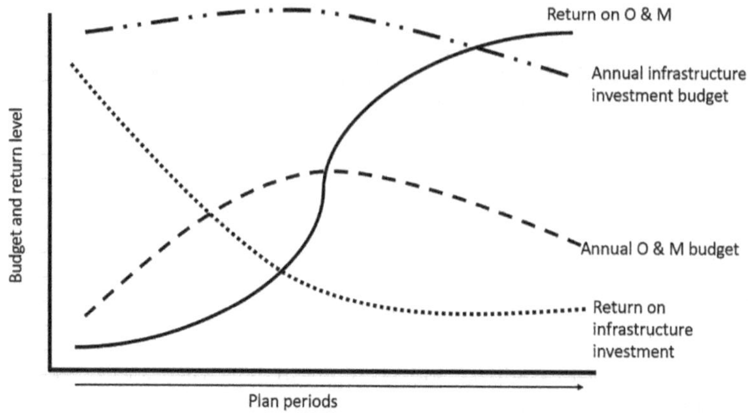

Fig. 14.1 Comparison of India's irrigation investment and return levels

from supply and demand side, and b) more focussed interventions that are relevant for improving water use efficiency sustainably in the future.

Comparing to a reasonable global good practice with that in India, the infrastructure stock peaks fast and then tails off in the subsequent periods But, as the infrastructure stock grows, the financial demands for maintaining and replacing them increase (Briscoe and Malik, 2006). In the Indian case, both the annual irrigation investment budgets and the O&M budget allocation grow but at different levels. Regarding the return to investment, the return to O&M investment is higher over the years than the capital investment on infrastructure development indicating the capital investment in the later plan periods goes on developing projects in marginal (less potential) areas where the return to investment is comparatively low (Fig. 14.1). As a consequence, project performance is poor in terms of poor water quality and highly varying supplies over both space and time. It is implied that India's development of water infrastructure has not been accompanied by an improvement in the governance of water resources and water (O&M) services. Hence, to improve the overall performance of the irrigation projects, management interventions through modernization programmes are highly warranted.

14.2 Need for Water-Related Interventions

As indicated in the earlier chapters, demand for water in the agriculture, domestic, industrial, environmental, and energy sectors will be increasing significantly in the next decades. Domestic, energy, and industrial water demand will increase more rapidly than demand from agriculture. Agriculture will still remain the sector with most use indicating the need for better interventions to improve water use efficiency.

The three crucial natural assets viz., water, land, and biodiversity will be supplemented by manmade assets like inputs, credit, marketing, and processing. On the other hand, inefficiency, scarcity, and poor governance are the three major liabilities resulting in poor performance of the irrigation sector. Hence the interface between the assets and liabilities should be carefully examined and interventions need to be introduced. Considering the importance of irrigated agriculture and its role in meeting the SDGs (zero hunger, nutrition security, income security, gender equality, clean water and sanitation, climate action, etc.) there is an urgent need to transform irrigation sector in the country through paradigm shift utilizing the complementarities between development, management, and governance phases of the irrigation sector. This calls for appropriate supply and demand-side interventions to be implemented through enabling policies. This chapter summarizes the problems and challenges faced by the irrigation sector and suggests interventions accordingly.

14.3 Major Issues and Possible Interventions

(a) Creation of new major and medium projects

Irrigation coverage still accounts for only about 54% of net sown area. This can be attributed to: the slow expansion of area under major and medium irrigation projects because of the difficulties in completing new/ongoing projects that give maximum multiple benefits; difficulties in land acquisition, which is long process subject to protracted litigation and delays in payment of compensation for the land. Besides, the challenges in inter-basin water transfers and lack of investments are also hampering the water management infrastructure improvement at the state level. In many states, the irrigation charges as well as their collection rates are low resulting in poor accountability (Shah, 2013).

Addressing groundwater irrigation challenges and sustainable use of groundwater resources will be one of the key drivers of agricultural production in the future. By 2050, the share of non-renewable groundwater use will be 30% compared to 25% at present. Investments in small-scale structures for groundwater recharge in locations where there are no adverse impacts on downstream users and sustainable development of groundwater in areas where it is adequately available are important future measures that need to be focused.

Expansion of micro irrigation adoption and changing cropping pattern could compliment groundwater use and offer significant opportunities for increasing water use efficiency and reducing overexploitation. Spreading awareness and competence in water saving technologies must be a high priority in the future programmes.

(b) Modernization of existing Irrigation projects and groundwater

A concerted effort is required to expedite modernization of ongoing projects whose water use efficiency is a mere 30–40%. Performance improvement of the existing projects even with increased budget allocation has not translated into "on time" and "on demand" availability for farmers, which needs to be improved. In future, funds must be allocated across projects according to their likelihood of creating additional irrigation potential (i.e., either additional area-wise or season-wise), adaptation to climate extremities, and vulnerabilities of the regions.

In the case of irrigation tanks, the major issues with tank irrigation in the country are: gradual declining of tank irrigated area over years; decrease in carrying capacity of tanks mainly due to encroachment and siltation; erratic rainfall pattern that has resulted in uncertain tank storage and crop failure; decrease in tank multiple uses resulting in disappearance of various tank-based local institutions; poor O&M budget and non-functioning of the WUAs and inadequate tank rehabilitation programmes ignoring the tank cascade approach.

In the case of groundwater irrigation the major challenges are: depletion of water resources within aquifers due to increase in the number of wells and their increasing abstraction capabilities; availability of cheap/subsidized/free electricity that has shifted farmers' dependency from surface water to groundwater; lack of regulations to control the groundwater extraction; the perverse groundwater–energy nexus; poorly performing watershed development activities in recharging the groundwater and lastly the climate change impacts in groundwater recharge.

Cross-learning of the successful cases of canal modernization and groundwater irrigation practices should be encouraged between the different states. New research and development activities must be promoted to strengthen the operation and better maintenance of the surface and groundwater systems.

(c) Wastewater treatment and reuse

Given the rate of urban population growth, the increasing amounts of wastewater discharge in India should be seen as an opportunity to improve the reuse and accordingly improve facilities to treat sewage water in the future. Secondary treated wastewater can be used to recharge the aquifer systems to benefit irrigated agriculture. As discussed in Chap. 6, already several successful cases of wastewater reuse are happening in India, and up-scaling programmes should be planned on priority basis. A successful case of wastewater treatment and reuse is that of Singapore city, where wastewater is recovered and treated to drinking water quality. Treated wastewater presently meets 30% of Singapore's water needs, with plans to triple volumes by 2060 (IIASA 2016). Lessons from such cases could provide a good learning for the major cities in India. Advances in energy and nutrient recovery should be properly utilized so that decentralized wastewater treatment and reuse could be financially viable in the future.

(d) Inter-basin water transfers

Inter-basin water transfers could be justifiable on the grounds that the burgeoning water demand could be met and a more assured surface water supply in major canal systems is ensured for the future. However, implementation of the inter-basin transfers is subject to several socio-political, technical, hydrological, financial, and environmental issues, which need to be addressed very carefully. For example, many of the interstate rivers lead to disputes in sharing the water among the riparian states involving substantial costs and delays. A viable mechanism is needed to deal with the interstate disputes and their settlements. A scientific approach in reworking the actual water demand of the basins should be considered. Unless, there is a consensus, interlinking of rivers is not going to be easy, though specific links have already been developed. Given the climate change impacts on the changing river flows, the river water treaties with neighbouring countries must also be re-examined and amended accordingly to sustain the water supplies.

(e) Climate-smart technologies and enhancing water productivity

Enhancing the water productivity in irrigated systems: A modest increase (1% annually) in the current water productivity could meet the additional consumptive water demand for grain crops in the future. With a 1.3% annual increase it could in fact eliminate the consumptive water demand of all crops in the future (Upali et al. 2007). Thus, it is important to identify regions with low and high crop yields, and low and high potential for increasing water use efficiency. Investment in research and extension and other physical interventions should be provided for increasing both yield and water use efficiency in these regions. Already various government initiatives are targeting an increase in overall water use efficiency of irrigation projects by 20% and this will have a major impact in meeting the future water demand (GOI 2008, 2017).

Enhancing the crop productivity in rainfed systems: India receives about half of its annual rainfall in 15–20 days causing intense floods and prolonged droughts. Given the average annual rainfall of about 1200 mm, India's major concern is how to manage the water supplies to minimize the recurrent water stress. In fact, this water stress is primarily a blue water issue, and opportunities still exist for improving management of rainfed areas that are mostly benefitted from the green water (Rockström et al. 2009). This means conserving rainwater in soil (green water) followed by harvesting rainwater (blue water), recharging groundwater (Pathak et al. 2009), and making its efficient use for crop growth by minimizing wasteful evaporation losses (Rockström and Falkanmark, 2000). In this context, improvements in agriculture and water management practices in the hill regions are also gaining importance in improving crop productivity.

Thus, if the existing rainfed and irrigated land could be made more productive, additional land and consequent additional irrigation water requirement would be negligible, thereby contributing for the reduction in water required. This can be complemented with the use of improved crop varieties, which can give higher yields using comparatively less water.

Integrated watershed management emphasizing on efficient use of green water (soil moisture), rainwater harvesting and its efficient use through drip, sprinklers must be promoted. Though the initiatives have started through country-wide drip irrigation schemes, the application on field is still below the potential. Information on when and how much water is needed and to be applied will help in precision management of irrigation, which is possible now through digital solutions/apps and drone technology.

The complementarities between rainfed and irrigated agriculture should be explored in meeting the future water demand in India. Use of green water can dominate food production where the consumptive use of green water is almost four times higher than that of blue water (Rockström et al. 2009). Hence increasing the productivity and production in rainfed systems will minimize the future water demand in irrigated agriculture. IWRM adoption can be better attempted by convergence of several departments related to agriculture, irrigation, drinking water, sanitation, and rural development, for e.g., the GoI has initiated the convergence of programmes through *Jal Shakti* Ministry which must be taken forward vigorously. However, in many regions, IWRM couldn't be successful because the policies/interventions suggested failed due to lack of cooperation between farmers, civil society, and governments (Shah 2023).

With climate change and frequent droughts and floods, development and adoption of the climate-smart technologies both in the irrigated and rainfed typologies must be another priority for the future. Climate-smart agriculture (CSA) technologies and precision agriculture, adopting market-based measures like carbon finance to promote green house gas mitigation and increase farmers' income are important measures to be planned and implemented. Furthermore, revamping of the existing agricultural support policies towards sustainability is also warranted (Gulati and Thangaraj 2023).

(f) Reworking the crop water requirements

Water allocations and crop budgets in the irrigation projects are based on the crop water requirements data or norms developed in the past. Use of climatological approach for re-estimation of crop water requirements adopting modern approaches incorporating the weather parameters will be relevant in making water allocation decisions in the future.

(g) Capacity building and out reach

One of the major constraints in technology adoption by farmers in India is lack of knowledge about technologies. About 59% of farmers do not get any extension support, only 11% is getting support from government programmes and the remaining 30% of the farmers get support from other farmers, media, and private agencies (GoI 2013). This lack of access to knowledge and training had severely limited their ability to increase productivity and income and reduced vulnerability. This is because the public agricultural extension system is loaded with multiple responsibilities such as the distribution of subsidies, inputs, etc. Also there is limited time available for the extension staff to get the updated skills in latest agricultural practices and technologies (Glendenning et al. 2010). Researchers and extension staff must adopt demand-driven approach for providing the integrated solutions to farmers' complex issues through building public–private partnerships and digital tools to improve extension services.

(h) Crop diversification and markets

Crop diversification towards less water intensive crops is being debated for decades without any visible impact. At present farmers follow the minimum support price (MSP) option to produce what they like to produce, which normally results in growing water-demanding crops like paddy and sugarcane. For example, growing paddy in Punjab and sugarcane in Maharashtra with groundwater irrigation results in overexploitation of scarce water resources. Crop diversification is possible provided farmers are assured of effective marketing channels for the non-water intensive crops with guaranteed price. In the recent years, Government of India (GoI) has proposed "*One district one product*" (ODOP) concept, which facilitates the complementarities of speciality of the product, skill development, capacity building, value addition, and better linkages with the markets. Farmer Producer Organisations (FPOs) were initiated by government all over the country. But they need to be further strengthened in the knowledge delivery systems, value-chain, and marketing facilities. "*One District One Product*" (*ODOP*) through empowered FPOs can ensure right price for the farmers and consumers as well as create rural employment which can facilitate crop diversification and ultimately less water demand for crop production.

(i) Digital agriculture

Awareness about digital agriculture is spreading particularly among the young farmers involved in agriculture. In this context, an integrated platform to provide one-stop advisories as well as inputs and sale of produce using innovations is gaining importance. As discussed in earlier chapters, the main bottlenecks for scaling up new technologies/products with marginal and small farms are lack of timely and reliable information on marketing channels and prices. To keep pace with the current challenges and aspirations of the new generation farmers, information and communication technology (ICT), remote sensing (RS), GIS, artificial intelligence (AI), cloud computing, drones and robots, nanotechnologies can offer wide opportunities towards digital agriculture. Several states have introduced agricultural technology platforms such as e-agri, e-NAM (electronic national agricultural market) portals, etc. for creating awareness and use of the new technologies. This is also an area where Public–private Partnerships can play a key role in the implementation of holistic and integrated platform for transforming rural areas. Compartmentalization needs to be overcome by strengthening data sharing mechanisms, ensuring quality and suitable control measures to avoid misuse of data to be used on the digital platform.

(j) Good governance

It is timely that a new comprehensive water law should be planned and suitably enacted for integrated water resources development and management to overcome the fragmented approach to water as multiplicity of prevailing laws relating to water has resulted in problems of the enforcement. Autonomous Water Resources Regulatory Authority at state level (similar to Electricity Regulatory Authority) should be constituted to look into all issues related to water. This will help move away from a narrow engineering-construction perspective towards a more multidisciplinary understanding of water (Shah 2013). To proceed along these lines, effective and efficient water governance is most important in today's context, which will support regular review and amendment of policies, laws, and regulations. Further, the previously proposed as well as existing water policies, their status and the weaknesses must be improved as needed. It is expected that good irrigation governance significantly improves the performance of the irrigation systems in each state and paves the ways for meeting the future water supply and demand. Hence, formulation of a novel institutional arrangement or altering the existing institutional arrangement is urgently needed. Some of the key components of the water governance are already explained in Chap. 13.

14.4 Inter-Linkage Between Investment (Development), Management, and Governance Phases

Even though India has now become a major global economy power, its water planning and management are still struggling to sustain. Consequently, the complexities and magnitudes of water development and management in the country have steadily increased over the past four decades. As indicated in Chap. 1, the inter-linkages between the three phases are strong. The development phase has most of the initial investments made under the five-year plans. In the management phase there was a paradigm shift towards managing the created potential where several interventions have been successfully tested in the past with varying levels of success. The lessons learned have necessitated the importance of governance phase which encompasses the converging roles of different institutions with transparency and accountability. It is to be remembered that governance phase can't be isolated from the investment and management phases. So, in the future of the irrigation sector, the expected contribution of these three phases must be in the ratio of 30: 30: 40 compared to what we had in the past with two phases viz., development and management phases (i.e., 40:60) where the governance is embedded within the management phase itself.

Looking at the supply side, climate change is playing havoc in the recent years where the flooding and prolonged droughts are common now and will be expected to continue in the future as well. It is expected that the overall supply may decrease by 10–15% from the current level under varying levels of rainfall. On the demand side, the problems associated with the management of the water resources are also increasing resulting in overall low water use efficiency and productivity per unit of water. The interventions broadly discussed above could pay the way for identifying specific supply and demand-side interventions to bridge the supply–demand gap.

14.5 Matrix of Interventions

A quick survey was conducted to get the expert views and suggestions on the type of interventions needed and their ranking. A total of 65 experts were contacted with varying backgrounds such as: engineering, agriculture, rural development; research, administrators, policy makers, NGOs, lawyers both at national and international levels. The matrix of interventions and the mean ranking of various interventions are given in Table 14.1.

Looking at the role of different interventions and the possible water saving, it is expected that supply-side interventions could help bridge the supply–demand gap by about 40% and the demand-side (mostly agricultural-related) interventions could bridge the gap by 60%. The 2030 Water Resources Group study has indicated that the agricultural interventions can collectively close the gap by 80% and that includes both irrigated and rainfed crop production measures and the remaining 20% gap could be bridged mostly through the rehabilitation of existing irrigation works and completion of pending projects (2030 Water Resources Group 2009). Again, within supply-side interventions, new investment in infrastructure will account for 30%, modernization of the existing projects will account for 40%, adoption of success cases account for 10%, and others such as wastewater reuse, watersheds & groundwater recharge, R&D activities, etc. about 20%.

In the case of demand-side interventions, adoption of technologies account for 30%, crop diversification 10%, crop water requirement re-estimation, hill agriculture development, micro irrigation, etc. 20%, better governance 30%, and others about 10% (Figs. 14.2, 14.3 and 14.4). Since the governance phase of the water sector is also embracing the investment and management phases, it is understood that good water governance will play a key role in the future of the water sector in India where it is expected to bridge the supply–demand gap by about 60%. Hence future policy prescriptions can target for good water governance, which will be a necessity for India.

Given the list of the above interventions, the marginal cost curves can provide a microeconomic analysis of the cost and potential water saving that could be possible. The technology cost curve also provides a benchmark of existing technologies and their potential cost to deliver additional water (2030 Water Resources Group 2009). The ranking of few of the above interventions was compared with the cost curves and found to be matching with the unit cost but the potential water saving under each intervention was slightly varying and this might be due to the assumptions made on the quantum of water saving.

14.6 Implementation of Interventions—River Basin Focus

Past experiences show that in spite of the implementation of several modernization interventions, the performance of the projects is still poor indicating the need for relooking the approach to be used while introducing these interventions. River basin

Table 14.1 Matrix of interventions and their rankings

Intervention	Rank (1–2)	Interventions	Rank (1–10)	Remarks
Supply side	2	*Interventions types*		
		1. Developing new irrigation projects (a) Major and medium (b) Minor tanks and small storage ponds	3	
		2. Modernizing/rehabilitation existing projects (a) Major and medium (b) Minor tanks—urban and rural tanks, (c) Rejuvenation of open wells and water harvesting structures (d) drainage improvements	1	
		3. Recharge aquifers through (a) Taming floods (b) Rainwater harvesting structures (check dam, percolation ponds, etc.) (c) Managed aquifer recharge	2	
		4. Wastewater recycle & reuse	4	
		5. Underground piped irrigation network	8	
		6. Adoption of national & international best cases & methods (e.g., canal automation scheduling, automatic weather stations, conjunctive use & bench marking, etc.)	5	
		7. Inter-basin water transfers	7	
		8. Water treaties with neighbouring countries	10	
		9. Improving policy framework taking into account the nexus angle	6	
		10. New R & D activities	9	
Demand side	*1*	*Intervention types*	*Rank (1–10)*	
		1. Climate-smart agriculture (CSA) practices/ technologies (farm level): (a) soil, crop, water (b) transformative investment	2	
		2. Project level interventions: (a) crop diversification, (b) micro irrigation in canal commands (c) Reworking crop water requirements under changing climate	3	
		3. Good governance (institutions, laws, taxes, subsidies, incentives, pricing of water)	1	
		4. Digital agriculture, precision agriculture using geospatial and ICT tools, RS, GIS Development of DSS and mobile apps, agri-portals, etc	5	

(continued)

Table 14.1 (continued)

Intervention	Rank (1–2)	Interventions	Rank (1–10)	Remarks
		5. Improved marketing, insurance products	4	
		6. Exploiting hill agriculture potential	10	
		7. Strengthening rainfed agriculture	9	
		8. Awareness, capacity building, and training of farmers and field staff	6	
		9. Empowering entrepreneurship in agriculture including new generation farmers; public–private partnership business models	8	
		10. New R & D activities	7	

Note The rankings differ according to the experts and their background and only the mean ranks are given in the table

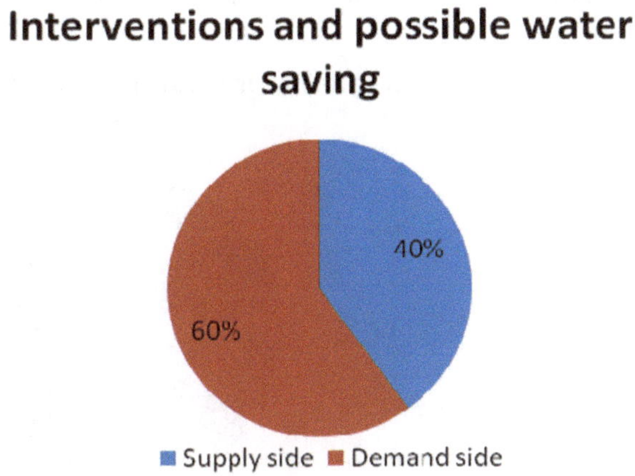

Fig. 14.2 Interventions and possible water saving

structure facilitates the identification of issues in the head, middle, and tail regions of the basin. For example, the head region of the basin may be affected by major land-use changes resulting in more land slides, heavy runoff, sedimentation during heavy rains affecting the reservoir storage and river flows. In the middle regions it could affect water distribution and water quality issues, and the tail region could be affected by water scarcity, land submergence, and drainage during heavy rains. Hence the interventions outlined in the above matrix can be examined for their suitability to different segments of the river basin and implemented accordingly.

Given the above matrix, a detailed list of general and focussed interventions along with their purpose, period of coverage (short term and long term) was prepared and presented below (Tables 14.2 and 14.3). Even though most of the general interventions are looking familiar in the water sector, the significance of these interventions is different in terms of their (a) specific focus, (b) actual implementation attempted

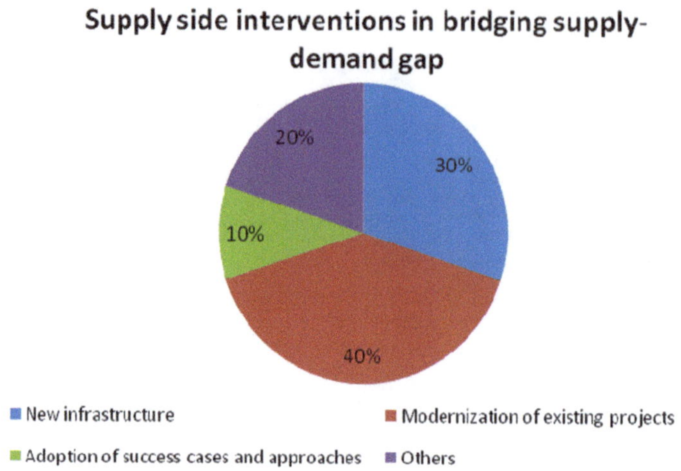

Fig. 14.3 Supply-side interventions in bridging supply–demand gap

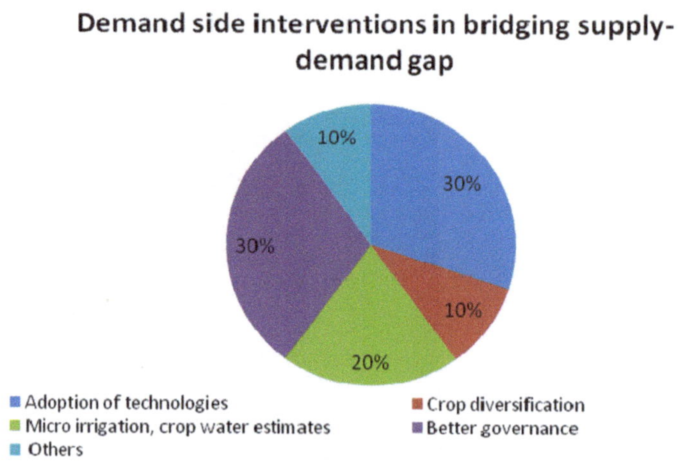

Fig. 14.4 Demand-side interventions in bridging supply–demand gap

in the past, and (c) their potential to mitigate the future supply–demand gap. Further, the scope of these interventions to implement them in the short term or long term is also indicated. Hence this matrix could be useful to policy makers and planners in selecting the appropriate type of interventions needed.

Table 14.2 Water supply-side interventions and their potential to reduce supply–demand gap

Broad interventions	Specific interventions needed	Level of interventions already attempted	Potential for reducing S–D gap	Long term	Short term	Remarks
1. Developing new Irrigation projects (a) Major and medium projects (b) Minor irrigation tanks (c) Underground piped irrigation network (d) Command area development works	Major & medium projects: Creation of new projects including green infrastructure projects in locations where hydrological and ecological and socio-economic and environmental parameters are favourable without affecting the functioning of the existing projects In the case of minor irrigation tanks, Eastern India provides more opportunities for small tanks and ponds with multiple uses. In other regions, new tanks can be planned where flooding due to climate change induced heavy rainfall and favourable terrain characteristics without affecting the functioning of the existing structures Piped water networks are possible where land acquisition for channels is a problem Include CAD works from the very beginning as an integral part of the project formulation and implementation Creating additional storage in locations where there is potential and where the value of water is high Adopt reservoir management strategy to improve adequacy, and efficiency and reduce losses	(a) High (b) Low (c) Low (d) High	(a) Medium (b) High (c) Medium (d) Medium	(a) X (b) X (c) X	(d) X	

(continued)

Table 14.2 (continued)

Broad interventions	Specific interventions needed	Level of interventions already attempted	Potential for reducing S–D gap	Long term	Short term	Remarks
2. Modernizing existing projects (a) Integrated basin management and reservoir management strategy to improve adequacy, efficiency and reduce losses canal automation with measuring devices (b) Converting canal to micro irrigation projects (c) Drainage improvements (d) Water accounting, Bench marking, RS & GIS	Stabilizing the storage and irrigated area through modernization of major and medium projects adopting latest modernization techniques and practices. This involves both structural and non-structural interventions. Agricultural sector interventions will be playing a major role with the adoption of climate-smart soil, crop, and water management practices. Use of ICT techniques for canal automation. Bench marking should be incorporated in the modernization plans. The SIMP approach can be examined and followed up according to the project type. Improving project efficiency by 20% Projects that qualify for the conversion of surface to micro irrigation-based irrigation should be selected and MI systems should be installed. Already few projects are showing promising results in the country In the case tanks, stabilizing the created storage through tank modernization involving catchment treatment, desilting, canal lining and on-farm water management can be done. Tank cascade or tank chain approach can be followed in tank modernization with participatory approach. Supply channels connecting the tanks should be rehabilitated to facilitate water transfer to downstream tanks during periods of high inflows. Urban tanks have different hydrology than rural tanks and need rehabilitation in improving the storage capacity and controlling water pollution Initiate bench marking of irrigation projects. Apply RS/GIS for water balance and improving water productivity. Adopt reservoir management strategy to improve adequacy, efficiency and reduce losses. Water accounting using remote sensing (Water Accounting Plus, etc.)	(a) Medium (b) Low (c) Low (d) Low	(a) High (b) High (c) Low (d) Medium	(a) X (b) X	(a) X (c) X (d) X	

(continued)

Table 14.2 (continued)

Broad interventions	Specific interventions needed	Level of interventions already attempted	Potential for reducing S–D gap	Long term	Short term	Remarks
3. Recharge aquifers and groundwater development (a) flood water management (b) rainwater harvesting (c) Rejuvenation of open wells and water harvesting structures (d) Sustaining groundwater resources	"Underground Taming of Floods for Irrigation" (UTFI) which involves diverting high water flows from rivers or canals at times when these flows pose flood risk and recharging the groundwater via village ponds or small dams that are modified for this purpose Rejuvenation of open wells, village ponds, percolation ponds, and check dams can be done through periodical desilting and catchment, management. Managed aquifer recharge concept can be adopted by linking with the water harvesting structures. Breaking the groundwater–energy nexus In the case of abandoned dug wells, recharge mechanisms should be established through channelling the rainfall runoff into the wells through filter pits Increasing water availability in normal years and mitigation of fluctuations in surface water supply in drought years, conjunctive use of surface and groundwater. Priority in exploitation of groundwater resources in unexploited regions; control of growth of wells in over-exploited. Policies to prevent overexploitation, promote incentives for micro irrigation in over-exploited regions	(a) Low (b) Medium (c) Low (d) Low	(a) High (b) Medium (c) Low (d) Low	(d) X	(a) X (b) X (c) X (d) X	
4. Hill area agri and water development (a) Water storage structures (b) Soil and water conservation practices (c) Modern irrigation practices	Planning, designing, and implementation of soil and water conservation structures, storage structures like check dams, farm ponds, check dams and percolation ponds, diversion weirs, etc Providing supplemental irrigation through drip and sprinkler irrigation	(a) Low (b) High (c) Medium	(a) High (b) High (c) High		(a) X (b) X (c) X	

(continued)

Table 14.2 (continued)

Broad interventions	Specific interventions needed	Level of interventions already attempted	Potential for reducing S–D gap	Long term	Short term	Remarks
5. Wastewater treatment and reuse (a) Agriculture use (b) Industry use	Efforts to be made for the full utilization of the created treatment potential by relaxing the constraints already identified Removing pollutants from wastewater and reusing it for different purposes depending on the treatment level. Wastewater treatment policies need to be revisited. Model policies like Gujarat's wastewater policy can be examined for replication in all the states. Intensification of wastewater-based irrigation in peri-urban areas should be planned. Business models should be promoted	Govt depts., water utilities, Industries, Environmental NGOs. Funding agencies, private institutions, Experts & stakeholders	(a) Low (b) High	(a) X	(a) X (b) X	
6. Desalination	Removing the salts from saline water to produce freshwater. Plants can be installed in coastal areas to meet the domestic demand and reduce the burden on irrigation projects RO units are now available even to use at village and home level. The cost and benefits of these interventions should be worked out	Low	Low	X		May take more time to implement at macro level
7. Inter-basin transfers	Moving water from water-abundant regions to water-scarce regions through manmade conveyance canal, storage systems. Large-scale water transfers between basins, increasing water demand could be met and reliable surface water supply is available. But, the extent of these water transfers depends on the improvement in crop water productivity and sustains the supplies to other sectors along with other socio-political, environmental, technical, and financial issues	Low	High	X		Subject to socio-political, interstate river disputes, politics, and environmental issues. May take a longer period to implement

(continued)

Table 14.2 (continued)

Broad interventions	Specific interventions needed	Level of interventions already attempted	Potential for reducing S–D gap	Long term	Short term	Remarks
8. Water treaties with neighbouring countries	Climate change-induced variation in river flows be monitored. Negotiating with neighbouring countries for water sharing and efficient utilization	Low	Medium	X		Subject to regional politics, govt policies, etc
9. Improving policy framework taking into account the nexus angle Energy management Transformative policies	Existing policies related to agriculture, irrigation, and power sectors need review and amendment for convergence on efficient water and energy use. Separation of feeders between agricultural use and domestic use. Allocation of electricity units each month and cash incentive if the use of electricity is below that allocation & the farmers get cash incentive on each unit "saved", Solar systems—grid-connected solar pump systems & solar power as remunerative crop, cheaper electric pumps or solar pumps in Eastern India. Policy reforms on solar	Low	Medium	X	X	
10. Advanced national-level climate change prediction models for developing future water scenarios, forecasting, etc	Already several prediction models are being developed and tested periodically. Given the change in climate parameters, advanced prediction models need to be used to predict exactly the rainfall pattern, floods, etc. for preparing the adaptation strategies in different regions of the country	Low	Medium		X	

Table 14.3 Water demand-side interventions and their potential to reduce supply–demand gap

Broad interventions	Purpose and specific interventions needed	Level of interventions already attempted	Potential for reducing S–D gap	Long term	Short term	Remarks
1. Adoption of climate-smart technologies (a) Soil-related, Crop-related, Water-related (b) Micro irrigation (c) Transformative CSA policies	Increase in water use efficiency and water productivity through efficient soil, crop, and irrigation technologies. The list of such technologies is given in Chap. 7 Crop-related: crop residue management, conservation tillage, mulching, change in crop pattern and cropping intensity. Use of drought-tolerant and early-maturing varieties. SRI practices Water-related: irrigation scheduling, deficit irrigation, water saving and best management practices such as drip and sprinkler. AWD practices Capping irrigation supply by limiting total seasonal ET and adjusting cropping to suit. Reduce groundwater abstraction by shifting paddy cultivation from the summer months to the monsoon season to reduce the irrigation and evapotranspiration. Reducing groundwater irrigation CSA technologies prioritization, financial outlay plan, convergence pathways, and institutional arrangements for implementation and monitoring. Transformative policies to strengthen CSA	(a) Low (b) Medium (c) Low	(a) High (b) High (c) Medium	(c) X	(a) X (b) X	
2. Reworking crop water requirements under changing climate	Use of climatological approach for re-estimation of crop water requirements for all crops adopting Fenman–monteith method incorporating all the weather parameters using IMD data sets covering all the regions in the country	Low	High		X	

(continued)

Table 14.3 (continued)

Broad interventions	Purpose and specific interventions needed	Level of interventions already attempted	Potential for reducing S–D gap	Long term	Short term	Remarks
3. Good governance (a) Institutions Laws, regulations Taxes, subsidies, incentives, pricing (b) Transformative Policies	Strengthening of RBOs, WUAs, FPOs, and other institutions working on water in coordination with Govt departments. Enactment of laws to address mainly water pollution, groundwater overexploitation, encroachment of water bodies, catchment area, etc. Revision of water taxes and improving cost recovery mechanisms, targeting the subsidies to needy farmers through better subsidy disbursement models, pricing of water in multiple uses, etc. in consultation with beneficiary groups. Introducing transformative policies to sustain climate-smart practices, finance and water governance. Conflict resolution mechanisms, enforcement mechanisms, networking and coalitions, capacity building, social and community support	(a) Medium (b) Low	(a) High (b) Medium	(a) X	(a) X (b) X	
4. Reworking O&M and subsidies	The O&M charges attributing to the irrigation components can be separated from other components such as domestic, industry, power, fishery, etc. using the separable costs and remaining benefits method. The subsidy attributed to irrigation can be worked out using the revised O&M cost of irrigation and cost recovery from water charges collection	Low	Medium		X	
5. Crop diversification towards less water intensive crops	Encourage MSP for less water intensive crops; cash incentives to crop diversification such as maize, pulses, minor millets, vegetables, etc. Include crops with more micronutrient composition for nutritional security; Establish efficient marketing channels and value-added products from non-rice cereals Vertical farming allied to solar power generation for vegetable cropping	Medium	Medium			

(continued)

Table 14.3 (continued)

Broad interventions	Purpose and specific interventions needed	Level of interventions already attempted	Potential for reducing S–D gap	Long term	Short term	Remarks
6. Strengthening crop productivity in rainfed agriculture	Using the river basin as the basis, improving the crop productivity in the rainfed agriculture by adopting climate-smart practices like soil and moisture conservation- contour bunding, land levelling, terracing, etc. Use of farm ponds for supplemental irrigation. Use of improved germplasm, modern equipment, etc	Medium	High		X	
7. Digital agriculture, use of ICT tools, Development of DSS	information and communication technology (ICT), remote sensing (RS), GIS, artificial intelligence (AI), cloud computing, drones and robots, nanotechnologies, mobile apps for judicious agricultural water management, introduction of agricultural technology platforms such as e-agri, e-NAM (electronic national agricultural market) portals, etc. for creating awareness and use of technologies Monitoring, forecast, and warning systems, expert know-how, simulation models, decision support systems, farm advisory	Medium	Medium		X	
8. Public–private partnership business models	Development and promotion of public–private partnership models both in agricultural and irrigation sectors such as seed production, agricultural machinery, input supplies, micro irrigation, irrigation scheduling, drainage, canal automation, water distribution, wastewater treatments, IT-based solutions, automatic weather stations, crop insurance products, data generation, etc	Low	Medium	X	X	

(continued)

Table 14.3 (continued)

Broad interventions	Purpose and specific interventions needed	Level of interventions already attempted	Potential for reducing S–D gap	Long term	Short term	Remarks
9. Awareness, capacity building, and training of farmers and field personnel	Adopt demand-driven approach rather than supply-driven approach for providing the integrated solutions to farmers' complex issues through building public–private partnerships among the needed institutions/organizations. change of mind-set of all actors from policy makers, research managers, administrators, editors of scientific journals, scientists to promote demand-driven research and extension for the small farm-holders. Programmes on awareness creation through different communication channels coupled with capacity building and training programmes for specialized topics to different stakeholders	Medium	Medium		X	
10. Improved marketing, crop insurance products	Identification and promotion of effective marketing channels for different products; improving the performance of FPOs, reworking on MSP concept to cover more crops, introduction of weather-based crop insurance for irrigated and rainfed crops Develop rural manufacturing to take people off the land and thus increase average farm sizes Moving from subsistence to productive irrigation. Organise "farming clusters" with neighbouring farmers grouping together to consolidate their landholdings to enable cropping in large blocks	Medium	Medium	X	X	

14.7 India's Water Future—Bridging the Gap

The supply–demand gap will be increasing from about 13% in 2025 to about 59% in 2050. Domestic, energy, and industrial water demand will further grow more rapidly than agricultural water demand, although agriculture sector will remain the dominant water demand sector. It is also expected that due to climate change, the global net irrigation requirements will increase by 3.5–5% by 2025, and 6–8% by 2075. In Indian situation, it is expected that the net irrigation requirements may increase by another 6% and gross irrigation requirements by about 10%. If this is considered, the supply–demand gap may still be high by 5–10%. However, there are possibilities to bridge the gap through focussed interventions in the future rather than generalized interventions done in the past as shown in the tables above. Proper implementation of the interventions through targeted programmes will also help in achieving the reliable annual water supply from the current 744 bcm to the attainable 1122 bcm level in the long run.

In addition to the interventions outlined above, few other important strategies that may be seemingly possible but not included in the supply–demand calculations are:

(a) **Desalination**: It is seemingly a possible strategy to meet the potable water demand in the adjacent cities and villages particularly close to coastal areas. It is an expensive and energy-intensive option because the cost can be passed on to the consumers. Practical issues like capital cost, size and scale of the project, operating and maintenance costs, pricing, cost recovery, etc. may constrain the large-scale development and management in the country. The major environmental concerns on desalination projects relate to the safe disposal of the brine, huge energy used in the process and CO_2 emissions. Desalination is generally not an attractive and affordable option for agriculture because of the high cost and large volume of water required for crop production. Already several small units are operating in different locations but with lot of operational and management constraints. Hence, their choice depends more on the financial trade-offs. This option is used in many developed countries. However, when desalination costs are dropped significantly due to technological advances such as use of renewable energies, this may be an option in the long run.

(b) **Inter-basin water transfers**: Even though inter-basin water transfers or national river linking is looking impressive and optimistic, it has seemingly a lot of limitations. The extent of these inter-basin water transfers depends upon the extent to which India can improve its crop water productivity using existing measures and sustain the supplies to other sectors as well. The implementation of the inter-basin transfers is subject to several socio-political, technical, hydrological, financial, and environmental issues which need to be addressed very carefully through in-depth location-specific and systematic studies.

Further, many of the interstate rivers have varying levels of legal disputes involving substantial costs and delays. The tribunal decisions often fail to consider the latest scientific understanding in the water sector. A viable mechanism is needed to deal with the emerging interstate disputes and preferably should be taken out of the purview of Courts. A scientific approach in reworking the actual water demand of the basins will support such an initiative. Few inter-links have already been initiated in few rivers (such as Ken-Betwa Link Project) showing some positive signals. The Ken-Betwa Link Project received the nod of the Union Cabinet in December 2021, as a part of government's plan to connect rivers to enable water from surplus areas to address needs in drought-prone regions. However, unless all the concerned states come to a consensus, holistic interlinking of rivers in the country will be really challenging. Further, given the climate change impacts on the changing river flows, it is also equally important that the river water treaties with neighbouring countries need to be re-examined and negotiated to sustain the river supplies.

These two supply-side measures (national river linking and desalination) are all on the more expensive part of the interventions and their choice requires financial trade-offs by decision-makers and funding agencies.

14.8 Counting What Matters Next

The lessons learned have necessitated the importance of Governance phase, which encompasses the converging roles of different institutions with transparency and accountability. Hence, in the future, the expected contribution of the three phases (i.e., investment, management, and governance) in improving the overall irrigation sector performance must be in the ratio of 30: 30: 40.

Changing climate in terms of decrease in future water supply and increase in future water demand warrants that India's water future in terms of water security needs to be ensured through time tested interventions outlined above. Past experiences show that in spite of the implementation of several modernization interventions, the performance of the projects is still poor indicating the need for relooking the approach to be used while introducing these interventions. River basin structure facilitates the identification of issues in the head, middle, and tail regions of the basin. Once the river basins become "closed river basins", the supply-side (engineering) interventions couldn't support to address the water's future issues in a sustainable way and demand management aspects will be important. Hence instead of the broad interventions just followed in the past, more specific/focused interventions outlined in the tables above can be examined for their suitability to different segments of the river basin and implemented accordingly. Further, it is important to select a portfolio of implementable policies and programmes supported by well-functioning water institutions (governance) in order to achieve sustainable outcomes. All of these interventions entail financial and social costs that need to be considered when addressing the India's water future in a changing climate. Therefore, the interventions suggested above focus mainly on the need for increased investment in strengthening and managing the irrigation facilities, developing climate-resilient crop varieties and creating awareness about climate change impacts and efficient water use.

While focusing on the above interventions, it is equally important to monitor the progress made from time to time. In this context, impact assessment of the infrastructure projects is warranted which can minimize the cost of wrong decisions or cost of not doing the right ones. E.g., the Dam Rehabilitation and Improvement Project (DRIP) is a timely initiative by the CWC involving the World Bank and all the stakeholders. A recent development in the area of canal irrigation modernization in India is the establishment of Support for Irrigation Modernization Program (SIMP) by the Asian Development Bank in association with CWC. The Composite Water Management Index (CWMI) recently developed by Niti Aayog, GOI helps the states to appropriately plan and implement the needed interventions. Further, adoption of Water Auditing and Water Auditing plus approaches with assured/refined data with science-based understanding can help in prioritizing the strategies to address India's water future in a better way.

References

Water Resources Group (2009) Charting our water future-economic frameworks to inform decision-making. The World Bank, Washington DC

Briscoe J, Malik RPS (2006) India's Water Economy-Bracing for a turbulent future. The World Bank, South Asia Division, Washington DC

Glendenning CJ, Babu SC, Asenso-Okyere K (2010) Review of the agricultural extension in India, IFPRI Discussion Paper 01048. International Food Policy Research Institute, Washington DC

GoI (2013) Key indicators of household consumption expenditure in India, 2011–12, Report No. NSS KI (68/1.0), National Sample Survey Organization, Ministry of Statistics and Programme Implementation, Government of India

Government of India (2008) Integrated water resources development and management, Central Water Commission, New Delhi

Government of India (2017) Operational guidelines of per drop more crop (micro irrigation) component of PMKSY. Ministry of Agriculture & Farmers Welfare, New Delhi

Gulati A, Thangaraj P (2023) El Niño and weather vagaries: deconstructing historical trends. Agri-Food Trends Anal Bullet (AF-TAB) 3(1)

International Institute for Applied Systems Analysis (IIASA) (2016) Water Futures and Solutions: Asia 2050 Knowledge and Innovation Support for the Water Financing Program of the Asian Development Bank (RETA 6498) Laxenburg, Austria

Pathak P, Sahrawat KL, Wani SP, Sachan RC, Sudi R (2009) Opportunities for water harvesting and supplemental irrigation for improving rain-fed agriculture in semi-arid areas. In: Wani SP, Rockström J, Oweis T (eds) Rainfed agriculture: unlocking the potential. Comprehensive assessment of water management in agriculture series. CAB International, Wallingford, UK, pp 197–221

Rockström J, Falkenmark M (2000) Semiarid crop production from a hydrological perspective: gap between potential and actual yields. Crit Rev Plant Sci 19(4):319–346

Rockström J, Falkenmark M, Karlberget L (2009) Future water availability for global food production: the potential of green water for increasing resilience to global change. Water Resour Res 45, W00A12. https://doi.org/10.1029/2007WR006767

Shah T (2023) Water-Energy-Food-Environment (WEFE) nexus in action: global review of precepts and practice. Draft, IWMI, Anand, India

Shah M (2013) Water: towards a paradigm shift in the twelfth plan. In: Economic and political weekly, vol XLVIII, no 3

Upali A, Shah T, Turral, H Anand BK (2007) India's water future to 2025–2050: business-as-usual scenario and deviations. Research Report 123, International Water Management Institute, Colombo.

FSC
www.fsc.org
MIX
Papier | Fördert gute Waldnutzung
FSC® C083411

Kolibri 360 GmbH
Ferdinand-Jühlke-Straße 7
99095 Erfurt, Deutschland
produktsicherheit@kolibri360.de